Emanuel Mendes da Costa

Historia Naturalis Testaceorum Britanniæ, or the British Conchology

Containing the Descriptions and Other Particulars of Natural History of the Shells of Great Britain and Ireland

Emanuel Mendes da Costa

Historia Naturalis Testaceorum Britanniæ, or the British Conchology
Containing the Descriptions and Other Particulars of Natural History of the Shells of Great Britain and Ireland

ISBN/EAN: 9783744657365

Printed in Europe, USA, Canada, Australia, Japan

Cover: Foto ©berggeist007 / pixelio.de

More available books at **www.hansebooks.com**

Historia Naturalis Testaceorum Britanniæ,

OR,

THE BRITISH CONCHOLOGY;

CONTAINING

The DESCRIPTIONS and other Particulars of NATURAL HISTORY of the SHELLS of GREAT BRITAIN and IRELAND:

Illustrated with FIGURES.

In ENGLISH and FRENCH.

By EMANUEL MENDES da COSTA,

Member of the IMPERIAL CÆSAREAN ACADEMY NATURÆ CURIOSORUM, by the Name of PLINY IV. and of the BOTANIC SOCIETY of FLORENCE.

LONDON: Printed for the AUTHOR; and Sold by Messrs. MILLAN, B. WHITE, ELMSLEY, and ROBSON, Booksellers.

M.DCC.LXXVIII.

Historia Naturalis Testaceorum Britanniæ,

OU,

LA CONCHOLOGIE BRITANNIQUE;

CONTENANT

Les DESCRIPTIONS & autres Particularités d'HISTOIRE NATURELLE des COQUILLES de la GRANDE BRETAGNE & de l'IRLANDE:

Avec FIGURES en taille douce.

En ANGLOIS & FRANCOIS.

Par EMANUEL MENDES da COSTA,

Membre de l'ACADEMIE IMPERIALE CESARIENNE NATURÆ CURIOSORUM, sous le Nom de PLINE IV. & de la SOCIETE BOTANIQUE de FLORENCE.

LONDRES: Imprime pour l'AUTEUR: Chez Messrs. MILLAN, B. WHITE, ELMSLEY, & ROBSON.

M.DCC.LXXVIII.

To Sir ASHTON LEVER, Knight.

SIR,

THE noble, magnificent, and curious MUSEUM you have collected at *Leicester-House*, for the inspection of the Public, is universally allowed to be a NATIONAL HONOUR. The approbation of our SACRED SOVEREIGN, a PRINCE PATRON and ENCOURAGER of the SCIENCES and ARTS, and the satisfaction of all ranks of people who resort to view it, are indubitable marks of the success of your efforts. The Perfection of the Specimens; the Elegance and Order of their Arrangement; the Choice and Variety of the Objects, collected from all parts of the known world; at once declare your Munificence in an unbounded Expence, and your unwearied Assiduity in the STUDY of NATURE. A glorious study! that leads the human mind into a truly religious contemplation of the numberless, stupendous, and incomprehensible Works of the Awful and Eternal DIVINITY.—As a person who has always attached himself to this instructive and delightful pursuit, permit me thus publicly to offer my acknowledgment, by dedicating this work, my BRITISH CONCHOLOGY, to your Patronage; and I shall esteem myself happy if it meets with your Approbation and Protection. My earnest wishes are, that the Almighty may

grant

grant you many years of health and ſucceſs in your noble purſuit, for the encouragement of NATURAL HISTORY.

I have the Honour to be,

With great reſpect,

SIR,

Your very obliged

And obedient humble ſervant,

EMANUEL MENDES da COSTA.

London, 22d June, 1778.

PREFACE.

I BEG leave, in the following pages, to submit to the Candid Public my *British Conchology*, or Natural History of such Shells as are Natives of these Kingdoms; in the execution of which I have given all the attention in my power, and sollicited all the informations I was able to procure.

Every author, whatever be the work he is occupied about, not only regards the subject of his pursuits as important himself, but wishes it to be viewed in the same light by his readers, the more to engage their curiosity, and interest themselves in what he may have to offer.

There are indisputably many branches of Natural History, which, consider'd in a *political* light, demand the highest regard, as great *national advantages* spring from them; such are *mines, collieries, quarries, &c.* yet these are always more the speculations of *interest* than of *curiosity*.—But I do not, however partial I may be to the subject I have adopted, offer it to the public notice as of any such *importance*. It is one of the many pursuits reserved for minds at ease; for minds disengaged from the tumult of business, and both disposed and at leisure to contemplate that immense variety of beauty which Nature hath scatter'd round us.

 The

PREFACE.

JE prens la liberté de soumettre au Public Candide, dans ces pages suivantes, ma *Conchologie Britannique*, ou l'Histoire Naturelle de telles Coquilles qui sont Natives de ces Royaumes; dans l'accomplissement de laquelle j'ay donné toute l'attention en mon pouvoir, & sollicité toutes les informations que j'etois capable de procurer.

Chaque auteur, quel que soit l'ouvrage dont il est occupé, non seulement considere le sujet de ses poursuites comme important à lui même, mais il desire qu'il soit envisagé par ses lecteurs dans la même veue, à mieux engager leur curiosité, & à les interesser dans ce qu'il veut offrir.

Il y a incontestablement plusieurs branches de l'Histoire Naturelle, qui, considerées dans une veue *politique*, demandent la plus importante consideration, comme des grands *avantages nationaux* en naissent; tels sont les *mines* des *metaux* & de *charbon*, les *carrieres, &c.* cependant ceux ci sont toujours plus les speculations de l'*interet* que de la *curiosité*—Mais je ne veux, toutefois partial que je suis au sujet que j'ay approprié, l'offrir à l'avis public comme de telle *importance*. C'est une des differentes poursuites reservée pour les esprits à leur aise; pour les esprits degagés du tumulte des affaires, & aussi disposés & à loisir à contempler cette varieté immense de beauté que la Nature a parsemée tout autour de nous.

Le

The world is but too much inclined to treat with levity thoſe *ſtudies* which do not lead to its riches or preferments; and, while the generality of mankind have only theſe objects in view, it is no wonder that the *enthuſiaſm* of the *Naturaliſt* is ſo often *ridiculed*, who can deſert the beaten road of competition, where all are preſſing forward, and turn aſide into the ſilent paths of life, in queſt of nothing better than *contemplative pleaſure*. The *Naturaliſt*, however, never fails to attain the *point* he aims at, while he beholds the *many*, who have ſhot their *ſallies of wit* after him, perpetually *miſſing* theirs.

Le monde eſt que trop incliné à traiter avec legereté ces *etudes* qui n'aboutiſſent aux richeſſes ou aggrandiſſements; &, pendant que le peuple en general a ſeulement ces objets en veue, ce n'eſt pas une choſe merveilleuſe que l'*enthouſiaſme* du *Naturaliſte* eſt ſi ſouvent *ridiculiſé*, qui peut s'ecarter du grand chemin de brigue, ou touts s'avancent ardemment, & tourner à part dans les ſentiers paiſibles de la vie, en recherche de rien mieux que le *plaiſir contemplatif*. Le *Naturaliſte* cependant ne manque jamais de atteindre le *but* qu'il viſe, pendant qu'il voit le *grand nombre* de ceux, qui ont tiré leurs *traits de raillerie* contre lui, perpetuellement *manquant* leur coup.

Nature is an *immenſe volume*, which the *Great Creator* hath thrown open to man; on every *page* his *omnipotence* is recorded, and in every *page* may be traced his *providential care* for the *order* and *preſervation* of the creatures he has formed.—To inveſtigate with accuracy the ſcenes of beauty and wonder that ſurround us, and to trace the nice links which chain together the *Animal*, the *Vegetable*, and the *Mineral* kingdoms, would, even imperfectly as we might do it, aſk a far larger portion of life than is allotted to the race of man; for *Natural Hiſtory* is like the *innumerable veſſels* of the human *body*, whoſe larger parts are diſcernible, but which ſhoot out and ramify into ſuch an infinitude of ſmaller ones, as muſt for ever *elude* our *ſtricteſt reſearches*.

La Nature eſt un *volume immenſe*, que le *Grand Createur* a ouvert à l'homme; dans chaque *page* ſon *toute puiſſance* eſt enregitrée, & dans chaque *page* ſa *providence ſoigneuſe* pour l'*ordre* & la *conſervation* des etres qu'il a formé ſe peut tracer.—Pour rechercher avec exactitude les ſcenes de beauté & d'admiration qui nous environnent, & pour tracer les chainons fins qui lient enſemble les regnes *Animal*, *Végétal*, & *Minéral*, demanderoit, même d'une maniere imparfaite comme nous le ferions, une portion d'exiſtence beaucoup plus grande que eſt determinée au genre humain; car l'*Hiſtoire Naturelle* reſſemble aux *vaiſſeaux innombrables* du *corps* humain, dont les grandes parties ſont faciles à voir, mais qui pouſſent & ramifient dans une telle infinité de plus petites, que il faut pour toujours *eviter* nos *recherches* les plus *exactes*.

Though to ſearch deep into the whole ſyſtem of Nature is a ſtudy to which our faculties are inadequate, yet detached parts of this ſtupendous ſyſtem have employed

Quoique à examiner profondement dans le ſyſteme entier de la Nature eſt une etude à laquelle nos facultés ſont imparfaites, cependant des parties ſeparées de ce ſyſteme

ployed both the thoughts and pens of learned men, in different times and countries.—That part of it which relates to *Conchology* hath been ably treated of by many, though no *author* has in any distinct work confined himself solely to the *Conchology of these kingdoms*, which must be an object more particularly worthy the inquiry and attention of those who are the inhabitants of them.

Shells are certainly a beautiful part of the works of Nature: we observe a vast variety in their *genera*, and those branching into an innumerable variety of *species*; we see in them the most *splendid colours*, the most *perfect symmetry*, and cannot but wonder at all this seeming waste of beauty lavish'd on the depths of the sea, and only accidentally brought to the face of day. We are in general indebted for most of the shells as yet discover'd to the winds and tempests, which throw them up on the different coasts of the world, and *probably know not*, nor *ever shall*, the *thousandth* part of what the waters of the world conceal.—What would be our astonishment, could the beds and caverns of the ocean be thrown open to our view! or could we ever be inform'd of the use and destination of all those multitudinous little animals whose habitations are so eagerly coveted by the Naturalist, and treasured up with so much expence by the curious!—Then would be explain'd to us that astonishing variety which shells exhibit in their external forms; why some are *radiated* with such *mathematical exactness*, and some so *lavishly adorned*; why some are *armed* at *all points*, as if their existence was a *state* of *perpetual warfare*, and others apparently *defenceless*; why some are surrounded

systeme etonnant ont employé les pensées & les plumes des sçavants, en differents pais & differents tems.—Cette partie qui regarde la *Conchologie* à eté traité avec habileté par plusieurs, bien que aucun *auteur* s'est borné uniquement dans un ouvrage distinct à la *Conchologie de ces royaumes*, qui doit etre un objet plus particulierement digne de la recherche & de l'attention de ceux qui en sont les habitans.

Les coquilles sont certainement une partie elegante des ouvrages de la Nature: nous observons une varieté prodigieuse dans leurs *genres*, & ces genres poussent dans une varieté innombrable des *especes*; nous en voyons les *couleurs* les plus *eclatantes*, & la *symmetrie* la plus *parfaite*, & ne pouvons que admirer à toute cette magnificence, en apparence dissipée, prodiguée sur les profondeurs de la mer, & seulement par accident porté à la face du jour. Nous sommes generalement endetté pour la plupart des coquilles aux vents & aux tempetes, qui les jettent sur les differentes cotes du monde, & *probablement nous ne connoissons*, ni *même connoitrons nous à jamais*, la *millieme* partie de ce que les eaux de l'univers cachent.—Quel seroit notre surprise, si les lits & les cavernes de l'Ocean seroient ouverts à notre veue! ou si nous pourrons à jamais etre informé de l'usage & de la designation de touts ces petits animaux si abondants, dont les domiciles sont avidement desiré par les Naturalistes, & thesaurisé avec tant de depense par les Curieux!—Alors nous seroit expliqué cette varieté etonnante que les coquilles montrent dans leurs formes exterieures; pourquoi les unes sont *radiées* avec une telle *precision mathematique*, & d'autres si *prodigalement embellies*; pourquoi quelques unes sont

rounded with a *texture* that one might conceive capable of *reſiſting* every *danger*, while others can ſtand the conflicts of the troubled ſea in *dwellings* ſo *delicate* and ſo *ſeemingly inſecure*.—But the Almighty hath drawn a veil over all this knowledge. *He* hath *diſcover'd enough* to ſhew us in this, as well as in every other part of *his works*, the veſtiges of his *creative power*; all *beyond* was *unimportant* to *our happineſs*.

Having ſaid thus much to juſtify the laudableneſs of this ſtudy, I ſhall only add, that while we obſerve the amuſements of the world pall the appetites of thoſe who have purſued them ardently, we ſtill find the Naturaliſt daily more enraptured of the objects to which he devotes his attention. The truth is, that *Natural Hiſtory* not only gives *pleaſure* to, but *enlarges* the *mind* it occupies.—A *cabinet of ſhells* is a *volume* of fine wrote *ſermons*; thoſe who *read* them *attentively* will find their *morals improved* by the *peruſal*.—The *elegance* of their *forms* may inforce our *notice*, the *beauty* of their *colours* may ſollicit our *admiration*; but the *philoſophic eye* penetrates much farther; it *diſcriminates* every *minute variation*, not only *ſurveys* the *texture*, but *explores* the *anatomy*, and charm'd with the effect, *darts* forward to the *cauſe*, and, as Mr. *Pope* has elegantly expreſs'd it,

"Looks thro' Nature up to Nature's God."

ſont *partout armées*, comme ſi leur exiſtence etoit un *etat* de *guerre perpetuelle*, & d'autres en apparence *ſans defenſe*; la cauſe que quelques unes ſont couvertes d'une *tiſſure* que l'on concevroit capable de *reſiſter* à chaque *danger*, pendant que d'autres peuvent braver les chocs de l'ocean agité dans des *habitations* ſi *delicates* & en *apparence* ſi *mal aſſeurées*.—Mais le Tout-Puiſſant à tiré un voile ſur toute cette connoiſſance. *Il* a *decouvert aſſes* pour nous faire voir que en celle ci, pareillement comme en toutes autres parties de *ſes ouvrages*, les veſtiges de ſon *pouvoir creatif*; tout *au delà* n'etoit d'*importance* à *notre felicité*.

Ayant tant dit pour juſtifier que cette etude eſt digne de louange, j'ajouterai ſeulement, que pendant que nous obſervons les amuſements du monde degoute les appetits de ceux qui les ont pourſuivis ardemment, nous trouverons toujours le Naturaliſte de jour en jour plus en extaſe des objets aux quels il devoue ſon attention. La verité eſt, que l'*Hiſtoire Naturelle* ne rend ſeulement du *plaiſir*, mais *aggrandit* l'*eſprit* de celui qui en fait ſon etude.—Un *cabinet de coquilles* eſt un *volume* de *ſermons* parfaitement bien ecrits; ceux qui les *liſent attentivement* trouveront leurs *morales perfectionnées* en la *lecture*.—L'*elegance* de leurs *formes* peut contraindre notre *remarque*, la *beauté* de leurs *couleurs* peut ſolliciter notre *admiration*; mais l'*œil* du *philoſophe* penetre beaucoup plus; il *diſtingue* chaque *variation menue*, non ſeulement il *examine* la *tiſſure*, mais il *recherche* l'*anatomie*, & charmé de l'effet, il *lance* à reconnoitre la *cauſe*, &, comme M. *Pope* s'exprime avec elegance,

"Il parcourt à travers la Nature juſques au Dieu de la Nature."

It may not be amiſs to take notice of thoſe *authors* who have touched on the ſubject of *Britiſh Conchology* in their *writings*. Of theſe, the *accurate* and *judicious* Dr. *Liſter* ſtands the *firſt*, who a century ſince, *viz.* in 1678, publiſhed his *Hiſtoria Animalium Angliæ*. He moſtly confines himſelf to the ſhells of the northern counties, and exhibits *ſixty-three ſpecies*, of which he has given very circumſtantial deſcriptions, and excellent delineations. It is but a *tribute* juſtly due to his *memory*, to ſay it is a *moſt admirable work*; and I publicly acknowledge that I have adopted it as a model for my own.

The Doctor's number of Engliſh ſpecies was certainly very ſcanty; but he remedied this, in ſome manner, by his *Hiſtoria Conchyliorum*, which he publiſhed fourteen years after, in which he added an *A.* for *Anglia*, to ſeveral ſpecies he had *newly diſcovered.*—*Merret*, in his *Pinax Rerum Naturalium Britannicarum*, enters but little on the ſubject; and *Petiver*'s Account of Britiſh Shells is not much more than looſe catalogues and a few figures.

The topographical Britiſh authors who have publiſhed the Natural Hiſtory of particular counties, as *Sibbald*, *Wallace*, *Martin*, *Leigh*, *Morton*, *Dale*, *Borlaſe*, *Wallis*, *Smith*, and *Rutty*, *&c.* have all ſpoken of the ſhells of the diſtricts they were deſcribing; and the celebrated Mr. *Pennant*, in his 4th volume of *Britiſh Zoology*, has profeſſedly enter'd on the *Britiſh Conchology*.

In the work now before the reader, I beg leave to ſay that I have intirely follow'd

Il ne ſera hors de propos de notifier ces *auteurs* qui ont touché ſur le ſujet de la *Conchologie Britannique* dans leurs *ecrits*. De ceux ci, le *limé* & *judicieux* Dr. *Liſter* range le *premier*, qui un ſiecle paſſé, *ſçavoir*, en 1678, publia ſon *Hiſtoria Animalium Angliæ*. Il ſe borne, pour la plupart, aux coquilles des comtés du nord, & propoſe *ſoixante-trois eſpeces*, deſquelles il a donné des deſcriptions fort en detail, & de tailles douces excellentes. C'eſt ſeulement un *tribut* avec juſtice deu a ſa *memoire*, de dire que l'*ouvrage eſt tres admirable*; & j'avoue publiquement que je l'ai adopté comme un plan pour le mien.

Le nombre des eſpeces Angloiſes que le Docteur a propoſé fut certainement petit; mais il a remedié cela, en quelque maniere, dans ſon *Hiſtoria Conchyliorum*, qu'il publia quatorze années apres, dans lequel il ajoute une *A.* pour *Anglia*, à pluſieurs eſpeces qu'il avoit *nouvellement decouvert*. —*Merret*, dans ſon *Pinax Rerum Naturalium Britannicarum*, ecrit tres peu ſur le ſujet; & le Recit de Coquilles Britanniques que *Petiver* nous donne, n'eſt pas beaucoup mieux que des catalogues volantes & quelques figures.

Les auteurs topographes Britanniques qui ont publié l'Hiſtoire Naturelle de quelques comtés particuliers, comme *Sibbald*, *Wallace*, *Martin*, *Leigh*, *Morton*, *Dale*, *Wallis*, *Smith*, & *Rutty*, *&c.* ont touts recité les coquilles des comtés qu'ils ont decrit; & le celebre M. *Pennant*, dans ſon 4e. tome de ſa *Zoologie Britannique*, a ſpecialement traité ſur la *Conchologie Britannique*.

Dans cet ouvrage à preſent devant le lecteur, qu'il me ſoit permis de dire, que j'ai

low'd the fyftem already laid before the public in my *Elements of Conchology*.

As it is neceffary to give fhells fome *trivial names* for diftinction fake, I have, in doing it, always endeavour'd to form the *denomination* on fome *idea* arifing from the *fhape*, *texture*, or *colour*, &c. but when no fuch *correfpondent circumftances fuggefted a name*, the choice of one neceffarily became *arbitrary*.

I much doubt whether my defcriptions may not be *fometimes* taxed with *prolixity*; but the object of Natural Hiftory will, I hope, excufe it: *Precifion*, not *Elegance*, is required.

I have quoted all the *fynonyms of authors at large*; they are lefs liable to *error* than *quotations only of figures or of pages*, and elucidate the authors themfelves. Thefe *quotations* are placed according to the *order* of *time* when the *authors refpectively flourifhed*; but this work being a *Britifh Natural Hiftory*, the *Britifh* authors are placed before thofe of *foreign* countries.

Attention has been paid to note the chief *places* where each *fpecies* is found, except when the fhell is very *common*; and to *many* I have added the other *countries* of which the fame fpecies is a *native*, that a comparative view may be formed of the various climes in which the fame fpecies exift: a *particular* which I judged would be both *curious* and *inftructive*.

Another *circumftance* to be premifed is, that I have been very *cautious* in *fixing* the *fpecies* of *thefe kingdoms*. Authors are liable to be impofed on: thus Sir *Robert Sibbald* had the *pearly* or *Eaft-Indian nautilus* fent him from the *Weftern Iflands* of *Scotland*; Dr. *Plot* was impofed on even by an *Oxford*

j'ai entierement fuivi le fyfteme déjà fou-mis au public dans mes *Elemens de la Conchologie*.

Comme il eft neceffaire de donner aux coquilles quelques *noms de guerre* pour les diftinguer, j'ai effayé toujours de former ces *noms* fur quelque *idée* de la *forme*, la *tiffure*, ou la *couleur*, &c. mais quand aucunes *circonftances correfpondentes fuggeroient un nom*, le choix alors neceffairement devenoit *arbitraire*.

Je doute beaucoup fi quelques unes de mes defcriptions ne feront *quelquefois* taxées d'etre *prolixes*; mais l'objet de l'Hiftoire Naturelle j'efpere m'excufera fur ce point: la *Precifion*, & non l'*Elegance*, eft requifé.

J'ai cité touts les *fynonymes des auteurs amplement*; ils font moins expofé à l'*erreur* que les *citations feulement des figures ou des pages*, & expliquent les auteurs mêmes. Les *citations* font arrangées felon l'*ordre* du tems quand les *auteurs refpectifs* ont *fleuri*; mais cet ouvrage etant une *Hiftoire Naturelle Britannique*, les auteurs *Britanniques* font placés devant ceux des pais *etrangers*.

Une *attention* a ete prife à noter les *lieux* principals ou chaque *efpece* fe trouve, excepté quand la coquille eft fort *commune*; & à *plufieurs* j'ay ajouté les autres *pais* defquels la même efpece eft une *native*, que une vue comparative peut etre formée des differents climats dans lefquels cette même efpece exifte: un *particulier* que j'ay jugé feroit auffi *curieux* que *inftructif*.

Une autre *circonftance* à etre remarqué eft, que j'ai eté fort *circonfpect* en *fixant* les *efpeces* de *ces royaumes*. Les auteurs font fujet à etre dupé: ainfi Sir *Robert Sibbald* à receu le *nautile nacré* ou *des Indes Orientales*, des *Ifles Hebrides* de l'*Ecoffe*; le Dr. *Plot* fut dupé même par un *Profeffeur de*

Oxford Professor, in his *curious land snail* of *Cornbury Park*, in *Oxfordshire*, of which he has given a figure; and Dr *Lister* was most probably imposed on by the fishermen of *Scarborough*, in the *strombiformis bicarinatus*, described No. 64.—Other like instances occur, even without any design of deceit. I have received a fine *volute* from *Scarborough*; *rhombi* from other *English* coasts; and the *grimace buccinum* from the shore of *Sussex*. My conduct in such cases has been to *reject* all *single instances*, and admit none but such as were *determinable*, either by *repeated observations*, or the *quantity* of the species *found*; for *single examples* are not *positive proofs*, they may *happen casually*.

I have described the shells from the objects themselves, except in *six instances*, where I could not procure the originals to compleat the series; in which cases I have borrowed them from authors of veracity; and the reader will find *those species distinguished* by *Roman characters*, and the *authors* quoted *verbatim*.

No expence has been spared in the engraving and colouring the plates, and I flatter myself they will meet with the public approbation.

The species of *British Univalves* here set forth are *eighty-six*. I am very sensible that several species of shells, yet unknown to me, remain to be discovered in these kingdoms. Should any Ladies or Gentlemen, curious in these pursuits, be kind enough to communicate to me any new observations or discoveries, I shall with all due thanks acknowledge the honour they confer on me, and, if of any number, they shall hereafter be printed in the same *form* with this

de Oxford, dans son *limaçon terrestre curieux* de *Cornbury Park*, au *comté* de *Oxford*, de lequel il a donné une figure; & le Dr. *Lister* probablement en fut imposé par les pecheurs de *Scarborough*, dans la *strombiformis bicarinatus*, decrite No. 64.—D'autres tels exemples se presentent, même sans aucun dessein de faire accroire. J'ai receu une belle *volute* de *Scarborough*; des *rhombi* d'autres cotes d'*Angleterre*; & le *buccin la grimace* de la cote de *Sussex*. Ma conduite en tels cas à eté de *rejetter* touts les *exemples simples*, & de admettre que ceux qui etoient *determinés*, ou par des *observations reiterées*, ou par la *quantité* de l'espece qui se *trouvoit*; car des *exemples simples* ne sont des *preuves positives*, ils peuvent *arriver par accident*.

J'ai decrit les coquilles des objets mêmes, excepté en *six exemples*, ou je ne pouvois procurer les originaux à faire la suite complette, en ces cas je les ai tiré des auteurs de credit; & le lecteur trouvera *ces especes distinguées* par des *caracteres Romains*, & les *auteurs* cités *mot à mot*.

Je n'ay epargnè aucune depense dans les gravures & pour les enluminer, & je me flatte qu'elles seront approuvées du public.

Les especes des *Univalves Britanniques* ici decrites sont *quatre vingt six*. Je suis fort sensible que plusieurs especes de coquilles, qui me sont jusques à present inconnues, restent à etre decouvertes dans ces royaumes. Si les Dames ou Messieurs, curieux dans ces poursuites, voudroient etre si affables à me communiquer des observations nouvelles ou decouvertes, je ferai tres reconnoissant de l'honneur qu'ils me font, &, si en quelque nombre, elles

this *work*, as an *Appendix*; for I think to renew *editions*, for the sake of a *few additions*, is an *unjust tax* on *literature*.

It only now remains for me, before I conclude this Preface, to return my most grateful acknowledgments to my *subscribers* and *encouragers* of this work, and in *particular* to the following Ladies and Gentlemen, who generously granted me free access to their elegant and well-chosen collections, furnished me with specimens, and communicated many and very interesting observations, as *George Keate*, Esq; and his *Lady*—*John Fothergill*, M. D.—Mr. *Ingham Foster*—*Martyn Fonnereau*, Esq; —Counsellor *Thomas Griffin*—*Philip Rashleigh*, Esq;—Mr. *Samuel Platt*—*William Watson*, jun. M. D. and his *Lady*, of *Bath*—*William Cuming*, M. D. of *Dorchester*—*Richard Pulteney*, M. D. of *Blandford*—*Richard Hill Waring*, Esq; of *Leeswood* in *Flintshire*—and *Walter Synnot*, Esq;

EMANUEL MENDES da COSTA.

London, 22 June, 1778.

elles seront ci après imprimées dans la même *forme* que cet *ouvrage*, comme un *Supplement*; car je pense de augmenter les *editions*, à l'egard de *quelques additions*, est une *taxe injuste* sur la *litterature*.

Il me reste seulement, avant que de finir cette Preface, de rendre avec beaucoup de gratitude mes reconnoissances aux *personnes* qui ont *souscrit* & *protegé* cet ouvrage, & *specialement* aux Dames & Messieurs suivants, qui genereusement m'ont permis un acces libre à leurs collections elegantes & bien choisies, m'ont fourni des echantillons, & m'ont communiqué plusieurs observations tres interessantes, comme *George Keate*, Esq; & Madame *Keate*—*Jean Fothergill*, M. D.—Mons. *Ingham Foster*—*Martyn Fonnereau*, Esq;—le Conseiller *Thomas Griffin*—*Philip Rashleigh*, Esq;—Mons. *Samuel Platt*—*Guillaume Watson*, jun. M. D. & Madame *Watson*, de *Bath*—*Guillaume Cuming*, M. D. de *Dorchester*—*Richard Pulteney*, M. D. de *Blandford*—*Richard Hill Waring*, Esq; de *Leeswood* au *comté* de *Flint*—& *Walter Synnot*, Esq;

EMANUEL MENDES da COSTA.

A Londres, ce 22 Juin, 1778.

A Natural History of British Shells.

Histoire Naturelle des Coquilles Britanniques.

PART I. Simple Univalves. Univalvia non Turbinata.

PREMIERE PARTIE. Univalves Simples.

GENUS I. Patella seu Lepas; the *Limpet, Flither*, or *Pap Shell.*

An univalve shell, simple, or in no wise spiral, and of a conical shape. The animal, a *Slug*.

GENRE I. Le Lepas.

Une univalve simple, en aucune maniere spirale ou contournée, de figure conique. L'animal est une Limace.

* FLUVIATILES. River.

* FLUVIATILES.

I.

P. Fluviatilis. River. Tab. II. fig. 8. 8.

I.

Lepas Fluviatile. Pl. II. fig. 8. 8.

PATELLA *integra, exigua, fusca, fragilis, vertice inflexo.* Tab. 2. fig. 8. 8.

Patella fluviatilis, fusca, vertice mucronato, inflexoque.—List. Hist. Anim. Angl. p. 151. tit. 32. tab. 2. fig. 32.

Patella fluviatilis, exigua, subflava, vertice mucronato, inflexoque.—List. Hist. Conchyl. tab. 141. fig. 39.

Morton, Northamptonshire, p. 417.—Anonymous Conchology, p. 25. tab. 4. fig. 17.

Patella Lacustris. Lake.—Pennant, Brit. Zool. IV. No. 149.

Patella fluviatilis, fusca, vertice mucronato, incurvo, inflexoque.—Gualt. Ind. Conchyl. tab. 4. fig. B.

Calyptra.—Klein, Ostracol. p. 118. § 292. No. 3.

Lepas fluviatile.—Argenville, Conchyl. I. p. 372. tab. 31. fig. 1. 2. 3. II. p. 329. tab. 27. fig. 1. & p. 73. tab. 8. fig. 1.

Patella testa membranacea ovali, mucrone reflexo.—Linne, Faun. Suec. I. p. 369. No. 1292. II. No. 2200.

Lacustris. Patella, testa integerrima ovali membranacea, vertice mucronato reflexo.—Linne, Syst. Nat. p. 1260. No. 769.

The *shell* is thin, almost membranaceous, of a brownish *colour* like horn, semitransparent, and brittle. The most general *size* is that of a small pea.

The *shape* very conical, on a large oval *base* or *aperture*, commonly ⅙ of an inch in *width*; but the largest measure better than ¼ of an inch. In *heighth*, or from the *base* to the *vertex*, about ⅛ of an inch. The *vertex* is very sharp pointed, and somewhat hook'd or turn'd, and the *margins* are level and smooth.

The *outside* is circularly wrinkled; the *inside* smooth.

It is *found* in most of the rivers of Great-Britain, adhering to aquatic plants, stones, &c.

Obs.—This species is also found in most of the rivers and lakes of Europe. The shells, when *dead*, become white. Towards the latter end of September, Dr. Lister observed them to couple, and to fix their spawn plentifully on the stones and other bodies in the rivers: it was in small gelatinous globules, in each of which he could easily distinguish many small shells.

** MA-

La *coquille* est mince, presque membraneuse, d'une *couleur* brune, comme de la corne, demitransparente et fragile. Generalement environ la *grandeur* d'un petit pois.

La *forme* est fort conique, sur une *base* ou *ouverture* large et ovale, generalement ⅙ d'un pouce en *largeur*; mais les plus grands mesurent au dela de ¼ d'un pouce. En *hauteur*, ou de la *base* a la *pointe* ou *bouton du sommet*, environ ⅛ d'un pouce. La *pointe* ou *bouton du sommet*, est fort pointu & crochu, et les *bords* sont unis et de niveau.

L'*exterieur* est circulairement ridè, et l'*interieur* lisse.

Cette coquille se *trouve* dans presque toutes les rivieres de la Grande Bretagne, fixée sur les plantes aquatiques, les pierres, &c.

Obs.—Cette espece se trouve aussi dans presque toutes les rivieres et lacs de l'Europe. Les coquilles, quand *roulees*, deviennent blanches. Vers la fin de Septembre, le Dr. Lister a observé qu'elles s'accouploient, et fixoient leur fray en abondance sur les pierres & autres corps dans les rivieres: c'etoit des petits globules gelatineux, dans chacun desquels il pouvoit facilement distinguer plusieurs petites coquilles.

** DE

** MARINÆ. SEA. | ** DE MER.

II.

P. VULGARIS, COMMON. Tab. I. fig. 1. 1. 2. 2. 8. 8.

II.

LE LEPAS COMMUN. Pl. II. fig. 1. 1. 2. 2. 8. 8.

Patella integra ex livido cinerea, ſtriata. Tab. 1. fig. 1. 1. 2. 2. 8. 8.

Patella ex livido cinerea ſtriata.—Liſt. Hiſt. Anim. Angl. p. 195. tit. 40. tab. 5. fig. 40.

Conick Limpet 1[a].—Grew. Muſ. Reg. Soc. p. 140.

Phil. Tranſ. at Large, No. 222. p. 321. art. VIII. — Muſ. Sibbald. p. 125. 1.— Merret, Pin. Rer. Nat. Brit. p. 193.—Leigh, Lancaſhire, p. 139. tab. 3. fig. 7.—Wallace, Orkneys, p. 41.—Martin, Weſtern Iſles of Scotland, p. 145, 162, 293, & alibi.— Dale, Harwich, p. 378.—Borlaſe, Cornwall, II. p. 274.—Smith, Cork, II. p. 318.— Rutty, Dublin, I. p. 383.—Wallis, Northumberland, I. p. 400.—Anonymous Conchol. p. 8. tab. 1. fig. 1. 1. 3.

Patella vulgata. Common.—Pennant, Brit. Zool. IV. No. 145. tab. 89. fig. 145.

Patella integra.—Klein, Oſtracol. p. 115. § 283. No. 10.

Patella limbo integro, apice ſatis acuto, ſtriata, ſtriis craſſis, nodoſis, inæqualibus, ſubalbida, colore rubiginoſo aliquo modo faſciata.—Gualt. Ind. Conchyl. tab. 8. fig. Q. et An. fig. D. G. H? et tab. 9. fig. A. E?

Lepas.—Argenville Conchyl. Zoomorp. p. 21. tab. fig. 1. 2. 3.

Patella teſta ovata quatuordecim angulis exarata.—Linne, Faun. Suec. I. p. 369. No. 1291. II. No. 2199.

Vulgata. Patella teſta ſubangulata, angulis quatuordecim obſoletis margine dilatato acuto.— Linné, Syſt. Nat. p. 1258. No. 758.

The *ſhell* is thick and ſtrong, yet always horny or ſemitranſparent, of various *ſizes, ſhapes* and *colours,* ſome very *copped* or *conic,* others *merely conical,* and others nearly *flat.* Nevertheleſs, tho' ſo very different in appearance, they are only varieties of the ſame ſpecies.

The *heighth,* or from the *baſe* to the *vertex,* of the copped ones, is from near one inch to one inch and a half. The *diameter* of the *baſe* or *aperture,* from one inch and a half to two inches. The *aperture* is not circular, but narrow towards one end, or is of an oblong ovoid *ſhape.*

La *coquille* eſt epaiſſe & forte, mais cependant toujours demitranſparente ou comme de la corne, de differentes *grandeurs, formes* et *couleurs,* quelqu'unes fort *elevèes* ou *coniques,* d'autres *ſimplement coniques,* et d'autres preſque *plattes.* Neantmoins, quoique ſi differentes en apparence, elles ſont ſeulement des variètès de la même eſpece.

L'*hauteur,* ou de la *baſe* au *ſommet,* de celles qui ſont fort elevèes, eſt de pres d'un pouce a un pouce et demi. Le *diametre* de la *baſe* ou *ouverture,* d'un pouce et demi a deux pouces. L'*ouverture* n'eſt pas *circulaire,* mais etrecie vers une extremitè, ou eſt d'une figure ovoide oblongue.

The *vertex*, or *top* is situated towards the narrow side, or not central; is obtuse, and like a small bead or grain.

The very *copped* or *conic* ones, I imagine to be the *perfect shells*, or of *full growth*; the *conical*, to be growing shells, or that have not attained to maturity; and the *flattish* and very fine *coloured* ones to be quite *young shells*.

The *outside* is generally cover'd with filth, balani, and other accidental bodies, under which lies the *epidermis*, or *cuticle*, thin, coarse, membranaceous and blackish. The *shell* is *dull*, or without *gloss*; some, but few, are merely striated, or with very fine longitudinal *ridges*, which do not even notch or cut the margins; but the greatest number is thick set with large and prominent *ridges*, never any determinate number (tho' *Linne*, in his *Synonym*, hints them to be *fourteen*) and other thin intermediate ones, which altogether jagg or dentate the *margins* deeply. The *conical* shells have nearly the same structure, but the *flat* and *small coloured ones* have the ridges most generally very broad, prominent and irregular, which slash or indent their *margins* very much, and render their *contour* very irregular.

The *colours* on the *outside*—The *vertex*, and some space round it, is generally whitish or yellowish, seldom livid. The ground is of a livid greenish ashen colour, with yellowish, ambry, or dull bluish longitudinal rays. Sometimes the *ridges*, for a little way down from the vertex, are *red*, or *reddish*, and lower to the margins *white*, and sometimes of other colours. Also *circular*

Le *sommet*, ou la *pointe*, n'est pas central, mais situé vers l'extremité etroite, il est obtus, et comme un petit grain de chapelet.

Celles qui sont fort *elevées* ou *pointues*, je m'imagine être les coquilles *parfaites*, ou *adultes*; les *coniques*, des coquilles qui sont *moins avancées en age*; & les *plattes*, & celles qui sont elegamment *colorées*, des coquilles tout a fait *jeunes*.

L'exterieur, ou la *robe*, est generalement couvert de faleté, glands de mer, & autres corps casuels, au dessous desquels se trouve l'*epiderme*, ou la *peau*, mince, grosse, membraneuse, & noiratre. La coquille est *ternie*, ou sans *lustre*; qu'elqu'unes, mais tres peu, sont simplement striées, ou à *côtes* peu elevées fines et longitudinales, sans même denteler les bords; mais le plus grand nombre est garnie des grandes *côtes*, & faillantes, sans nombre determinè (quoique *Linne*, dans son *Synonyme*, paroit les fixer a *quatorze*) et d'autres intermediates, qui toutes ensemble debordent et decoupent la *base* fortement. Les coquilles *coniques* ont a peu près la même structure, mais les coquilles *plattes*, & les petites qui sont *colorées*, ont les côtes generalement fort larges, faillantes et irregulieres, qui decoupent ou dentelent grandement leur *base*, et rendent leur *contours* fort irreguliers.

Les *couleurs* de l'*exterieur* ou la *robe*.—Le *sommet*, & quelque espace a l'environ, est generalement blanchatre ou jaunatre, rarement livide. Le *fond* est de couleur livide verdatre cendrée, avec des raies longitudinales jaunatres, couleur d'ambre, ou bleuatres ternes. Quelquefois les *côtes* sont, pour un petit espace en bas du sommet, *rouges* ou *rougeatres*, & encore plus

circular bands or *belts*, run round the middle of many ſhells, livid, whitiſh, yellowiſh, or of other tints, which form very agreeable variegations.

plus bas ou aux bords *blanches*, & auſſi quelquefois d'autres couleurs. Des *bandes circulaires*, ou *ceinturons*, environnent le milieu de pluſieurs ; livides, blanchatres, jaunatres, ou d'autres teintures, qui forment un effet fort agreable.

The *inſide* is ſmooth, and of a fine gloſs or poliſh, with a kind of pearly luſtre; the *centre*, or *ſeat*, is always opake, white, ambry, or ſaffron colour'd; the *ground* finely rayed, from the center to the circumference, with numerous livid bluiſh or dark brown rays, and ſaffrony ſpots, and ſometimes the margins are ſpotted alternately white and dark brown, the latter ſpots proceeding from the brown ridges on the upper ſide; but all are very beautiful.

L' *interieur*, ou le *dedans*, eſt liſſe, d'un beau luſtre ou poli nacré; le *centre* eſt toujours opaque, blanc, couleur d'ambre ou de ſafran; le *fond* elegamment rayè, du centre aux bords, par des rayes nombreuſes livides bleuatres ou brunes foncées, & de taches couleur de ſafran, quelquefois les bords ſont tachés alternativement de taches blanches & brunes foncées, ces derniers taches procedent des côtes ſur le deſſus; mais toutes ces bigarrures ſont tres belles.

Beſides the above appearances, many and very different *varieties* occur. I have ſeen *limpets*, when much toſs'd on the ſhores, or *dead ſhells*, quite milk white and opake; others, apparently *live* or *freſh ſhells*, entirely of a yellow colour; and of the ſame colour, variegated with orange ſpots or rays intermixed with the livid ones; and likewiſe with rays of brown, white, &c.

Outre ces apparences, il y a pluſieurs autres & tres differentes *varietes*. Je les ai vues, quand beaucoup roulées ſur les rivages, ou *coquilles mortes*, tout a fait blanches et opaques; d'autres, en apparence des *coquilles fraiches* ou *vives*, totalement d'une couleur jaune, & de la même couleur variée de taches ou des rayes couleur d'orange entremelées avec des raies livides, et quelquefois des raies blânches, brunes, &c.

The *ſmall beautiful flat ſhells* are called by the Engliſh, *auriculas*, and by the French, *ſoucis*, or *marygolds*, from their reſemblance to thoſe *flowers*. They indeed ſeem quite a different ſpecies, from their *flatneſs*, *ſmallneſs*, and *elegance of colours*; but as I cannot find that they have any diſtinguiſhing *character* or *criterion*; that a gradual tranſition of *colours* and *ſhapes* may be traced from the ſmalleſt to the largeſt; and that they are always found together in the *ſame places*; are reaſons ſufficient to determine

Les *petites coquilles plattes et colorées*, ſont nommées par les Anglois, des *auricula* ou *oreilles d'ours*, et par les François, *ſouci*, de leur reſſemblance a ces fleurs. Elles paroiſſent a la verite totalement une eſpece differente, a cauſe d'ètre ſi *plattes*, *petites*, et pour *l'elegance de leur couleurs*; mais je ne peus trouver aucun caractere diſtinctif; comme une tranſition graduelle de *couleurs* et *formes* ſe peut tracer de la plus petite a la plus grande; et comme elles ſe trouvent toujours enſemble dans les

determine them only *young ſhells* of the *ſpecies.*

These great differences have induced authors to propoſe ſhells of this *kind* for other different *ſpecies*. Thus, the

les *mêmes lieux*, ſont des raiſons ſuffiſantes a les determiner etre ſeulement de *jeunes coquilles* de cette *eſpece*.

Ces grandes differences ont perſuade les Auteurs a propoſer ces coquilles pour d'autres differentes *eſpeces*. Ainſi, le

1. *Patella maculoſa fere ſtriata, modo levior A.*—Liſt. Hiſt. Conchyl. tab. 535. fig. 14.
2. *Patella ſtriis rugoſis aſpera, apice, acuminato.*—Borlaſe Cornwall, p. 275. tab. 28. fig. 3.
3. *Lepas ſouci.*—D'Avila Cab. p. 84. No. 29.
4. *Auricula. Patella, integra media, & parva, eleganter colorata.*—Anonym. Conchol. p. 17. tab. III. fig. 10.
5. *Patella depreſſa.*—Pennant, Brit. Zool. IV. No. 146. tab. 89. fig. 146.

Are all *ſhells* of this *ſpecies*.

It is a very common ſhell on all the *Britiſh* and *Iriſh* coaſts. *Martin*, in his *Orkneys*, p. 225. notes that the largeſt limpets he ever ſaw were at *St. Kilda.*

In the *Orkneys* and *Ireland*, they are eat abundantly by the poor, but elſewhere they are accounted coarſe food, and uſed only by the fiſhermen for baits.

It is alſo a common ſpecies on all the *European* ſhores.

Sont toutes de *coquilles* de cette *eſpece*.

Le lepas eſt fort commun ſur toutes les côtes de la *Grande Bretagne* et *l'Irlande*. *Martin*, *Orcades*, p. 295, obſerve que les plus grands lepas qu'il a jamais vu etoient a *St. Kilda*.

Dans les *Orcades* et en *Irlande*, le menu peuple ſe nourrit frequemment de ces lepas, mais ailleurs ils ſont contè une nourriture groſſiere, & ſeulement uſés par les pêcheurs pour des amorces.

C'eſt auſſi une eſpece de coquille commune ſur toutes les côtes de l'*Europe*.

III.

P. Parva, Small. Tab. VIII. fig. 11.

Patella integra, parva, sublævis, albescens, radiis rubentibus, tab. 8. fig. 11.

The *shell* small, *size* of a pea, rather thin and semitransparent, dull, or of no gloss, of a depressed conic *shape*; the *vertex* placed much on one side, and blunt; the *base* or *aperture* large and circular, and the *margins* level and smooth.

Outside whitish, near smooth, having only a few fine whitish prominent streaks that run longitudinally, or from the top to the bottom, and is coloured with broad intermediate rays, also longitudinal, and some circular bands of a very pale or dull washy red.

Inside white and smooth, but without any gloss.

I received several from the coasts of *Dorsetshire*, but never from any other of the *British* shores; consequently they are scarce.

It is a *non-descript species*, for I cannot find it proposed by *Linné*, or any other *conchologist*.

III.

Le petit Lepas. Pl. VIII. fig. 11.

La *coquille* petite, de la *grandeur* d'un pois, plutôt mince et demitransparente, ternie, ou sans lustre, d'une *figure* conique applatie; le *sommet* est situé beaucoup vers une extremité, & obtus; la *base* ou l'*ouverture* est grande & circulaire, & les *bords* sont de niveau & unis.

L'*exterieur* est blanchatre, et presque lisse, ayant seulement quelques lignes blanches fines qui sont saillie et courent longitudinalement, ou du sommet au bord, & est coloré avec des rayes larges intermediattes, longitudinales, outre quelques bandes circulaires d'une couleur rouge lavêe ou pale.

Le *dedans* blanc & lisse, mais sans aucun lustre.

J'ai recu plusieurs de ces lepas de la côte du comté de *Dorset*, mais jamais d'aucune autre côte de la *Grande Bretagne*; consequemment elles sont rares.

Cette *espece* est une *non-decrite*, car je ne le peus trouver proposée par *Linné*, ou aucun autre *Auteur*.

IV.

P. Coeruleata, Blue rayed. Tab. I. fig. 5. 6.

IV.

Le Lepas à Lignes bleues. Pl. I. fig. 5. 6.

Patella integra, parva, lævis, cornea, cæruleis lineis elegantissime insignita, tab. 1. fig. 5. 6.

Patella minima lævis pellucida, aliquot cæruleis lineis eleganter insignita.—List. Hist. Conchyl. tab. 543. fig. 27.

Patella minor, fusca, tenuis, umbone nigro ad extremitatem anteriorem detruso, tribus inde lineis cæruleis per dorsum decurrentibus pulchre distincta.—Wallace, Orkneys, p. 41.

Small

Small oval limpet, thin and transparent, with blue lines, of the Western Isles of Scotland, Sir R. Sibbald, Phil. Transf. at Large, No. 222. art. VIII. p. 321.—And *Thin patella streaked blue,* Martin, Western Isles, p. 38.

Patella Anglica parva, prætenuis, cymbuli formis, lineis cæruleis guttatis.—Muf. Petiver, Cent. 8. p. 68. No. 725.

Patella minima lævis, ovata, pellucida, cæruleis à quatuor ad novem lineolis elegantissime insignita.—Borlase, Cornwall, p. 276. tab. 28. fig. 1.

Blue rayed limpet. Patella integra exigua Levis, cornea, cæruleis lineis insignita.—Anonym. Conchol. p. 20. tab. 4. fig. 4. superior.

P. Pellucida. Transparent.—Pennant, Brit. Zool. IV. No. 150. tab. 90. fig. 150.

Calyptra.—Klein, Ostracol. p. 118. § 292. No. 6.

Pellucida. Patella testa integerrima, obovata, gibba, pellucida, radiis quatuor cæruleis.—Linné Syst. Nat. p. 1260. No. 770.

Lepas d'eau douce demi-ovoide transparent, a trois lignes bleues.—D'Avila, cab. I. p. 428. No. 962.

These above synonyms are for the *young shells*, and the following are for the *old* ones.

Ces premiers synonimes sont pour les *jeunes coquilles*, et les suivants pour les *coquilles vielles.*

Lepas seu patella altera minor, vertice lævi, tanquam adunco, minus rugosa; et *Patella altera minor ipso vertice Levi & quasi adunco.*—List. App. ad Hist. Anim. Angl. in Goedartio suo de Insectis, p. 38. tab. 2. fig. 10. 10.

Patella Lævis fusca.—List. Hist. Conchyl. tab. 542. fig. 26. 26.

Harthpool horn limpet. Patella, cornea minor.—Petiver, Gazophyl. tab. 75. fig. 3.

Patella radiata inflato apice postulato & albescente.—Borlase, Cornwall, p. 276. tab. 28. fig. 6.

Anonym. Conchol. p. 20. tab. 4. fig. 4. inferior.—Pennant, Brit. Zool. IV. tab. 90. fig. Muta & Inferior, which is his *P. lævis*, or *Smooth*, No. 151.

Patella Levis fusca, in acutum mucronem veluti conum desinens.—Buonanni, Muf. Kircher. p. 437. No. 29. fig. 29. & Ricreazione, fig. 29.

Calyptra.—Klein, Ostracol. p. 118. § 292. No. 4. tab. 8. fig. 6. 7.

This *species* of limpet, in its states of *young* and *old*, is so extremely different, that it seems *two distinct species*; and for such, authors have both figured and described them, as may be seen by the synonyms quoted above.

Cette *espece* de lepas, dans ses etats de *jeune* & *adulte*, est si extremement differente, qu'elle paroit etre *deux especes distinctes*; & pour telles les auteurs les ont decrites & figurées, comme il se peut voir par les synonymes ci dessus cités.

Though

Quoique

Though so extremely different in their *two states*, yet the shells all preserve their *peculiar character* of the *blue lines*. Their *gradations* from the *young* state to the *old* are easily to be traced, and they are always met with in the *same places*.

Quoique si extremement differentes dans leur *deux etats*, cependant les coquilles conservent toujours leur *caractere singuliere* des *lignes bleues*. Leur *degres graduels* de *jeune* a l'état *adulte* ou *vieux*, est aisé a tracer, & elles sont toujours trouvées dans les *mêmes lieux*.

In the *young state*, the *shell* is extremely thin and horny, of a very *conic* shape, and of the *size* of a pea. The *aperture* or *base* is ovoid, the *margins* are smooth and level, and the *vertex* is sharp, and bends towards one side.

Dans l'*etat jeune*, la *coquille* est extremement mince, et comme de la corne, d'une forme tres *conique*, et de la *grandeur* d'un pois; l'*ouverture* ou la *base* est ovoide, les *bords* sont unis et de niveau, et le *sommet* est pointu et recourbé vers un côté.

The *outside* is perfectly smooth, exactly like horn, and beautifully decorated with some thin *lines* or *streaks* of a fine *bright blue* colour, generally four or five in number; but, according to Borlase, they extend to nine. These blue lines run parallel, and near each other, and tend from the vertex down the back to the margins.—The *inside* is horny and smooth.

L'*exterieur* est totalement lisse, et ressemble à de la corne; il est elegamment orné de quelques *lignes* ou *rayes* fines, d'une belle couleur *bleue claire* et *eclatante*, generalement il y a quatre ou cincq de ces lignes; mais, selon Borlase, elles s'augmentent jusques a neuf. Ces lignes bleues courent paralleles, proches l'une de l'autre, et tendent du sommet, le long du dos, aux bords.—L'*interieur* est comme de la corne et lisse.

In the *old state*, the *shell* becomes very thick, and four or five times as large, but yet retains a semi-pellucidity, and somewhat horny or ambry appearance.

Dans l'*etat vieux*, la *coquille* devient fort epaisse, et quatre ou cinq fois plus grande, mais neantmoins elle retient une demi-transparence, et quelque ressemblance a de la corne ou de l'ambre.

Some are conic, others more depressed, or near flat; all have the *blue streaks* more or less, which run very diverging or distant from each other, and besides have many transverse or circular wrinkles. The *aperture* is slightly elliptical, and though the *margins* are very smooth, yet in most they do not lie level or even.

Quelqu'unes sont coniques, d'autres ont peu de convexité, ou sont presque plattes; mais toutes ont les *lignes bleues* plus ou moins, qui sont distantes l'une de l'autre, et outre elles ont plusieurs rides circulaires ou transversales. L'*ouverture* est quelque peu elliptique, & quoique les *bords* sont fort unis, cependant dans plusieurs ils ne sont point de niveau.

The *vertex*, in these old ones, is very prominent and peaked, but lays flat-ways,

Le *sommet*, dans ces vielles coquilles, est fort saillant & pointu, mais couché,

or not erect, and at about two-thirds of the length of the ſhell.

The *outſide* is pale or livid, horny or ambry; the *vertex* is white, and the *blue lines* are not of ſo fine and bright a colour, but moſt generally are duſky.

The *inſide* is whitiſh, horny or ambry, with ſome pearlineſs, or an opal-like glare, which ſometimes is ſo fine as to reflect and play in variety of colours, like an *opal*, or like *mother of pearl*.

It is found in ſome plenty on the coaſt of *Cornwall*, as at the *Land's End*, *Falmouth Harbour*, *Whitſand Bay*, &c. On the *Dorſet* coaſt, near *Weymouth*, not ſo frequent, and are found adhering to the roots of algæ. Dr. *Liſter* and *Petiver* had their ſhells from *Hartlepool* in *Durham*, and *Huntley-Nab* in *Yorkſhire*; *Martin* and Sir *Robert Sibbald*, from the *Weſtern Iſles* of *Scotland*, particularly on the weſtern coaſt of the Iſle of *Harries*; and *Wallace* mentions them to be a ſhell of the *Orkneys*.

Linné recites them to be *Norwegian* and *Mediterranean* ſhells.

Obs. 1.—*D'Avila*, from its thinneſs, miſtakingly ranks it as a *freſh-water ſhell*.

Obs. 2.—Former conchologiſts, as *Bonanni*, &c. have denied the *bright blue* colour to be ever found in *ſhells*. *Linné* judiciouſly remarks, that this ſhell proves the axiom falſe; however, it certainly is the only inſtance againſt it, yet diſcovered.

P. Fis-

ou non erigé, et ſitué a environ les deux tiers de la longeur de la coquille.

L'*exterieur* eſt pale, livide, ou comme de la corne, ou de l'ambre; le *ſommet* eſt blanc, et les *lignes bleues* ne ſont pas d'une couleur ſi belle et eclatante, mais generalement plus obſcures.

Le *dedans* eſt blanchatre, ou comme de la corne, ou de l'ambre, avec un luſtre de nacre, ou d'opale, qui quelquefois eſt ſi beau que de reflechir et briller d'une varieté de couleurs, comme l'*opale*, ou la plus belle *nacre*.

Cette eſpece ſe trouve en quelque abondance ſur la côte de *Cornwall*, comme au *Land's End*, le *Port de Falmouth*, la *Baie de Whitſand*, &c. Sur la côte de *Dorſet*, près de *Weymouth*, mais non pas ſi communes, ou elles ſe trouvent attachées aux racines des algues. Le Dr. *Liſter* & *Petiver* les ont recus de *Hartlepool* en *Durham*, et de *Huntley-Nab* au comté de *York*; *Martin*, & le Sieur *Robert Sibbald*, des *Iſles Occidentales d'Ecoſſe*, particulierement de la côte occidentale de l'*Iſle de Harries*; et *Wallace* obſerve que c'eſt une coquille des *Orcades*.

Linné les propoſe comme des coquilles *Mediterranées* & *Norwegiennes*.

Obs. 1.—*D'Avila*, parceque la coquille eſt ſi mince, erronement la range comme une *coquille d'eau douce*.

Obs. 2.—Les conchologiſtes precedents, comme *Bonanni*, &c. ont denié que la couleur *bleue eclatante* ſe trouvoit jamais ſur les *coquilles*. *Linné* remarque la deſſus avec jugement, que cette coquille prouve que l'axiome eſt faux; cependant, il eſt certain que ceſt le ſeul exemple au contraire, encore decouvert.

Lepas

V.

P. FISSURA. SLIT.

Tab. I. fig. 4.

V.

LEPAS à ENTAILLE.

Pl. I. fig. 4.

Patella integra parva, alba, cancellata, fissura notabili in margine. Tab. 1. fig. 4.

Patella exigua alba, cancellata, fissura notabili in margine.—List. H. Conch. tab. 543. fig. 28.—Petiv. Gaz. tab. 75. fig. 2.—Anon. Conch. p. 20. tab. 4. fig. 2.

P. fissura. Slit.—Penn. Brit. Zool. No. 152. tab. 90. fig. 151.

Patella integra clathrata.—Klein, Ostrac. p. 116. § 283. No. 7.

Fissura. Patella testa ovali, striata, reticulata, vertice recurvo, antice fissa.—Lin. Syst. Nat. p. 1261. No. 778.

Lepas d'eau douce reticulé, avec une petite fente ou entaille.—D'Avila, Cab. I. p. 428. No. 962.

Lepas minima, vertice adunco, obsolete flavida, fissura notabili in margine distincta.—Martini Conch. p. 145. fig. 109. 110.

The *shell* is wholly white, and without any gloss or polish, generally thin and semitransparent, of a *conic* shape, very elevated; the *size* of a large pea. The *vertex* or *top* is not central, but large, sharp, and hooked backwards.

The *base* or *aperture* is oval; the *margins* level, and very finely crenated or notched.

The *outside* is thickly striated or ridged from the top to the margins, or longitudinally, as also transversely, so as to form a curious cancellated or lattice work. On the fore part, in the middle of the margin, is a strait regular *fent* or *fissure*, tending upwards on the shell for about one quarter of an inch, from which (as being an unerring or peculiar *character* of the *species*) it is called the *Fissura* or *Slit* Limpet.—The *inside* is smooth.

It is found in plenty on the coast of *Cornwall*; and on the coast of *Devonshire*, particularly at *Barnstable*.

La *coquille* est totalement blanche, et sans aucun lustre, generalement mince & demi-transparente, de forme *conique*, fort elevée, de la *grandeur* d'un grand pois. Le *sommet* n'est pas central, mais grand, pointu, & crochu en arriere.

La *base* ou l'*ouverture* est ovale; les *bords* de niveau, & bellement crenelés ou entaillés.

L'*exterieur* est orné des stries ou côtés fort serres, longitudinales, du sommet aux bords, comme aussi des transversales, de sorte que de former un ouvrage reticulé ou en reseau fort curieux. Au devant dans le milieu du bord se trouve une *entaille* ou *fente*, droite & reguliere, qui s'eleve environ un quart d'un pouce, de laquelle (comme etant un caractere particulier & fixe de cette espece) il se nomme le *Fissura* ou *Lepas à entaille*.—Le *dedans* est lisse.

Cette coquille se trouve en abondance sur la côte de *Cornwall*; et aussi sur la côte du comté de *Devon*, particulierement a *Barnstable*.

Linné notes this ſpecies to be found at *Algiers*; and I have ſeen very fine ones from *Falkland's Iſland*, in the *Atlantic Seas*.

Obs.—I cannot help remarking Linné's error, who ranks this *ſhell* with the *patellæ apice perforato*; neither can I paſs unnoticed *D'Avila*'s miſtake, who ranks it as a *river ſhell*.

Linné dit qu'on trouve cette eſpece a *Algiers*; et j'ai vu de tres belles coquilles de l'*Iſle de Falkland* ou *Malouine*, dans la mer *Atlantique*.

Obs.—Je ne ne peus que remarquer l'erreur de *Linné*, qui range cette *coquille* parmi ſes *patellæ apice perforato*; ni puis je paſſer ſous ſilence l'erreur de *D'Avila*, qui la range parmi les *coquilles d'eau douce*.

VI.

P. Pileus Morionis Major.

Large Fool's Cap.

Tab. I. fig. 7. 7.

VI.

Lepas Bonnet de Dragon.

Pl. I. fig. 7. 7.

Patella integra, albeſcens, ſtriata, vertice ſpirali, intus roſacea. Tab. 1. fig. 7. 7.

Patella rugoſa alba, vertice admodum adunco. The fool's cap of a pale bloſſom colour.—Borlaſe, Cornw. II. p. 276. tab. 28. fig. 4.

P. pileus morionis major. Patella integra media, albeſcens, ſtriata, vertice ſpirali, intus roſacea.—Anon. Conch. p. 25. tab. 4. fig. 18.

P. Hungarica, bonnet.—Penn. Brit. Zool. No. 147. tab. 90. fig. 147.

Patella alba ſtriata, vertice intorto.—Muſ. Kircher, p. 437. No. 23. fig. 23. & Buon. Ric. fig. 23.

Patella vertice intorto, major, toto dorſo ſerico fuſco indumento ad marginem uſque contecta, & villoſa; intus cavitate acuta, intorta, inſignita, ex albo purpureo colore maculata.—Gualt. I. Conch. tab. 9. fig. W.

Lepas cabochon et *lepas bonnet de dragon.*—Argenv. Conch. II. p. 189 & 383. tab. 2. fig. R. & tab. 1. fig. A.

Ungarica. Patella teſta integra conico acuminata ſtriata vertice hamoſo revoluto.—Lin. Syſt. Nat. p. 1259. No. 761.

Lepas bonnet de dragon.—D'Avila, Cab. I. p. 86. 87. No. 32. 34.

Lepas ſtriata, extus albida, intus roſacea, vertice adunco, cavitate ſimplici.—Martini Conch. p. 143. tab. 12. fig. 107. 108.

The *ſhell* is rather thin and ſemipellucid, when full grown, of the *ſize* of a walnut; but ſmall ones, of the ſize of filberds, are moſt commonly met with. It is covered with an *epidermis* or *cuticle*, eſpecially on the

La *coquille* eſt plûtôt mince & demi-tranſparente, la *grandeur* quand adulte eſt celle d'une noix; mais des petites coquilles, de la grandeur des noiſettes, ſont plus communes. Elle eſt couverte d'un *Epi-*

the margins, a thin and tough ſkin ſet with a pile of ſhort, thick, brown, ſilky filaments.

The *ſhape* is conic, very elevated or copped; the *baſe* or *aperture* is round; and the *margins* are not quite level, being always ſomewhat ragged or uneven.

The *vortex* turns ſpirally, and falls very low backwards, whereby the ſhell is ſo greatly fore-ſhortened on the *ſide* it hangs over, that it is near perpendicular, and only about three-eighths of an inch in length, while the other *ſide* is ſo lengthened as to meaſure near two inches.

The *outſide* has no gloſs, is whitiſh, and very thickly ſtriated with fine longitudinal furrows from the top to the margins; the ſurface is not even, but always has rugged ſpaces and depreſſions, like bruiſes, and towards the margins it is alſo wrought with circular or tranſverſe wrinkles.

The *inſide* is ſmooth, finely glazed or gloſſy white, with a beautiful roſy bluſh or carnation colour.

This ſpecies is only found on the *Corniſh coaſt*, and even is very ſcarce there, being moſt generally dredged ſome miles from the ſhore; for the ſhell is ſo thin, that it will hardly bear rolling from its native ſpot to the beach. It is generally found affixed on a ſpecies of *eſcallops*, called *frills* in Cornwall.

Linné ſays, it is alſo a ſhell of the *Mediterranean ſea.*

Epiderme, eſpecialement ſur les bords, une peau mince et coriace compoſée d'un poil de filamens, courts, gros, bruns, et comme de la ſoye.

La *forme* eſt conique, et fort elevée; la *baſe* ou l'*ouverture* eſt rond; et les *bords* ne ſont pas tout a fait de niveau, etant toujours dechirés ou inegaux.

Le *ſommet* tourne en ſpirale, & courbe ſi bas en arriere, qu'il raccourcit la coquille tant de cet *coté* qu'il eſt preſque perpendiculaire, & ſeulement environ trois huitiemes d'un pouce en longeur, pendant que l'autre *coté* eſt ſi allongé qu'il a près de deux pouces en longeur.

L'*exterieur* n'a point de luſtre, il eſt blanchatre, et ſtriè par des canelures fines longitudinales tres ſerrées ou du ſommet aux bords; la ſurface n'eſt pas unie, car elle a toujours des eſpaces raboteux et des enſoncements, comme des contuſions, et vers les bords elle eſt auſſi ridée par des rides circulaires ou tranſverſales.

Le *dedans* eſt liſſe, d'un beau luſtre blanc de lait, coloré ou teint d'une belle couleur de roſe ou d'incarnat.

Cette eſpece ſe trouve ſeulement ſur la côte de *Cornwall*, et même eſt fort rare la, etant pour la pluſpart pechées a quelques lieues de la côte; car la coquille eſt ſi tendre qu'elle ne peut ſoufrir d'être roulée de ſa place natale aux rivages de la mer. Elle eſt generalement fixée ſur une eſpece de *peigne*, nommés *frills* en Cornwall.

Linné dit que c'eſt auſſi une coquille de la *Mer Mediterranée.*

VII.

P. Larva, Reticulata. Reticulated Masque Limpet.

Tab. I. fig. 3.

VII.

Lepas *Masque (ou percè en dessus)* **Treillis.**

Pl. I. fig. 3

Patella parva cinerea, vertice perforato. Tab. 1. fig. 3.

An. Lift. H. Conch. tab. 527. fig. 2?

P. Græca. Striated.—Penn. Brit. Zool. No. 153. tab. 89. fig. 153.

Anon. Conch. tab. 7. fig. 15.

Patella clathrata.—Klein Oftrac. p. 116. § 284. No. 2.

An *Patella nimbosa.*—Lin. Syft. Nat. p. 1262. No. 781?

The *shell* is very thick for its fize. It is fmall, not much exceeding three quarters of an inch in *length*, and above half an inch *broad*, *conic* and *elevated*, above one quarter of an inch in height, and of an *oblong shape*.

The *outside* dull, wrought all over with longitudinal and tranfverfe thick prominent ridges or ribs, fo as to form a very curious deep cancellated or lattice work. The *vertex* inclines to one end, or is not central, and is *perforated* by an oblong hole, rather larger than the head of a common pin. The whole *shell* of this fide is dirty afhen brown.

Infide. The *aperture* is oblong; the *margins* level and notched or crenated within; the whole fmooth and glazed, and of a livid colour.

Several of this fpecies, about this fize, have been fifh'd up near *Weymouth*, in *Dorfetshire*, but are rare; and I could never learn that they inhabit any other of the *British coafts*.

This fpecies is pretty common in the *West Indies*.

La *coquille* eft fort epaiffe pour fa grandeur. Elle eft petite, peu plus que trois quarts d'un pouce en *longeur*, et plus d'un demi pouce en *largeur*, *conique* et *elevée*, plus d'un quart de pouce en *hauteur*, et d'une *forme oblongue*.

L'*exterieur* eft terne, & a côtes groffes & faillantes ou en vive arrete longitudinales & tranfverfales, de forte a former un ouvrage reticulé ou a treillis fort curieux & profond. Le *fommet* incline vers une extremite, ou n'eft point central, et eft *percé* d'un trou oblong, quelque chofe plus grand que la tete d'un epingle commune. Toute la *coquille* de cet côté eft de couleur brune cendrée fale.

Le *dedans.* L'*ouverture* eft oblongue; les *bords* de niveau & crenéles en dedans; le tout liffe & luftré, & d'une couleur livide.

Plufieurs de cette efpece, environ cette grandeur, ont ete pechées pres de *Weymouth*, au comté de *Dorfet*, mais elles font rares; & je n'ay jamais oui dire qu'elles etoient habitantes d'aucunes autres *côtés Britanniques*.

Cette efpece eft affés commune dans les *Indes Occidentales*.

GENUS

GENRE

GENUS II.

Haliotis, the Sea Ear.

An univalve shell, flattish, almost open, nearly simple, having only a single spire turned in at one end, and a row of orifices its length. The animal, a *Slug*.

GENRE II.

Haliotis, Oreille de Mer.

Coquille univalve, applatie, presque ouverte, a peu pres simple, ayant qu'une seule spirale tournée à une extremité, & une rangee de trous disposés sa longeur. L'animal est une *Limace*.

* MARINÆ. Sea.

VIII.

H. Vulgaris, Common.

Tab. II. fig. 1. 2.

* MARINÆ. Marines.

VIII.

Oreille de Mer commune.

Pl. II. fig. 1. 2.

Auris marina quibusdam: Patella fera Rondoletii, λεπὰς ἀγεια Aristotelis; mother of pearl, Anglice.—List. H. An. Angl. p. 167. tit. 16. tab. 3. fig. 16.

Auris marina major, latior, plurimis foraminibus eorumve vestigiis ad 40 *circiter conspicua, clavicula elata.*—List. H. Conch. tab. 611. fig. 2.

Auris marina Anglica nobis.—Grew, Mus. p. 139.—Mus. Sibbald, 130.—Mus. Peti. C. 9 & 10, p. 81. No. 801.—Anon. Conch. tab. 8. fig. 12.

Haliotis tuberculata, tuberculated.—Penn. Brit. Zool. No. 144. tab. 88. fig. 144.

Patella major. s. fera.—Gesn. Aquat. 807. 808.—Belon, Aquat. 395. *le grand bourdin*. Aldrov. Exsang. 551. fig. 1. 2.—Buon. Ric. p. 141. fig. 10. 11.—Klein Ostrac. p. 19. § 52. No. 3. *Auris striata.*—Argenv. Conch. I. p. 245. tab. 7. fig. A. et F. et II. p. 195 & 23. tab. 3. fig. A. et F. et tab. 1. fig. C. D. *Oreille marine.*

Auris marina, major profunde sulcata, magis depressa, fusco colore obsito, intus argentea.—Gualt, Ind. Conch. tab. 69. fig. I.

Haliotis striata rugosa.—Lin. Faun. Suec. I. p. 379. No. 1326. & II. No. 2198.

Tuberculata. Haliotis, testa subovata, dorso transversim rugoso tuberculato.—Lin. Syst. Nat. p. 1256. No. 741.

This *shell*, like the *limpets*, when living, adheres to the rocks. They are always covered with filth, vermiculi, balani, and other small shells; under this filth it has a filmy or thin tough blackish *epidermis* or *cuticle*.

The

Cette *coquille*, comme les *lepas*, quand vivante, s'attache aux rochers. Elles sont toujours couvertes de saletés, vermisseaux, glands de mer, et autres petites coquilles; au dessous de cette croute sale est son *epiderme* ou *peau*, mince, membraneuse, coriace & noiratre.

La

The *ſhell* is thick, ſtrong and opake, from three to four and a half inches *long*, and from two to three and a half inches *broad*, *flattiſh*, or but *little convex* on the *upper ſide*. On the *under ſide*, it is quite patulous, or *wide open*.

The *ſhape* is an oblong oval, ſomewhat like to an *human ear*, from whence it has obtained its *name*; and, like a turbinated univalve, it turns ſpirally at one end.

The *outſide* is of different ſhades, of a fine reddiſh brown colour, in a mottled or variegated manner; often theſe variegations are whitiſh, and ſometimes formed in waves of a greeniſh colour, and is very agreeable or pretty. *One ſide*, lengthwiſe, riſes gradually convex quite to the row of orifices, which forms a ridge or edge, and from thence ſlightly ſlopes down into a *ſide*, or *wall*, quite to the margin. The *whole ſhell* is wrinkled tranſverſely or acroſs, the wrinkles ſtrong and pretty deep. At *one extreme* it turns in a *ſpire*, and from this ſpire the *row of orifices* begins, which runs in a curve line the length of the ſhell to the oppoſite end, not along the middle of the ſhell, but much on one ſide; at the beginning the *holes* appear like ſmall bliſters, and gradually augment in ſize till the oppoſite margin, where they are large and open: theſe *holes* are from *thirty-five* to *forty* in number, and are all cloſed, or like bliſters, till to the latter *ſeven* or *nine*, which are large, and quite pierced or open. When the outer coat is taken off, the *ſhell* is of the ſame fine pearly ſubſtance as withinſide.

The

La *coquille* eſt epaiſſe, forte & opaque, de trois juſques a quatre et demi pouces en *longeur*, et de deux a trois et demi pouces en *largeur*, *applatie*, ou peu convexe en *deſſus*. En *deſſous* elle eſt preſque entierement *ouverte*.

La *forme* eſt ovale oblongue, quelque choſe reſſemblante a l'*oreille humaine*, d'ou elle a pris ſon *nom*; & comme une univalve contournée, elle tourne en ſpirale a une extremite.

L'*exterieur* eſt de differentes nuances, d'une belle couleur rougeatre brune, d'une maniere variolée ou variée; ſouvent ces bigarrures ſont blanchatres, & ſouvent formées en ondes d'une couleur verdatre, & eſt agreable ou fort joli. *Un côté*, en longeur, s'eleve convexe par degrè juſques a la rangée de trous, qui forme un bord ou vive arrete, & de la il va en pente a l'autre *côté*, comme une *muraille*, juſques au bord. Toute la *coquille* eſt ridée a travers, par des rides fortes & aſſés profondes. A une *extremité* elle tourne *ſpiralement*, & de cette ſpirale commence la *rangée de trous*, diſpoſes ſur une ligne courbe la longeur de la coquille juſques a l'extremité oppoſée, non pas au milieu de la coquille, mais beaucoup d'un côté; au commencement les *trous* paroiſſent comme des petites puſtules, & par degres augmentent en grandeur juſques au bord oppoſé, ou ils ſont grands & ouverts: ces *trous* ſont au nombre de *trente-cinq* a *quarante*, & touts ſont fermés, ou comme des puſtules, juſques au derniers *ſept* ou *neuf*, qui ſont grands, & tout a fait percés ou ouverts. Quand la *coquille* eſt depouillée de ſa robe, elle eſt de la même belle ſubſtance de nacre que en dedans.

Le

The *inside* is pearly, and of great ſplendor, concave, or like a diſh; the fore and end margins are quite ſharp; the back or further margin riſes high, and near ſtrait like a wall, and at its top forms a thick, ſtrong, large and flat ledge. It is this wall which is the boundary of the inſide concavity, and at the end of which the ſpire is turned, and is very viſible; the row of holes here, within-ſide, runs as it were in a ſlight furrow or gutter.

This *ſhell* is found in great abundance on the coaſts of the iſle of *Guernſey*. It is alſo found at *Seaford*, near *Newhaven*, at *Eaſtbourn*, and along that *eaſtern coaſt* of *Suſſex*; and likewiſe, according to Mr. *Pennant*, is frequently caſt upon the *ſouthern coaſts* of *Devonſhire*.

Obs.—Dr. *Liſter*, from its ſpiral turn at one end, ſays *this ſhell* is wrongly ranked as a *ſimple univalve*; he therefore places it, in his *Hiſt. Anim. Angl.* as a very wide or open *ſnail* or *cochlea*.

GENUS

Le *dedans* eſt nacre, du plus bel orient, concave, ou comme un plat; le bords de devant et de la fin ſont tout a fait tranchants; le bord exterieur eſt haut, & s'eleve preſque droit comme une muraille, et a ſon ſommet forme un bord, epais, fort, grand, & plat. Cet bord eſt la borne de la concavite interieure, et a la fin duquel la ſpirale eſt tournée, qui eſt fort viſible; & la rangée de trous ici, en dedans, eſt diſpoſée dans un leger conduit ou goutiere.

Cette *coquille* ſe trouve en grande abondance ſur les côtes de l'iſle de *Guernſey*. Elle ſe trouve auſſi a *Seaford*, pres de *Newhaven*, & a *Eaſtbourn*, & le long de cette *côte orientale* de *Suſſex*; pareillement, ſelon *Pennant*, elle eſt frequemment trouvée ſur la *côte meridionale* du comté de *Devon*.

Obs.—Le Dr. *Liſter*, a cauſe de la ſpirale qu'elle a, a une extremité, dit que cette *coquille* eſt erronnement rangée comme une *univalve ſimple*; il l'a pour cette raiſon placée dans ſon *Hiſt. Anim. Angl.* comme un *limaçon* ou *cochlea* etendu ou ouvert.

GENRE

GENUS III.

Serpula. The Worm Shell.

Univalves tubular, cylindrical, found ſingle or looſe, in cluſters and adhering to ſhells, ſtones, and other bodies, irregularly ſinuous by windings and twiſtings. The animal, a *Terebella*, Auger, or Whimble Worm.

GENRE III.

Serpula. Vermiculaire.

Univalves tubulaires, cylindriques, trouvées iſolées, en groupes & adherant a des coquilles, des pierres, et a d'autres corps, irregulierement ſinueuſes par leurs tours & entrelaſſements. L'animal eſt un *Terebella*.

* MARINÆ. Sea.

IX.

S. Vermicularis. Common.
Tab. II. fig. 5.

* MARINÆ. Marines.

IX.

Le Vermiculaire commun.
Pl. II. fig. 5.

TUBULI *in quibus vermes. Worm ſhells.*—Merret, Pin. p. 194.

Tubulus vermicularis minor oſtreorum, aliiſque teſtis adnaſcens. Small worm ſhells.—Dale Harw. p. 391. No. 1. and p. 455. No. 1.—Rutty, Dublin, I. p. 388.—Ellis, Corallines, tab. 38. fig. 2.—Anon. Conch. tab. 10. fig. 10. 13. 17. 18. & 19.

S. intricata. Complicated.—Penn. Brit. Zool. No. 157. tab. 91. fig. 157.

S. vermicularis. Worm.—Idem, Brit. Zool. No. 159. tab. 91. fig. 159.

Teſſon chargé de vermiſſeaux, et vermiſſeaux de mer.—Argenv. Conch. I. p. 84. tab. 5. fig. 1. et p. 352. tab. 29. fig. B. C. E.

Vermiſſeaux de mer.—Argenv. Conch. II. p. 197. tab. 4. fig. B. C. E. et p. 24. tab. 1. fig. L. M. N.

Concha æquilatera, in qua innumeri tubuli vermiculares nidificant, ut inde appareat quomodo hoc genus teſtaceorum omnes ſubmarinas plantas, productiones, ſaxa, & quiſquilias undique occupet. And *Tubuli marini irregulariter intorti, vermiculares, in congeriem ſimul uniti. Aliquando congeries hæc vermiculorum in talem & tantam molem coaleſcit, ut ſimilem obſervaverim, quæ tripalmarem diametrum habebat, & viginti tres libras pondere æquabat.*—Gualt. Ind. Conch. tab. 10. fig. P. T.

Dentalium teſta cylindracea inæquali flexuoſa contorta.—Lin. Faun. Suec. I. p. 380. No. 1328.

Intricata. Serpula teſta filiformi ſcabra tereti flexuoſa.—Lin. Syſt. Nat. p. 1266. No. 796.

Vermicularis. Serpula teſta tereti ſubulata curvata rugoſa.—Idem, p. 1267. No. 805.

The Cette

The *ſhell* is ſlender, rounded or *cylindric*, not tapering, but of equal *diameter*, from that of a thin packthread, through all the intermediate degrees, to that of a common writing quill.

It is of a teſtaceous ſubſtance, hard, and pretty thick, naturally *white* (if not accidentally ſoil'd) dull, or with no poliſh, generally ſmooth, except a few circular wrinkles.

The *inſide* is quite tubular, and ſomewhat glazed or gloſſy.

The *form* of theſe worm ſhells is inexpreſſibly irregular, from their various windings and twiſtings. They are found moſt generally in groups or cluſters, adhering to one another, to ſtones, ſhells, &c. *Single* or *looſe* ones are alſo met with, but ſeldom.

Theſe *ſhells* are very frequent on all the *Britiſh ſhores*.

Obs. 1.—*Linné* has augmented my *two ſpecies* of *ſerpulæ* (*this* and the *next*) into *four ſpecies*, without reaſon, as I cannot find any fixed characters to determine them as *different ſpecies*. *Pennant* has followed that celebrated author. The irregularity of the *forms* of ſerpulæ does not admit any fixed *character*, and the *work* on them is the only *arbiter*. On this *character* of the *work* I have formed my *two ſpecies*; yet, however, I cannot help obſerving, that worm ſhells of *both ſorts* are conſtantly found in the ſame cluſters or groups, without any other diſtinction; therefore even theſe are, perhaps, only a *variety*, inſtead of *a ſpecies*.

Obs. 2.—Many of the *looſe* or *ſingle ſerpulæ* are frequently found *ſpiral*, and ſome-

Cette *coquille* eſt deliée, arrondie ou *cylindrique*, & ne va point en appetiſſant, mais par tout eſt d'un *diametre* egal, des la *groſſeur* d'une ficelle, par touts les degrès intermediats, juſques a celle d'une plume commune a ecrire.

Elle eſt d'une ſubſtance teſtacée, dure, & aſſes epaiſſe, naturellement *blanche* (ſi par accident elle n'eſt point ſalie) terne, ou ſans luſtre, & generalement liſſe, excepté quelques rides circulaires.

Le *dedans* eſt tubulaire, et quelque peu luſtré.

La *forme* de ces tuyaux vermiculaires eſt extremement irreguliere, par ſes differents entortillements & ſinuoſités. Elles ſe trouvent le plus ſouvent en amas ou groupes, adherant les uns aux autres, & a des pierres, coquilles, &c. et auſſi *detachés* ou *iſoles*, mais rarement.

Ces *tuyaux vermiculaires* ſont fort communs ſur toutes les *côtes Britanniques*.

Obs. 1.—*Linné* a augmenté mes *deux eſpeces* de *ſerpulæ* (*celleci* & la *ſuivante*) a *quatre eſpeces*, ſans raiſon, car je ne peus trouver un caractere fixe a les diſtinguer comme des *eſpeces differentes*. *Pennant* a ſuivi cet auteur celebre. L'irregularité des *formes* des ſerpulæ n'admette point de *caractere* fixe, & le *travail* ſur ces coquilles eſt le ſeul *arbitre*. Sur cet *caractere* du *travail* j'ay formé mes *deux eſpeces*; cependant, je ne peus que remarquer que les tuyaux vermiculaires, des *deux ſortes*, ſont conſtamment trouvés dans les mêmes groupes ou amas, ſans aucune autre diſtinction; ainſi peut etre même ces deux ſortes ſont ſeulement *une varieté*, au lieu d'*une eſpece*.

Obs. 2.—Pluſieurs de ces *tuyaux ſolitaires* ou *detachés* ſont aſſes communement trouvés

ſometimes ſo long, regular, and perfect, that they emulate *turbinated ſhells*. The *iconical* conchologiſts, as *Gualtieri*, *Seba*, *D'Argenville*, &c. figure many ſuch, but they are only *luſi* of theſe kinds, and not *diſtinct ſpecies*. Theſe *luſi* of *ſpiral ſerpulæ* are not unfrequently met with on the *Britiſh ſhores*, with the others.

trouvés contournés en *ſpirale*, & quelquefois ſi longs, reguliers & parfaits, qu'ils reſſemblent a des *coquilles contournées en ſpirale*. Les *conchologiſtes* qui donnent des *figures*, comme *Gualtieri*, *Seba*, *D'Argenville*, &c. figurent pluſieurs tels vermiculaires, qui ſont ſeulement des *luſi*, & non des *eſpeces diſtinctes*. Ces *luſi* des *vermiculaires ſpirales* ne ſont pas rares ſur nos *côtes Britanniques*, avec les autres.

X.

S. Angulata. Angular.

Tab. II. fig. 9.

X.

Vermiculaire Angulaire.

Pl. II. fig. 9.

S. teſta tereti carinata, ſeu ſulcis elevatis exarata; *Angulata*. Tab. 2. fig. 9.

Anon. Conch. pl. xi. fig. 8. 9. 10. 16.

Serpula triquetra. Angular.—Penn. Brit. Zool. No. 156.

S. contortuplicata. Twined.—Id. No. 158. tab. 91. fig. 158.

Argenv. Conch. I. p. 352. tab. 29. fig. D. & II. p. 197. tab. 4. fig. D.

Triquetra. S. teſta repente flexuoſa triquetra.—Lin. Syſt. Nat. p. 1265. No. 795.—Faun. Suec. II. No. 2206.—Muſ. Reg. p. 698. No. 428.

Contortuplicata. S. teſta ſemitereti rugoſa glomerata carinata.—Id. p. 1266. No. 799.—Muſ. Reg. p. 698. No. 429.—Faun. Suec. II. No. 2205.

Une groſſe come chargée de vermiculaires blancs & triangulaires à côte longitudinale en vive arrête ſur le milieu du corps.—D'Avila, Cab. 1. p. 102. No. 65.

This *worm ſhell* aſſumes the ſame irregular *forms*, from its twiſtings, as the foregoing. *It* is alſo of the ſame *ſubſtance* and *colour*, and is moſt generally found intermixed with the foregoing in the ſame cluſters or groups, and adheres in like manner to ſhells, ſtones, &c. It is likewiſe found *ſpiral*, or in the *luſi* or *varieties*.

Cet *vermiculaire* prend toujours les mêmes *formes* irregulieres, cauſées par ſes ſinuoſites & entortillements, que le precedent. *Il* eſt auſſi de la même *ſubſtance* & *couleur*, & ſe trouve generalement entremelé avec le precedent dans les même groupes ou amas, & attaché de la même maniere a des coquilles, pierres, &c. Comme auſſi en forme *ſpirale*, & en *luſi* ou *variétés*.

The

Les

The *differences* are, it is most generally not so *large, thick, cylindric*, or *smooth*, and assumes a more *creeping form. It* always *tapers* and runs out to a very fine thread or extreme, and is constantly wrought with one or more *sharp prominent ridges*, not regular or perfect, that run its *length*; and by these ridges the *exterior surface* is rendered somewhat *angular*.

This *species* is alike frequent on all the *British shores*.

Obs.—This *fig.* 9. as also *fig.* 5, of the last *species*, shew both these kinds intermixed in the same groups.

Les *differences* sont, *il* n'est pas pour la plusspart si *grand, epais, cylindrique*, ou *lisse*, & s'approprie une *forme* plus rampante. *Il* diminue toujours en diametre, jusques a terminer dans une extremité comme un fil, & il est constamment travaillè en une ou plusieures *cotés* en *vive arrète*, ni regulieres ni parfaites, mais qui courent sa *longeur*; & par ces côtes la *surface exterieur* est rendue quelque chose *angulaire*.

Cette *espece* est aussi frequente sur toutes les *côtes Britanniques*.

Obs.—Cette *figure* 9. comme aussi *figure* 5, de *l'espece* precedente, montrent ces deux sortes entremelées dans les mêmes groupes ou amas.

XI.

S. Teredo. The Ship Worm.

Vermiculaire perceur, ou de Navire.

Serpula testa cylindracea flexuosa, lignum perforans. Teredo.

Anon. Conch. tab. 10. fig. 1. & 4.

Teredo navalis. Ship.—Penn. Brit. Zool. No. 160.

Rousset, *Observations sur les vers de mer, qui percent les vaisseaux, &c.* La Haye, 1733, avec fig. 8vo.

Dentalium testa membranacea cylindracea, ligno inserta.—Lin. Faun. Suec. I. p. 380. No. 1329.—& *Teredo intra lignum testa flexuosa.* Id. II. No. 2087.

Navalis. Teredo.—Lin. Syst. Nat. p. 1267. No. 807.

The *ship worm* is always *cylindrick*, and of a flexuous or waved *form*, but never cluster'd or twisted; for they are always single or seperate, each in its cell. *It* is *large* and *thick*, the most general *diameter* that of the little finger; but the *shell* is thin, smooth, or without any wrinkles or striæ, *exteriorly* white and dullish, but *interiorly* has a glaze or polish.

Le *vermiculaire perceur*, ou *de navire*, est toujours *cylindrique*, & d'une *forme* ondée ou en courbures, mais jamais entortillè ou en groupe; car ils se gardent seuls & separès, chacun dans sa cellule. *Il* est *grand* & *gros*, du *diametre* en général du petit doigt; mais la *coquille* est mince, unie, ou sans stries ou rides, *exterieurement* blanche & ternie, mais *interieurement* elle est lustrée ou polie.

The

Sa

The nature and effects of its piercing and eroding the timber of ſhipping, &c. is too fatally experienced.

I have not given a figure of this ſpecies, as its anatomy and entire natural hiſtory are ſo accurately ſet forth, and illuſtrated with figures, by *G. Sellius*, in his *Hiſtoria Naturalis Teredinis*, ſeu, *Xylophagi Marini*, printed at *Utrecht* in 1733, in *quarto*.

This *animal* is rather to be accounted an *exotic*, and imported to us from *other ſeas*.

Sa nature & effets de penetrer & ronger le bois de charpente ou de navire, ſont malheureuſement trop connus.

Je n'ai pas donné une figure de cette eſpece, comme l'anatomie & l'hiſtoire naturelle complete de ce ver, ſont ſi ſoigneuſement decrites, & embellies des tailles douces, par *G. Sellius*, dans ſon *Hiſtoria Naturalis Teredinis*, ſeu, *Xyphologi Marini*, imprimée a *Utrecht*, en 1733, en *quarto*.

Cet *animal* doit plutôt être contè comme *etranger*, & apporté des autres *parages*.

XII.

S. Spirorbis. Spiral.

Tab. II. fig. 11.

XII.

Vermiculaire Nautiloide.

Pl. II. fig. 11.

S. parva orbiculata & ſpirali, ammoniæ inſtar convoluta. Spirorbis. Tab. 2. fig. 11.

Vermiculus exiguus albus nautiloides, algæ fere adnaſcens.—Liſt. H. Conch. tab. 533. fig. 5. but in *Huddesford's edition* tranſpoſed to tab. 553.

Planorbis minimus, algis frequenter adnaſcens. Wrack Spangle.—Petiv. Gaz. tab. 35. fig. 8.

Very ſmall worm ſhells.—Dale, Harw. p. 391. No. 2. & p. 455. No. 2.

Depreſſed orbicular cochleæ on algæ.—Wallis, Northumb. I. p. 402. No. 41.

Anon. Conch. Tab. 11. fig. 6. 6.

Serpula ſpirorbis. Spiral.—Penn. Brit. Zool. No. 155. tab. 91. fig. 155.

Tubulus marinus irregulariter intortus, vermicularis, minimus, rugoſus, ammoniæ inſtar convolutus, plantis ſubmarinis adhærens, candidus.—Gualt. I. Conch. tab. 10. fig. O.

Spirorbis. Serpula teſta regulari, ſpirali, orbiculata, anfractibus, ſupra introrſum ſubcanaliculatis ſenſimque minoribus.—Lin. Syſt. Nat. p. 1265. No. 794.—Faun. Suec. II. No. 2204.

This *ſpecies* is very *ſmall*, and like a ſpangle, the largeſt not meaſuring above one-eighth of an inch over, *flat*, *ſpiral* on the ſame level or *helical*, like a cornu ammonis, and made up of two *wreaths* only; the

Cette *eſpece* eſt fort *petite*, & comme une paillette, la plus grande n'ayant au deſſus du huitieme d'un pouce de diametre, *platte*, contournée en *ſpirale* ſur un même plan ou en *helix*, comme une corne d'am-

the *wreaths* are smooth, very cylindrick or convex, and contiguous, but the *center* is hollow or umbilicated.

The *under side*, as they always adhere to other bodies, is flat by the adhesion.

The *shell* is strong and compact for the size, *white* and dull, or with no polish. *Each* is single or insulated, and perfect; for they are never complicated, joined, or laid on one another, but are set singly, and dispersed all over the leaves of the wracks, and other submarine plants, like so many spangles.

This *species* is found in abundance on most of our *British shores*, generally sticking to the sea plants, sometimes on shells and other bodies.

Obs.—Dr. *Lister*, in his original edition, rank'd this *shell* among the *worm shells*, (tab. 533. fig. 5.) calling it *Nautiloides*, only from its wreath'd form like to a *nautilus*; but his re-editor, the Reverend Mr. *Huddesford*, has been pleased to reverse the Doctor's *arrangement*, by transposing it to the *Nautilus family*, where it now is, (tab. 553.) and thereby fixes an error of arrangement on Dr. *Lister*'s memory, which that *excellent* and *accurate conchologist* was not guilty of.

d'ammon, & seulement a deux *replis* ou *revolutions*; ces *replis* sont lisses, fort cylindriques ou convexes, & contigus, mais le *centre* est vuide ou umbiliquè.

Le *dessous*, comme ces vermiculaires adherént toujours a d'autres corps, est plat par ce moyen.

La *coquille* est forte & compacte pour sa grandeur, *blanche*, ternie ou sans poli. *Chacune* est seule ou isolée, & parfaite; car elles ne sont jamais entrelassées, jointes, ou posées l'une sur l'autre, mais detachées, & dispersées par tout les feuilles de l'algue, ou autres plantes marines, comme autant des paillettes.

Cette *espece* se trouve en abondance sur toutes nos *côtes Britanniques*, plus communement attachée aux plantes marines, quelquefois aussi sur des coquilles & autres corps.

Obs.—Le Dr. Lister, dans l'edition originale, a rangé cette *coquille* entre ses *vermiculaires*, (pl. 533. fig. 5.) la nommant *Nautiloides*, seulement, a cause de sa forme contournèe comme un *nautile*; mais son re-editeur, le Rev. Monf. Huddesford, a plu de renverser l'*arrangement* du Docteur, en la transposant a la *famille* des *Nautiles*, ou elle se trouve a present, (sçavoir pl. 553.) & de cette maniere il fixe une erreur d'arrangement sur la memoire du Dr. Lister, dont cet *excellent* & *exact conchologiste* n'etoit pas coupable.

GENUS

GENRE

GENUS IV.

DENTALE, the TOOTH SHELL.

Univalve simple, or not spiral, slender, tubular, of a determinate curved conical shape, and open at both ends. The animal, an *Auger* or *Whimble Worm*.

GENRE IV.

LE DENTALE.

Univalve simple, ou point contournée en spirale, deliée, tubulaire, d'une forme determinée, conique & courbe, & ouverte aux deux extremités. L'animal est un *Terebella*.

* MARINÆ. SEA.

XIII.

DENTALE VULGARE. COMMON.

Tab. II. fig. 10.

* MARINÆ. DE MER.

XIII.

DENTALE COMMUN.

Pl. II. fig. 10.

DENTALE *læve albescens. Vulgare.* Tab. II. fig. 10.

Dentale læve album, altera extremitate rufescens.—List. H. Conch. tab. 547. fig. 2. Grew. Mus. p. 145.—Petiv. Gaz. tab. 65. fig. 9.—Anon. Conch. tab. 12. fig. 3. 4.

Dentale Leve, curvum album.—Borlase Cornw. p. 276. tab. 28. fig. 5.

Dentale, entalis, common.—Penn. Brit. Zool. No. 154. tab. 90. fig. 154.

Dentales, seu *Antales.*—Gesn. Aquat. 348.—Rondel. Pisc. 110.—Buon. Ricr. p. 141. fig. 9.

Tubulus marinus regulariter intortus, arcuatim incurvatus, & versus unam extremitatem acuminatus, dentalis dictus, lævis, candidus.—Gualt. I. Conch. tab. 10. fig. E.

Antales.—Argenv. Conch. I. p. 246. tab. 7. fig. K. II. p. 196. tab. 3. fig. K.

Dentalium testa subcylindracea lævi obliqua, hinc angustiore.—Lin. Faun. Suec. I. p. 379. No. 1327. II. No. 2201.

Entalis. Dentale testa tereti subarcuata, continua lævi.—Lin. S. N. p. 1263. No. 786. —Mus. Reg. p. 697. No. 427.

The *shell* is of a tapering, curved and conical *shape*, quite *tubular*, and *open* at both ends, from one and a half inch to two inches *long*, and somewhat above one eighth of an inch in *diameter*, at the top or

La *coquille* est d'une *forme* conique, courbée, & qui s'etrecit vers une extremité, totalement *tubulaire*, & *ouverte* aux deux bouts, d'un pouce & demi a deux pouces en *longeur*, & au dela d'un pouce en

or broadeſt part; from whence it tapers ſo much to the other end, as not to meaſure one third of that *girth*.

It is *ſtrong* and rather thick, *ſmooth*, ſomewhat gloſſy, and, when *live* or freſh, *white*, with a waſh or tinge of yellowiſh brown, eſpecially towards the lower end, or taper extremity. The *inſide* is white and gloſſy.

This *ſpecies* is found in plenty on many of the Britiſh ſhores, as the *Scilly iſlands*, *Cornwall*, *Devonſhire*, *Hampſhire*, &c.

en *diametre*, ou elle eſt plus large, & de là appetiſſe tant vers l'autre extremite, qu'elle ne meſure pas la troiſieme partie de cette *groſſeur*.

Elle eſt *forte* & plutôt epaiſſe, *liſſe*, quelque peu luſtrée, & quand *vivante*, *blanche*, avec un teint de brun jaunatre, eſpecialement vers le bout menu. Le *dedans* eſt blanc & luſtré.

Cette *eſpece* ſe trouve en abondance ſur pluſieurs de nos côtes Britanniques, comme aux *Iſles de Scilly*, en *Cornwall*, et aux comtes de *Devon*, *Hants*, &c.

PART

E

PART II.

Revolved Univalves, Univalvia involuta.

Univalves whose spires are latent within the body of the shell, and do not appear externally, so that they have no turban.

SECONDE PARTIE.

Univalves contournées en dedans.

Univalves roulées, ou dont les revolutions sont interieures ou cachées dans le corps de la coquille, ainsi qu'elles ne paroissent pas a l'exterieur, & n'ont point une clavicule.

GENUS V.

Bulla, the Dipper.

Univalve suboval, a little convoluted and umbilicated at the base, the aperture or mouth its length, very open or patulous, and the lips smooth or even. The animal, a *Slug*.

GENRE V.

Bulla, la Bulle.

Univalve a peu pres ovale, un peu contourné & umbiliquée a la base; l'ouverture ou la bouche, qui court sa longeur, est fort grande ou evasée, avec les levres lisses & unies. L'animal est une *Limace*.

* MARINÆ. Sea.

XIV.

Lignaria. Wood.

Tab. I. fig. 9.

* De MER.

XIV.

L'Oublie.

Pl. I. fig. 9.

B*ULLA major, leviter et dense tranverse striata. Lignaria.* Tab. 1. fig. 9.

Concha veneris major, leviter et dense striata.—List. H. Conch. tab. 714, fig. 71.

Concha veneris major lævis, lineis luteis et albis dense distincta, apertura longe inæqualis The larger striped concha veneris.—Borlase Cornw. p. 277. tab. 28. fig. 14.

Bulla lignaria, Wood.—Penn. Brit. Zool. No. 83. tab. 70. fig. 83.

Buon. Ricr. p. 175. fig. 3.

Bulla.

Bulla.—Klein, Ostrac. p. 82. § 222. No. 3.

Lignaria. Bulla testa obovata oblongiuscula transverse striata, vertice subumbilicato.—Lin. S. N. p. 1184. No. 379.

Oublie, ou papier roulé, tonne a bouche entiere.—D'Avila, cab. p. 206. No. 387.

The *size* of this *species*, at a medium, is from one and a half inch to two inches *long*, about an inch *broad* at the top, and about half an inch at the bottom or base, of an *oblong oval shape*, greatly narrowing towards the base or lower extremity, and very swell'd or ventricose.

The *shell* is thin, brittle, and semiperspicuous.

The *outside* has no gloss, and is finely and thickly *striated* across or transversely, of a pleasing light yellowish brown *colour*, thick set with white veins, which answer to the striæ, and much resembles a piece of very vein'd wood, whence *it* has obtained its *Latin* and *English trivial* names.

The *Inside.* The *body* of the shell on this side is an ovoid, about double the *size* of an olive stone, of the same *colour*, &c. as the back, which involutes or revolves within, but so loosely or open, that the *wreath* answering the *pillar* or *columella* is distinctly seen from the outside. The rest is taken up by the *aperture*, which is very patulous and *entire:* on the upper part it rises very high above the body, full half the length of the shell, is wide and arched, and from thence forms only a long and more narrow opening to the lower end. It is white, glossy, and striated, exactly answering to the outside striæ.

La *grandeur* de cette *espece*, à un milieu, est un pouce & demi a deux pouces en *longeur*, environ un pouce en *largeur* au haut, et environ un demi pouce au bas ou a la base, d'une *forme oblongue ovale*, se retrocissant beaucoup vers la base ou extremité inferieure, et tres bombée ou ventrue.

La *coquille* est mince, fragile, & demi-transparente.

L'*exterieur* n'a point de lustre, il est bellement *strié* transversalement d'une maniere tres serré, d'une *couleur* agreable jaunatre claire brune, avec des veines blanches tres serrées, qui repondent aux stries, & ressemble beaucoup a un morceau de bois fort veiné, d'ou cette *coquille* a obtenue ses *noms de guerre Latin* & *Anglois*.

Le *dedans.* Le *corps* de la coquille de ce coté est une ovoïde, environ deux fois la *grandeur* d'un noyau d'olive, de la même couleur, &c. que l'exterieur, qui fait ses revolutions interieurement ou en dedans, mais si lachement ou peu serrè, que la *revolution* qui repond a la *columelle*, est tres visible de l'exterieur. La reste est occupè par l'*ouverture* ou la *bouche*, qui est fort evasée & *entiere:* a la partie superieure elle s'eleve fort haut au dessus du corps, pleinement la moitie de la coquille, est fort large & en voute, & de là forme seulement une ouverture longue & etroite jusques a la partie inferieure. Elle est blanche, lustrée, & striée, exactement conforme aux stries exterieures.

The *outer lip* is thin and even from the turn of the arch to the body. On the *inner lip* it has a thick white gloſſy edge, which enlarges or ſpreads as the body involutes, and forms a white gloſſy and broad *pillar lip*.

The *baſe* is thick, for the ſhell thickens towards it, obtuſe, ſomewhat convoluted but not umbilicated, white and gloſſy.

This *ſpecies* is found on ſome of the coaſts of *Ireland*, and in *Cornwall*, but not very frequent. It is alſo ſometimes fiſhed up on the coaſts of *Devon* and *Dorſet*.

La *levre exterieure* eſt mince & unie dès le tour de la voute juſques au corps. Sur la *levre interieure* il y a un bord epais, blanc et luſtré, qui s'elargit ou s'etend a proportion que le corps fait ſes revolutions interieures, & forme la *levre* de la *columelle*, qui eſt blanche, luſtrée & large.

La *baſe* eſt epaiſſe, car la coquille s'epaiſſit vers là; elle eſt obtuſe, quelque peu contournée en ſpirale, mais non umbiliquée, blanche & luſtrée.

Cette *eſpece* ſe trouve ſur quelques côtes de l'*Irlande*, et en *Cornwall*, mais non pas frequemment. Elle ſe trouve auſſi quelquefois peché ſur les côtes des comtés de *Devon* et de *Dorſet*.

XV.

NAVICULA. PINNACE.

Tab. I. fig. 10.

XV.

BULLE GONDOLE.

Pl. I. fig. 10.

Bulla ovalis, fragilis et pellucida. Navicula. Tab. 1. fig. 10.

Bulla. Ampulla, obtuſe.—Penn. Brit. Zool. No. 84.

Nux marina umbilicata, minutiſſime per longitudinem ſtriata, ſubrotunda, ore admodum patulo, tenuis, fragilis, candida.—Gualt. I. Conch. tab. 13. fig. D D.

Hydatis. Bulla teſta rotundata, pellucida longitudinaliter ſubſtriata, vertice umbilicato.—Lin. S. N. p. 1183. No. 377.

Bulles d'eau blanches, papyracées. Tonnes à bouche entiere.—D'Avila Cab. p. 207. No. 389.

This *ſhell* is oval, exactly of the *ſhape* of the egg of a wren or other ſmall bird, and of the *ſize* of a filberd. It is extremely *thin*, very *brittle*, and *tranſparent*, but not pellucid or clear, being like water a little dirtied, not gloſſy, and near *ſmooth*, having only ſome few ſlight longitudinal wrinkles, and extreme fine tranſverſe ſtriæ, hardly perceivable.

It

Cette *coquille* eſt ovale, exactement de la *forme* de l'œuf d'un roitelet ou autre petit oiſeau, et de la *grandeur* d'une noiſette. Elle eſt extremement *mince*, fort *fragile*, et *tranſparente*, mais non pas parfaitement claire, etant comme de l'eau un peu ſale, point luſtreé, et preſque *liſſe*, ayant ſeulement quelques rides legeres longitudinales, et quelques ſtries tranſverſales, ſi fines, qu'elles ſont a peine viſibles.

Elle

It is very convex or swell'd on all sides. The *structure* the same as the foregoing, but does not narrow so much at one end. The *body* is large and swell'd, and of consequence the *aperture* is not so wide.

The *aperture*, on the upper part, does not rise much higher than the body, but is arched and *entire*; at the bottom it reaches somewhat lower, or beyond it. The *outer lip* is even; the *inner* and *pillar lips* the same, and have no glossy thick edge, as the foregoing.

The *base*, or lower end of the body, is strongly *umbilicated*.

These described *shells*, and also that of Pennant, and others I have seen, were all fish'd up at and near *Weymouth*, in *Dorsetshire*; but I never heard of them on any other British coast, consequently it is rare in our seas.

Neither Gualtieri or D'Avila mention their *natal* place; but Linné says they are *natives* of the *Mediterranean sea*, and notes it to be often, only of the *size* of a small pea.

Obs.—*Pennant* is extremely erroneous in surmising it to be but a young shell of the *B. Ampulla*, and in quoting that *Bulla* of Linné for this shell.

Elle est fort convexe ou ventrue de tous côtés. La *structure* est la même que la precedente, mais elle ne s'etrecit pas tant vers une extremité. Le *corps* est grand et bombée, & par consequence l'*ouverture* n'est pas si evalée.

L'*ouverture*, ou la *bouche*, ne s'eleve pas beaucoup plus haut que le corps, mais elle est en voute et *entiere*; au bas elle deborde quelque chose au de là du corps. La *levre exterieure* est unie; la *levre interieure* & de *la columelle* la même, & elles n'ont point le bords epais & lustrè, comme la precedente.

La *base*, ou bout du corps, est fortement *umbiliquée*.

Ces *coquilles* decrites, comme aussi celle de Pennant, & autres que j'ay vues, etoient toutes pechées a et près de *Weymouth*, au comté de *Dorset*; & je n'ai jamais entendu dire qu'elles se trouvoient sur aucunes autres côtes Britanniques, par consequent elle est rare dans nos mers.

Ni Gualtieri ou D'Avila font mention de leur lieu *natal*; mais Linné dit qu'elles sont *natives* de la *mer Mediterranée*, & remarque qu'elles sont souvent seulement de la *grandeur* d'un petit pois.

Obs.—*Pennant* erre extremement en la supposant etre seulement une jeune coquille du *B. Ampulla*, et en citant la même *Bulle* de Linné pour cette coquille.

XVI.

BULLA, the BUBBLE.

Tab. II. fig. 3.

XVI.

BULLE D'EAU.

Pl. II. fig. 3.

Bulla pellucida, fragilissima, tota hians, s. apertura amplissima. Tab. 2. fig. 3.

Bulla, patula. Open.—Penn. Brit. Zool. No. 85. tab. 70. fig. 85. A.

Nux marina transversim minutissime striata, ore omnium amplissimo, tenuissima, fragilissima, pellucida, candida.—Gualt. I. Conch. tab. 13. fig. E E.

Aperta. Bulla testa subrotunda pellucida, transversim substriata, tota hians.—Lin. S. N. p. 1183. No. 376.

Oublies blanches papyracées. Tonnes a bouche entiere.—D'Avila, cab. I. p. 207. No. 389.

The *shell* of the *size* of a filberd, somewhat glossy, brittle, extremely thin, quite pellucid or clear, and exactly resembles a *bubble* or *bladder of water*; nearly globose, but narrows much towards the lower end. The *outside* has some fine, almost imperceptible, longitudinal wrinkles, and the *inside* is set with like wrinkles.

The *aperture* is so extremely large, that the whole shell almost lies open to view. the *contour* of it is somewhat *oval*, the upper part very dilated, the lower narrow. The *body* resembles a small oval bead, scarcely measuring one quarter of an inch, and is like a flap just turned inwards, for it does not involute more than in one *single turn*; so that, properly speaking, it has neither *wreaths* nor *piller*, for the turn is so wide or slack that it is quite conspicuous to view, and thereby the whole *shell* seems almost *open* and *concave*. The *aperture* is entire; the upper part rises near two thirds above the body, and extends about one quarter of an inch below it. The *lips* are even; the *base* or end of the body has a slight wreath, but is not umbilicated.

All

La *coquille* est de la *grandeur* d'une noisette, quelque peu lustrée, fragile, extremement mince, tout a fait transparente ou claire, & ressemble exactement a une *bouteille d'eau*; presque ronde, mais s'etrecit beaucoup vers l'extremité inferieure. Le *dehors* à quelques rides fines longitudinales, à peine visibles, & le *dedans* est aussi garni de rides semblables.

L'*ouverture* est si evasée, que la coquille entiere reste presque ouverte a la vue; le *contour* est quelque chose *ovale*, la partie superieure fort evasée, l'inferieure etrecie. Le *corps* resemble a un petit grain de chapelet ovale, a peine de la longeur d'un quart de pouce, comme une oreille un peu contournè en dedans, car il ne contourne plus que dans une *seule revolution*; ainsi, a proprement parler, il n'a ni *revolutions*, ni *columelle*, car cette revolution est si lache ou large, qu'elle est tres visible, & par cet moyen toute la *coquille* paroit presque *ouverte* & *concave*. L'*ouverture* est entiere; la partie superieure s'eleve près de deux tiers au dessus du corps, & s'etend environ un quart de pouce plus bas. Les *levres* sont unies. La *base*, ou la *fin* du corps, a une revolution legere, mais elle n'est point umbiliquée.

Toutes

All the *ſhells* of this ſpecies, I know, were fiſh'd up near *Weymouth* in *Dorſetſhire*, and not any where elſe on the Britiſh coaſts; they are even not frequent, ſo that it ſeems a *rare*, as well as a *curious* ſhell.

Neither Gualtieri or D'Avila mention their *natal* places. Linné had his from the *Cape of Good Hope*. Said author *queries* whether it be a mere *variety* of his *bulla naucum*. It certainly is *not*, but a *diſtinct ſpecies*. I am inclined to believe it a ſhell, inhabitant of the *Mediterranean ſea*.

Toutes les *coquilles* de cette eſpece, que je connois, furent pechées près de *Weymouth*, au comté de *Dorſet*, & non en autre lieu des côtes Britanniques; elles ne ſont même pas communes, ainſi qu'elle paroit etre une coquille auſſi *rare* que *curieuſe*.

Ni Gualteri ou D'Avila font mention de leur lieu *natal*. Linné l'a reçu du *Cap de bonne Eſperance*. Le dit auteur fait queſtion, ſi ce n'eſt une ſimple *varieté* de ſon *bulla naucum*. Certainement elle n'eſt pas, mais une *eſpece diſtincte*. J'incline a croire que c'eſt une coquille, habitante de la *mer Mediterranée*.

XVII.

Cylindracea, Cylindric.

Tab. II. fig. 7. 7.

XVII.

Bulle blanche.

Pl. II. fig. 7. 7.

Bulla exigua cylindracea, lævis et nivea. Tab. 2. fig. 7.

Concha veneris, exigua, alba, vere cylindracea.—Liſt. H. Conch. tab. 714. fig. 70.

Bulla, cylindracea. Cylindric.—Penn. Brit. Zool. No. 85. tab. 70. fig. 85.

Oliva cylindrica, vel nucleus olivæ.—Klein, Oſtrac. p. 83. § 224. No. 5. tab. 5. fig. 99. a. b.

Porcellana integra admodum tenuis, fimbriata; dorſo pulvinato, candidiſſima.—Gualt. I. Conch. tab. 15. fig. 4.

An Spelta.—Lin. S. N. p. 1182. No. 372?

Pallida. Bulla teſta cylindrica, ſpira elevata acuta.—Lin. Muſ. Reg. p. 588. No. [illegible]

Pallida. Voluta teſta integra oblongo ovata, ſpira elevata, columella quadruplicata.—[illegible] S. N. p. 1189. No. 405.

The *ſhell* of a cylindric *form*, moſt general *ſize* that of a grain of wheat, but bulkier, and often of double or treble that bigneſs, ſtrong but rather thin, quite ſmooth, of a pure or milk white *colour*, and

La *coquille* eſt de *forme* cylindrique, la *grandeur* la plus generale celle d'un grain de froment, mais plus gros, et ſouvent de double ou triple cette grandeur, forte, mais plutot mince, totalement liſſe, d'une belle

and of an exquisite polish. The *inside* is also smooth and white, but not glossy.

The *aperture* is wide at top, arched, and entire; it thence narrows to the bottom, not linear, but continues pretty wide.

The *outer lip* is even and smooth; but the *inner* or *pillar lip*, at the beginning of the *pillar*, is wrought with four strong prominent pleats or wrinkles.

The *base*, or end of the body, extends somewhat below the aperture, is obtuse, and wrought with two faint wreaths.

This *shell* is found, but not in abundance, on the *western coasts* of England. Lister notes them from *Barnstaple* in *Devonshire*; and I have received them from *Cornwall*, and from about *Weymouth* on the *coast* of *Dorset*.

This *species* inhabits the *Mediterranean sea*, and is not unfrequent on the coasts of *Italy*.

GENUS

belle *couleur* blanche de lait, et d'un poli exquis. L'*interieur* est aussi lisse & blanche, mais point lustrè.

L'*ouverture* est evasée au haut, en voute, et entiere; de là elle s'etrecit jusques au bas, non pas etroite, mais continue asses large.

La *levre exterieure* est unie et lisse; mais l'*interieure*, ou de la *columelle*, au commencement de la *columelle*, est garnie de quatre rides ou plis forts & saillants.

La *base*, ou la fin du corps, deborde quelque chose au delà de l'ouverture, est obtuse, et a deux revolutions legeres.

Cette *coquille* se trouve, en abondance, sur les *côtes occidentales* de l'Angleterre. Lister les notifie de *Barnstaple*, au comté de *Devon*; & je les ai receu de *Cornwall*, & des environs de *Weymouth* sur le côte de *Dorset*.

Cette *espece* est habitante de la *Méditerranée*, et se trouve asses frequemment sur les côtes de l'*Italie*.

GENRE

GENUS VI.

CYPRÆA, the COWRY.

Univalve revolved or involuted, hemispherical, the aperture, on the flat part, the length of the shell, very narrow or linear, with both the lips tooth'd. The animal, a *Slug*.

GENRE VI.

CYPRÆA, la PORCELAINE.

Univalve roulée ou contournée interieurement, hemispherique; l'ouverture, sur la partie applatie, la longeur de là coquille, fort etroite ou en forme de fente, avec lès deux lèvres dentées. L'animal est une *Limace*.

* MARINÆ. SEA.

XVIII.

PEDICULUS, seu MONACHA.
The SEA LOUSE or NUN.
Tab. II. fig. 6. 6.

* MARINÆ. DE MER.

XVIII.

PORCELAINE POU DE MER.
Pl. II. fig. 6. 6.

CYPRÆA exigua transversim striata, maculæ fuscæ dorso inspersa. Pediculus seu Monacha. Tab. 2. fig. 6. 6.

Concha veneris exigua, alba, striata. Nuns.—List. H. An. Angl. p. 168. tit. 17. tab. 3. fig. 17.

Concha veneris exigua, striata, leviter admodum rufescens, in summo dorso integro, maculæ rufescentes.—List. H. Conch. tab. 707. fig. 57.

Grew. Mus. p. 138.—Mus. Petiv. p. 5. No. 17.—Leigh, Lancash. tab. 3. fig. 13. —Wallace, Orkn. p. 41.—Wallis, Northumb. p. 402. No. 40.

Concha veneris exigua purpurascens, striis minimis transversis, tribus maculis fuscis dorso inspersa. The purple spotted nuns, alias *cowrie.* And

Concha veneris minima nullis maculis insignita. The smallest nuns without spots.—Borlase Cornw. p. 277. tab. 28. fig. 12. 13.

Cypræa pediculus, common.—Penn. Brit. Zool. No. 82. tab. 70. fig. 82.

Porcellana elatior, pediculus.—Klein, Ostrac. p. 85. § 230. No. 13. c.

Pou de mer.—Argenv. Conch. I. p. 310. tab. 21. fig. L. II. p. 270. tab. 18. fig. L. and D'Avila, cab. p. 275.

Porcellana vulgaris, parva, globosa, striata, candida, dorso sinuato.—Gualt. I. Conch. tab. 14. fig. P. & tab. 15. fig. R.

Pediculus. Cypræa testa marginata transversim sulcata.—Lin. S. N. p. 1180. No. 364. —Mus. Reg. p. 582. No. 211.

F The La

The *shell* is thick and strong, very convex, about the *size* of a horse bean, striated circularly or across the back; the striæ fine, and like prominent threads, and run quite to the lips of the aperture on the under part; some of them are often furcated, or inosculate into one another.

La *coquille* est epaisse et forte, tres bombée, environ la *grandeur* d'une feve de cheval, striée circulairement ou a travers du dos; les stries sont fines, et comme des fils saillants, & courent jusques aux levres de l'ouverture en dessous: quelques unes d'eux sont fourchues, ou s'entrelassent l'une dans l'autre.

The *aperture* linear, both the *lips* dentated, and the *outer* one rises in a thick high border on the back.

L'*ouverture* est en forme de fente, les deux *levres* sont dentées, & l'*exterieure* s'eleve sur le dos dans un haut & gros bourrelet.

In *colour* it varies much; many are whitish, and others washy red or pale flesh, without any spots; but more generally they are of a pale red, with a tint of ashen, and have three spots in a row on the top or summit of the back. These *spots* are of a dark or deep ash colour, irregularly roundish, and the middle one is the largest.

En *couleurs* elle varie beaucoup; plusieurs sont blanchatres, autres rouge lavé ou de chair pale, sans aucunes taches; mais en general elles sont rouge pale, avec un teint cendré, & ont trois taches dans une rangée sur le haut ou sommet du dos. Ces *taches* sont d'une couleur obscure ou cendre foncée, irregulierement rondes, & celle du milieu est la plus grande.

It is a common *shell* on most of our shores, as *Yorkshire, Northumberland, Cheshire, Cornwall, Devonshire*, and *Sussex*, &c. also on the shores of *Scotland*; and, according to Wallace, are vulgarly called *John-a-Groat's buckies* in the *Orkneys*.

C'est une *coquille* tres commune sur la plupart de nos côtes, comme aux comtés de *York, Northumberland, Cheshire, Cornwall, Devon*, & *Sussex*, &c. aussi sur les côtes de l'*Ecosse*; &, selon Wallace, elles sont nommées vulgairement aux *Orcades John-a-Groat buckies*.

Obs. 1.—There is a *shell*, only a *variety* of the species, found in great plenty in the West Indies, and other parts, which differs from this in a *single circumstance*, *viz.* it has a furrow along the middle of the back, which this *British* kind is destitute of.

Obs. 1.—Il y a une *coquille*, une *varieté* seulement de cette espece, trouvée en tres grande abondance aux Indes Occidentales, & autres pais, qui differe de celle ci dans une *seule circonstance*, sçavoir, elle à un sillon longitudinal le long du milieu du dos, que la *Britannique* n'a pas.

Obs. 2.—Linné errs, in saying the *English* shells are only white, and without spots.

Obs. 2.—Linné se meprend, en disant que les coquilles *Angloises* sont seulement blanches, & sans taches.

PART III.

TURBINATED UNIVALVES.

UNIVALVIA TURBINATA.

Univalves ſpiral, whoſe ſpires are external and viſible, and form a turban.

TROISIEME PARTIE.

UNIVALVIA TURBINATA.

UNIVALVES CONTOURNÉES EN SPIRALE.

Univalves ſpirales, dont les revolutions ſont exterieures & viſibles, & forment une clavicule.

GENUS VII.

TROCHUS, the TOP.

Univalve conic or pyramidal, the aperture or mouth moſt generally ſubangular and flattiſh. The animal a *Slug*.

GENRE VII.

TROCHUS, le SABOT.

Univalve conique ou pyramidale, l'ouverture ou la bouche eſt plus generalement quelque choſe angulaire et applatie. L'animal eſt un *Limace*.

* TERRESTRES. LAND.

A

* TERRESTRES. TERRESTRES.

A

TROCHUS *terreſtris*, Liſteri.

Buccinum parvum, ſive trochilus ſylvaticus, agri Lincolnienſis.—Liſt. H. An. Angl. p. 123. tit. 9. ſine figura.—Phil. Tranſ. at large, No. 105. fig. 9.—Morton, Northampt. ch. 7. p. 415.

The *ſize* hardly of half a pepper-corn, *pellucid*, and of a yellowiſh *colour*. The *baſe* is flat; it ends in a pointed *tip*, like a trochus, and has ſix or ſeven *wreaths* or ſpires. Dr. Liſter found it in the moſs at the roots of the large trees in *Burwell woods*, in *Lincolnſhire*, but was rare.

La *grandeur* a peine celle du demi d'un grain de poivre, *tranſparente*, & de *couleur* jaunatre. La *baſe* eſt applatie; elle finit dans un *bout* pointu, comme un ſabot, et à ſix ou ſept *revolutions* ou orbes. Dr. Liſter les a trouvées dans la mouſſe aux racines des grands arbres dans *les bois* de *Burwell*, au comté de *Lincoln*; mais elle etoit rare.

Mr. Morton's *shell* agrees in all regards with that of Dr. Lister, except that his had no more than five *wreaths*, instead of six or seven, and the *live shell* is a little *bristly*; he found it in *Morsley wood*, in *Northamptonshire*, but it is extremely rare.

La *coquille* de Monf. Morton accorde en touts egards avec celle du Dr. Lister, excepté que le sien n'avoit que cincq *orbes*, au lieu de six ou sept, & la *coquille*, quand *vivante*, etoit un peu *berissée*; il l'a trouvé dans *le bois* de *Morsley*, au comté de *Northampton*, mais elle est extremement rare.

B

B

Trochilus terrestris Mortoni.

Trochilus exiguus quatuor spirarum elegantissime striatus.—Morton, Northampt. ch. 7. p. 416.

A very small *shell*, scarce the fourth part of a barley-corn in *size*, of the trochus *shape*, of a light brown *colour*, of four *wreaths*, fashioned as those of the trochi, with a pretty flat *base*, neatly streak'd or *striated*, transversely, or to the turn of the whirls, with deep striæ. The larger wreaths rise up into a sharpish ridge. He thought it a *non-descript*, and found it in the clefts of a sallow, in a thicket of sallows, near a pond in *Thorpe Mandeville* lordship, in *Northamptonshire*.

Une fort petite *coquille*, a peine la quatrieme partie d'un grain d'orge en *grandeur*, de la *forme* d'un sabot, d'une *couleur* brune claire, avec quatre *revolutions*, en forme de celles du sabot, & une *base* asses applatie, finement et bellement *striée* transversalement, ou selon les revolutions, par des stries profondes. Les plus grandes revolutions s'elevent en vive arrete. Il pense qu'elle est *une non-decrite*, & l'a trouvé dans les fentes d'un saule, dans un petit bois de saules, près d'un etang dans la seigneurie de *Thorpe Mandeville*, au comte de *Northampton*.

C

C

Trochus terrestris tertius.

Trochus, terrestris. Land.—Penn. Brit. Zool. No. 108. tab. 80. fig. 108.

Minute, conic, livid. A *new species* discovered in the mountains of *Cumberland*.

Menue, conique, livide. Une *nouvelle espece* decouverte dans les montagnes de *Cumberland*.

As Mr. Pennant notes this shell for a *new species*, I have ranked it as *such*, yet not without some doubts of its being of the same species as A.

Comme Monf. Pennant l'a notifié pour une *nouvelle espece*, j'e l'ai arrangé ici pour *telle*, non pas sans quelques doutes qu'elle est de la même espece que A.

** MARINÆ. Sea.
XIX.
Zizyphinus. Livid.
Tab. III. fig. 2. 2.

** MARINÆ. De Mer.
XIX.
Pl. III. fig. 2. 2.

Trochus pyramidalis imperforatus, lividus, rubro variegatus, limbo in summo quoque orbe circumdatus. Zizyphinus. Tab. 3. fig. 2. 2.

Trochus albidus maculis rubentibus distinctus, sex minimum spirarum.—List. H. An. Angl. p. 166. tit. 14. tab. 3. fig. 14.

Trochus pyramidalis variegatus, limbo angusto in summo quoque orbe circumdatus.—List. H. Conch. tab. 616. fig. 1.

Trochus ziziphinus, livid.—Penn. Brit. Zool. No. 103. tab. 80. fig. 103.

Wallace, Orkn. p. 40.—Smith, Cork, II. p. 318.—Wallis, Northumb. p. 402. No. 39.

Buon. Ricr. p. 193. fig. 93.—*Trochus fasciatus,* Klein, Ostrac. p. 24. § 65. No. H. tab. 2. fig. 36.

Trochus ore angusto, & horizontaliter compresso, lævis, zizyphino colore lucide depictus. And *Trochus ore angusto, & horizontaliter compresso, striis minimis circumdatus, ex subrosco & candido radiatim nebulatus, & in spirarum commissuris costula ex subrubro & candido tessellata cinctus.*—Gualt. I. Conch. tab. 61. fig. B. C.

Zizyphinus. Trochus testa imperforata conico livida lævi, anfractibus marginatis.—Lin. S. N. p. 1231. No. 599.—Faun. Suec. II. No. 2168.—Mus. Reg. p. 650. No. 336.

Culs de lampe de moyenne grandeur, lisses, marbrès de rouge et de violet, à orbes separès par un cordon.—D'Avila, cab. p. 127. No. 155.

The *shell* is strong and glossy, but rather thin, and about double the *size* of a filberd.

The *shape* is very pyramidal, from a broad base gradually tapering to a fine point or tip. The *base* is *not umbilicated,* but has only a long curved shallow cavity aside the inner lip.

The *base* is broad, slightly convex, finely and circularly striated, or according to the run of the whirls, and the striæ are not close set. The *colour,* when *alive,* is livid, or bluish ashen, with longitudinal red

La *coquille* est forte et lustrée, mais plutôt mince, et environ le double de la *grandeur* d'une noisette.

La *forme* est fort pyramidale, d'une base large s'appetissant par degre à un bout ou extremité fort pointue. La *base* n'est *point* umbiliquée, mais seulement à une cavite longue courbe & peu profond a coté de la levre interieure.

La *base* est large, peu convexe, legerement & circulairement striée, ou suivant le tour des revolutions ; les stries ne sont point serrées. La *couleur,* quand *vivante,* est livide, ou bleuatre cendrée, avec des rayes

red or brown rays; but, when *dead*, is whitish, with the same red rays.

The *mouth* is subangular and flattish; *within* slightly channelled by the outside striæ, else smooth and pearly. The *outer lip* even; the *inner* and *pillar lip* thick, broad, oblique, smooth, concave, and pearly.

The *body* and the *turban* have six *wreaths*, all channelled circularly with broad furrows, but not close set, and the *upper edge* of each wreath is distinguished by a remarkable broad smooth *border*.

The *colours* on the body and turban are the same as on the base, but stronger, and variegated in larger waves, so that it shews far more striking to the sight; but the broad border that surrounds the upper edge of each wreath is remarkably pretty, being more regularly chequered with the same colours.

The under *coat* of this shell is *mother of pearl*.

This *species* is common on most of our English shores, as *Sussex*, *Essex*, *Cornwall*, *Yorkshire*, *Lincolnshire*, &c. and the *Orkneys* and *Western Isles* in Scotland.

It is a common European and Mediterranean shell.

rayes longitudinales rouges ou brunes; mais, quand *morte*, elle est blanchatre, avec les mêmes rayes.

La *bouche* est un peu angulaire et applatie; en *dedans* legerement cannelé par les stries exterieures, autrement lisse & nacrè. La *levre exterieure* unie; l'*interieure* & *celle* de la *columelle* epaisse, large, oblique, unie, concave, et nacrée.

Le *corps* et la *clavicule* ont six *revolutions*, toutes cannelées circulairement par des cannelures larges, mais point serrées, & le *bord superieur* de chaque revolution est distinguè par une *bordure*, large & lisse.

Les *couleurs* de la robe & clavicule sont les mêmes que sur la base, mais plus fortes, & variées en ondes plus grandes, ainsi frappe plus la vue; mais la *bordure* large, qui regne autour du bord superieur de chaque revolution, est extremement jolie, etant plus regulierement marquetée des mêmes couleurs.

La seconde *robe* de cette coquille est de *nacre*.

Cette *espece* est commune sur la pluspart de nos cotes Angloises, comme *Sussex*, *Essex*, *Cornwall*, *Yorkshire*, *Lincolnshire*, &c. et aux *Orcades* et Isles *Hebrides* en Ecosse.

C'est une coquille commune de l'Europe & de la mer Mediterranée.

XX.

Papillosus. Shagreen'd.

Tab. III. fig. 3. 3.

XX.

Pl. III. fig. 3. 3.

Trochus pyramidalis umbilicatus ruber, series papillarum donatus. Papillosus. Tab. 3. fig. 3. 3.

An List. H. Conch. tab. 630. fig. 16. tab. 631. fig. 17. tab. 632. fig. 19?

An *young shells of the trochus maculatus.*—Lin. S. N. p. 1227. No. 580. & Gualt. I. Conch. tab. 61. fig. E?

Cul

Cul de lampe à flammes longitudinales, alternatives blanches & rouges, à stries granuleuses & umbiliqué: et *Cul de lampe marbré de blanc & de rouge, à stries circulaires granuleuses & tuberculeuses.*—D'Avila, cab. I. p. 127. 128.

The *shell* is rather thin, tho' strong, half the *size* of a walnut, of a very pyramidal or pointed *shape*, but not very high or lengthened.

The *base* is flattish, and finely wrought with circular ridges and furrows, but not very close set. The furrows are broad and smooth, but shallow; the ridges are narrow, not very prominent, and are finely cut across with numerous notches. The *colour* of this part is thickly speckled, in a chequer-like manner, with a light agreeable red on a whitish ground.

The *mouth* is high and round; within smooth and white, but not pearly.

The *outer lip* is even and thin; the *inner* and *pillar lip* is a mere white oblique edge, but from the top of it descends a wide and deep *cavity*, very smooth and white, at the bottom of which a tortuous *umbilicus* is seen.

The *body* and *turban* is thick set with circular ridges. The *ridges* are made up of *rows* of knobs or pimples, like those on shagreen skin, and size of small pins heads; but at the *edges* of each wreath they are more large and more prominent, so as to bump the edges much.

The *wreaths* are six, and are only separated by a slight furrow.

The *colour* of the turban is a pale reddish ground, with large longitudinal waves and shades of a deep or brownish red.

I received some few of these *shells* from *Cornwall* (in a great quantity of others, natives of that coast) by an intelligent gentleman

La *coquille* est plutôt mince, quoique forte, de la *grandeur* d'une noix, d'une *forme* fort pyramidale ou pointue, mais pas fort haute ou effilée.

La *base* est un peu applatie, & bellement travaillée en stries et sillons legers, peu serrés. Les sillons sont larges, unis, et peu profonds; les stries sont etroites, peu saillantes, & bellement decoupées a travers en entailleures nombreuses. La *couleur* de cette partie est extremement tachetée, en forme d'echiquier, de rouge clair agreable sur un fond blanchatre.

La *bouche* est haute & ronde, & en dedans lisse & blanche, mais non pas nacrée.

La *levre exterieure* est mince & unie; l'*interieure* & *celle* de la *columelle* est un simple bord blanc et oblique, mais du haut descende une *cavité*, large, profonde, fort lisse et blanche, au fond de laquelle se voit un *umbilic* tortueux.

Le *corps* & la *clavicule* sont a stries circulaires fort serrées. Ces *stries* sont composées de *rangées* de grains ou tubercules, comme ceux sur la peau chagrin, & de la grandeur de la tete d'une petite epingle; mais aux *bords* de chaque revolution ils sont plus grands & plus saillants, de sorte que de rendre les bords assés tuberculeux.

Les *revolutions* sont six, & separées seulement par un sillon leger.

La *couleur* de la clavicule est un fond pale rougeatre, avec des grands ondes & nuances longitudinales, d'une couleur brune-rouge chargée.

J'ay receu quelqu'unes de ces *coquilles* de *Cornwall* (dans une grande quantité d'autres, natives de cette côte) d'un amateur

gentleman of veracity and curiosity; but must own I have never met with this *species* since from any other *British coasts*.

teur tres curieux et de credit; cependant il faut avouer, que je n'ai jamais rencontré avec cette *espece* depuis, d'aucunes de nos *côtes Britanniques*.

XXI.

CONULUS. CONULE.

Tab. II. fig. 4. 4.

XXI.

Pl. II. fig. 4. 4.

Trochus pyramidalis parvus, imperforatus, anfractibus linea elevata interstinctis. Conulus. Tab. 2. fig. 4. 4.

Trochus pyramidalis parvus, ruberrimus, fasciis crebris exasperatus.—Lift. H. Conch. tab. 616. fig. 2.

T. conulus. Conule.—Penn. Brit. Zool. No. 104. tab. 80. fig. 104.

Conulus. Trochus testa imperforata conica, lævi, anfractibus linea elevata interstinctis.—Lin. S. N. p. 1230. No. 598.

The *shell* is thick and strong for its *size*, which seldom exceeds that of a cherry kernel. The *shape* is very pyramidal, and tapers to a sharp point.

The *base* is flat, *imperforated*, and circularly ridged and furrowed, of a pale ashen red *colour*. The *mouth* round and high, and white and smooth *within*, but not glossy.

The *body* and *turban* consists of six *wreaths*, also wrought with circular ridges and furrows as the base, and each wreath has a *prominent ridge* on its upper part, that distinguishes its separation from the following one. This ridge is prettily coloured with ashen and red, but more of the latter colour, and the *tip* or last wreath, which is generally of a very fine purplish rosy colour.

This *shell* is not uncommon on the shores of *Sussex*, and I have also received them from the *Devonshire* coast.

OBS.

La *coquille* est epaisse & forte pour sa *grandeur*, qui rarement surpasse celle d'un pepin de cerise. La *forme* est fort pyramidale, & appetisse dans une pointe aigue.

La *base* est applatie, *non umbiliquée*, & à stries & sillons circulaires, d'une *couleur* cendre rouge pale. La *bouche* ronde et haute, & en *dedans* blanche et lisse, mais non pas lustrée.

Le *corps* & la *clavicule* consiste de six *revolutions*, aussi à stries & sillons circulaires comme la base, & chaque revolution à un *bord saillant* a sa sommité, qui distingue sa separation de la suivante. Cet bord est joliment bigarré de cendre et rouge, mais plus de la derniere couleur, et la *pointe*, ou derniere revolution, est generalement d'une fort belle couleur pourpre rose.

Cette *coquille* n'est pas rare sur les côtes de *Sussex*, & je les ai aussi reçu des côtes du comté de *Devon*.

OBS.

OBS. 1. — *Linné*, from the prominent ridge on the whirls, as alſo from its being imperforated, ſays, *perhaps*, it is a *variety* in *ſmallneſs* of the *T. Zizyphinus*. The different growths of ſhells certainly muſt produce different ſizes; yet, with this ſeeming circumſtance, I totally *diſſent* from that celebrated profeſſor's *opinion*, and think it a *diſtinct ſpecies*.

OBS. 2. — *Linné* has alſo miſtakingly quoted the *Trochus, tit.* 15. *fig.* 15. of *Liſt. H. An. Angl.* for this ſhell; but *that* is the *T. lineatus, No.* 24, infra.

OBS. 1.—*Linné*, du bord ſaillant ſur les revolutions, comme auſſi parceque elle n'eſt pas umbiliquée, dit, que *peut etre*, c'eſt une *variete* en *petiteſſe* du *T. Zizyphinus*. Les differents accroiſſements de coquilles faut certainement produire differentes grandeurs; cependant, avec cette circonſtance apparente, je *differe* totalement en *opinion* du celebre profeſſeur, et la penſe etre une *eſpece diſtincte*.

OBS. 2.—*Linné* s'eſt egaré auſſi en citant le *Sabot, tit.* 15. *fig.* 15. de *Liſt. H. An. Angl.* pour cette coquille; mais *cela* eſt le *Sabot lineatus, No.* 24, infra.

XXII.

PARVUS. SMALL.

Trochus pyramidalis imperforatus, exiguus, albus, rubro undatim perbelle depictus. Parvus.

Trochus pyramidalis exiguus, rubris lineolis undatim depictus, Striatus.—Liſt. H. Conch. tab. 621. fig. 8.

An varietas *conuli pulcherrime variegati.*—Lin. S. N. p. 1230. No. 598?

This *ſhell*, in *ſize*, *ſhape*, *imperforation*, *mouth*, and *wreaths* or *whirls*, in no wiſe differs from the foregoing.

It is alſo finely *ſtriated* in a circular manner, or to the run of the whirls, but not ſo ſtrongly. It *differs* in not having the *prominent ridge* on the upper part of each whirl; in having a ſomewhat *pearly* under coat; in being *gloſſy* or poliſhed; and entirely in *colours*, the ground being white, beautifully mottled, waved or ſtreaked with red or fine dark brown, and *never* has the *roſy tip*.

Cette *coquille*, en *grandeur*, *forme*, *non umbiliquée*, *bouche*, & *revolutions*, differe nullement de la precedente.

Elle eſt auſſi *ſtriée* circulairement, ou ſelon le tour des revolutions, mais non pas ſi fortement. Elle *differe* en manquant le *bord ſaillant* ſur la ſommité de chaque revolution; en ayant quelque *nacre* au deſſous de la robe; en etant *luſtrée* ou polie; et totalement en *couleurs*, le fond etant blanc, elegamment tacheté, ondé ou rayé de rouge ou brun chargé, et *n'a jamais* la *pointe couleur de roſe*.

Tho' it might seem to be only a *variety*, yet I cannot but pronounce it a *distinct species*.

I have received these *shells* from the coasts of *Devonshire* and *Dorset*.

Quoiqu'elle peut paroitre seulement une *varieté*, cependant je ne peus que la prononcer une *espece distincte*.

J'ai reçu ces *coquilles* des côtes de *Devon* & *Dorset*.

XXIII.

CINEREUS. ASHEN.

Tab. III. fig. 5. 5.

XXIII.

Pl. III. fig. 5. 5.

Trochus pyramidalis umbilicatus, cinereus, lineis angustis nigrescentibus notatus. Cinereus. Tab. 3. fig. 5. 5.

Trochus pyram. parvus, ex viridi sive subcaeruleo variegatus, insigniter umbilicatus.—List. H. Conch. tab. 633. fig. 21.

An *smooth spire.* Grew. Mus. p. 132?—Dale Harw. p. 381. No. 3.—Wallace Orkn. p. 40.—Wallis Northumb. p. 402. No. 38.

An *trochus nostras lineis angustis & lineis latis oblique rufescentibus.*—Mus. Petiv. p. 88. No. 849. 850? et Gaz. tab. 101. fig. 12?

Trocho-cochlea undata & umbilicata.—Klein Ostrac. p. 42. § 119. No. 2.

The *shell* is thick and strong, of the *size* of a cherry, *shape* obtusely pyramidal, or not quite tapering to a point.

The *base* is very concave, with some circular furrows; the *mouth* roundish and capacious, *within* fine mother of pearl; the *outer lip* smooth and even; the *inner* or *pillar lip* has two jags or slight teeth, and two furrows crossing it transversely: from hence it widens, runs oblique, and forms a spacious cavity, at the bottom of which lies the *umbilicus*, deep, cylindric, and so hollow as to admit the head of a large pin. All this part is of a dark ash, greatly variegated with blackish lines or streaks, which run lengthways and across; but the beginning of the *umbilicus* is generally pearly, and of a fine light greenish colour.

The

La *coquille* est epaisse et forte, de la *grandeur* d'une cerise; la *forme* pyramidale obtuse, ou qui n'appetisse pas en pointe.

La *base* est fort concave, avec quelques sillons circulaires; la *bouche* quasi ronde & ample, en *dedans* d'une tres belle nacre; la *levre exterieure* lisse & unie; l'*interieur* ou de la *columelle* à deux entailleures ou dents legeres, et deux sillons qui la traversent: de là elle s'etend, court oblique, & forme une cavité etendue, au bas de laquelle se trouve l'*umbilic*, profond, cylindrique, & si creux que d'admettre la tété d'une grande epingle. Toute cette partie est de couleur cendre chargée, extremement bigarrée de lignes ou raies noiratres, transversales & longitudinales; mais le commencement de l'*umbilic* est generalement nacrè, & d'une belle couleur verdatre claire.

Le

The *body* and *turban* have five bellied or fwell'd *wreaths* or *whirls*, feparated by a very depreffed line; they are circularly *ftriated*, but faintly, and the *colours* are exactly the fame as on the bafe.

The outer coat being taken off, the whole fhell is fine *mother of pearl*.

This *fpecies* is common in *Suffex*, *Effex*, *Cornwall*, *Chefhire* and *Northumberland*, as alfo in the *Orkneys*.

It is likewife found in the Mediterranean fea.

Le *corps* & la *clavicule* ont cincq *orbes* bombés ou renflés, fepares par une ligne fort enfoncée; ils font *ftriés* circulairement, mais legerement, et les *couleurs* font de même que fur la bafe.

Etant depouillée de fa robe, toute la coquille eft d'une tres belle *nacre*.

Cette *efpece* eft commune en *Suffex*, *Effex*, *Cornwall*, *Chefhire*, & *Northumberland*, comme auffi aux *Orcades*.

Elle fe trouve pareillement dans la mer Mediterranée.

XXIV.

Lineatus. Streaked.

Tab. III. fig. 6. 6.

XXIV.

Pl. III. fig. 6. 6.

Trochus minor obtufe pyramidalis vel fubglobofus, umbilicatus, fubalbidus, ex nigro oblique lineatus. Lineatus. Tab. 3. fig. 6. 6.

Trochus crebris ftriis fufcis & tranfverfe & undatim difpofitis donatus.—Lift. H. An. Angl. p. 166. tit. 15. tab. 3. fig. 15.

Trochus parvus, ftriatus, undatim ex fufco denfe radiatus.—Lift. H. Conch. tab. 641. fig. 31.

T. cinerarius. Cinereous.—Penn. Brit. Zool. No. 106*.

Trochus ore ampliore, & fubrotundo, ftriatus, fubalbidus, ex nigro oblique lineatus.—Gualt. I. Conch. tab. 61. fig. N.

An *Varius. Trochus tefta oblique umbilicata convexa, anfractibus fubmarginatis.*—Lin. S. N. p. 1229. No. 589?

The *fhell* is very thick and ftrong, of the *fize* of a hazle nut, *fubglobofe* or obtufely pyramidal, or not produced to a fharp point.

The *bafe* is a little convex, *ftriated* very flightly in a circular manner, of a whitifh *colour*, prettily variegated with very fine lines or ftreaks of a pale red or dufky afhen; they run tranfverfe or acrofs the whirls in an oblique manner, but parallel and

La *coquille* eft tres epaiffe & forte, de la *grandeur* d'une noifette, *quafi ronde* ou pyramidale obtufe, ou qui n'appetiffe pas en pointe aigue.

La *bafe* eft un peu convexe, *ftrié* circulairement & finement, d'une *couleur* blanchatre, joliment variée de tres belles lignes ou rayes de rouge pale ou cendre obfcur; elles courent a travers les orbes obliquement, mais ferrées & paralleles. La

and close set. The *mouth* is roundish, smooth, whitish, and pearly *within*. The outer *lip* even; the *inner* or *pillar lip* thick and oblique, and aside of it is a narrow curved hollow, at the bottom of which lies the *umbilicus*, pretty deep, but very small.

The *body* and *turban* have five *whirls*, convex or swell'd, and are separated by a slight prominent edge and furrow; they are more strongly striated than the base, but in the same circular manner; and all this part is coloured and streaked exactly the same as on the base.

The coat being taken off, the shell is *mother of pearl*.

I have received this *species* from *Barmouth*, the coast of *Flintshire*, and *Anglesey* in *Wales*, *Ince* in *Cheshire*, *Exmouth* in *Devonshire*, and from the *Essex* shores.

La *bouche* est quasi ronde, lisse, blanchatre, & nacrée en *dedans*. La *levre exterieure* unie; l'*interieure* ou de la *columelle* epaisse & oblique, & a son coté se trouve une cavité courbe & etroite, au bas de laquelle est l'*umbilic*, qui est asses profond, mais tres petit.

Le *corps* & la *clavicule* ont cincq *orbes*, qui sont bombés ou renflés, & separès par un bord peu saillant & un sillon; ils sont striés plus fortement que la base, mais aussi circulairement; et toute cette partie est coloré & rayée exactement de même que la base.

La coquille, etant depouillée de sa robe, est de *nacre*.

J'ai reçu cette *espece* de *Barmouth*, des côtes de *Flint*, & de *Anglesey* en *Galles*, de *Ince* au comté de *Chester*, *Exmouth* au comté de *Devon*, et des côtes de *Essex*.

XXV.

TUBERCULATUS. KNOBBED.

Tab. III. fig. 1. 1.

XXV.

LA SORCIERE.

Pl. III. fig. 1. 1.

Trochus pyramidalis umbilicatus, anfractibus supra marginatis, infra nodulosis, albus, rubro variegatus. Tuberculatus. Tab. 3. fig. 1. 1.

Trochus maritimus nostras orbibus elatis.—Mus. Petiv. p. 88. No. 846.

Trochus acuminatus, orbibus striis transverse et undatim dispositis donatus. The wavy striated trochus, pearl coloured.—Borlase Cornw. p. 278. tab. 28. fig. 6.

T. magus. Tuberculated.—Penn. Brit. Zool. No. 107. tab. 80. fig. 107.

Sabot sorciere.—Argenv. Conch. I. p. 263 tab. 11. fig. S. II. tab. 8. fig. S.—D'Avila Cab. p. 123. No. 143.

Magus. Trochus testa oblique umbilicata convexa, anfractibus supra obtuse nodulosis.—Lin. S. N. p. 1228. No. 585.—Mus. Reg. p. 647. No. 332.

The *shell* is generally about the *size* of a walnut, thinnish, tho' strong, of a pyramidal,

La *coquille* est generalement environ la *grandeur* d'une noix, plutôt mince, quoique forte,

dal *form*, but flatted, or not very produced or long.

The *base* is near flattiſh, and ſtriated circularly, or according to the run of the whirls, with fine diſtant ſtriæ; the *mouth* roundiſh, very capacious, and *within* pearly; the *outer lip* thin and even; the *pillar lip* oblique, very thick and white. The *umbilicus* proceeds from a thick edged ſemicircular depreſſion, that riſes from the top of the pillar; it is round, capacious, and ſo very deep as to ſhew ſome of the lower interior whirls.

The *body* and *turban* have five *whirls*, ſeparated by a broad depreſſion; they are all convex or ſwell'd, and each is ſurrounded on the upper part by a very prominent edge, and round the bottom with a row of pretty large bumps or tubercules, which protuberate the ſaid lower edges greatly. This whole part is alſo circularly ſtriated as the baſe.

The *ſhell* is white, beautifully and thickly variegated with tranſverſe zigzag ſtripes and waves of fine red, which ſometimes degenerates into a dark browniſh red.

The outer coat being taken off, the ſhell is fine and *pearly*.

This *ſpecies* is found on ſeveral of our Engliſh ſhores; in *Suſſex* it is common; in *Dorſet*, *Devonſhire*, *Cornwall*, *Wales*, &c.

It is alſo an inhabitant of the Mediterranean ſea, and other parts of Europe.

XXVI.

forte, de *forme* pyramidale, mais applatie, ou point fort allongée.

La *baſe* eſt preſque applatie, & ſtriée circulairement ou ſelon le tour des orbes par des ſtries fines & diſtantes; la *bouche* eſt quaſi ronde, fort ample, & en *dedans* nacrée; la *levre exterieure* mince & unie; celle de la *columelle* oblique, fort epaiſſe, & blanche. L'*umbilic* procede d'un enfoncement à bord epais demi rond, qui s'eleve du haut de la columelle; il eſt rond, ample, & ſi profond que de faire voir quelqu'unes des baſſes revolutions interieures.

Le *corps* & la *clavicule* ont cincq *orbes*, ſeparés par un enfoncement large; ils ſont bombés ou renflés, & chacun eſt entouré ſur ſa ſommité par un bord fort ſaillant, et a l'entour de ſa partie inferieure par une rangée d'aſſes grands tubercules ou boſſes, qui rendent les dits bords inferieurs fort boſſus. Toute cette partie eſt ſtriée circulairement de même que la baſe.

La *coquille* eſt blanche, elegamment & grandement variée par des ondes & raies tranſverſales, en ziczac, d'une belle couleur rouge, qui degenere quelquefois en brun rouge chargé.

Etant depouillée de ſa robe, la coquille eſt d'une belle *nacre*.

Cette *eſpece* ſe trouve ſur pluſieurs de nos côtes Angloiſes; en *Suſſex* elle eſt commune; aux comtés de *Dorſet*, *Devon*, *Cornwall*, *Galles*, &c.

Elle eſt auſſi habitante de la mer Mediterranée, & autres parages de l'Europe.

XXVI.

XXVI.

Umbilicalis. Umbilical.

Tab. III. fig. 4. 4.

XXVI.

Pl. III. fig. 4. 4.

Trochus planior umbilicatus, undatim ex fusco pei belle radiatus. Umbilicalis Tab. 3. fig. 4. 4.

Trochus planior undatim ex rubro late radiatus.—Lift. H. Conch. tab. 641. fig. 32.

The umbilicated top fhell.—Dale Harw. p. 381. No. 4.

Trochus umbilicaris. Umbilical.—Penn. Brit. Zool. No. 106. tab. 80. fig. 106.

Trocho cochlea undata & umbilicata plana.—Klein Oftrac. p. 42. § 119. No. 1.

Cochlea marina terreftriformis, umbilicata, ftriata, aliquantulum depreffa, undatim ex rubro & fufco radiata.—Gualt. I. Conch. tab. 64. fig. C.

Cinerarius. Trochus tefta oblique umbilicata, ovata, anfractibus rotundatis.—Lin. S. N. p. 1229. No. 590.

The *fhell* is thick and ftrong, *fize* of a fmall grape, no ways pyramidal, but *flattifh*, the *body* being only fomewhat convex, and not lengthened.

The *bafe* is fet with fine diftant circular ftriæ. The *mouth* round and capacious; its *infide*, except a few flight ftriæ correfpondent to fome on the outfide, is fmooth and pearly; the *outer lip* even and fharp; the *inner* or *pillar lip* thick and oblique, and from the top of it arifes a pretty large oblique white cavity that leads to the *umbilicus*, which runs deep, and is capable of admitting a common pin's head.

The *body* and *turban* have five *whirls*, not produced or lengthened, but only juft rifing obtufe on one another, fo as to form a fmall convexity. They have the fame circular ftriæ as on the bafe, and are only feparated from one another by a furrow.

The fhell is *whitifh*, or fullied with afhen, very beautifully marked with purplifh

La *coquille* eft epaiffe & forte, de la *grandeur* d'un petit raifin, en aucune maniere pyramidale, mais quafi *applatie*, le *corps* etant feulement quelque chofe convexe & point allongé.

La *bafe* eft a ftries fines circulaires et diftantes. Le *bouche* ronde & ample; le *dedans*, excepté quelques ftries legeres correfpondentes a celles de la robe, uni & nacré; la *levre exterieure* unie & aigue; *l'interieure*, ou celle de la *columelle*, epaiffe & oblique, & de fon haut s'eleve une affes grande cavitè, oblique & blanche, qui tend a un *umbilic*, profond, & capable d'admettre la tété d'une epingle commune.

Le *corps* & la *clavicule* ont cincq *orbes*, non allongés, mais feulement couchés obtus, l'un fur l'autre, ainfi que de former une petite convexite; ils ont des ftries circulaires de même que la bafe, & font feulement feparés l'un de l'autre par un fillon.

La coquille eft *blanchatre*, ou obfcurie de cendre, elegamment marquée avec

purplifh brown oblique ftripes; they run lengthways of the fhell, are thick fet, and very regular, and by their obliquity form a kind of radiation. When nicely obferved, fome ftripes are furcated.

The outer coat being taken off, the fhell is of *mother of pearl.*

A beautiful and moft common fhell on all the Britifh fhores.

avec des raies obliques de couleur pourpre brune; elles courent la longeur de la coquille, font fort ferrées, & fort regulieres, et par leur obliquité forment une forte de rayonnement. Quand obfervè attentivement, quelques rayes font fourchues.

La coquille, etant depouillée de fa robe, eft de *nacre.*

C'eft une efpece tres elegante & tres commune fur toutes les côtes Britanniques.

FAMILY OF COCHLEÆ, or SNAILS.

Turbinated univalves, with a round *mouth*, or approaching thereto, and *intire*, being perfectly bordered or circumscribed. The animal, a *Slug*.

This *family* subdivides into *five genera*, viz. 1. The *Nerits*; 2. The *Helices*; 3. The *Snails*; 4. The *Turbo*; and 5. The *Strombiformis*, or *Long Snail*.

LA FAMILLE DE COCHLEÆ, ou LIMAÇONS.

Univalves contournées en spirale, avec une *bouche* ronde, ou approchante, et *entiere*, etant tout a fait bornée ou circonscrite. L'animal est une *Limace*.

Cette *famille* se soudivise en *cinq genres*, qui sont, 1. La *Nerite*; 2. Le *Helix*; 3. Le *Limaçon*; 4. le *Turbo*; et 5. Le *Strombiformis* ou *Limaçon effilé*.

GENUS VIII.

NERITA. NERIT.

Cochleæ, or *snails*, full bodied, or near *globose*; the *turben* flat and level with the body, or never much produced; the *mouth* semicircular, for the *pillar* is oblique, and diametrically crosses it.

GENRE VIII.

NERITA. La NERITE.

Cochleæ, ou *limaçons*, ventrus ou presque *ronds*; là *clavicule* applatie & de niveau avec le corps, ou jamais beaucoup allongée; la *bouche* demi ronde, car la *columelle* est oblique, & la traverse diametralement.

* FLUVIATILES. RIVER.

XXVII.

FLUVIATILIS. RIVER.

Tab. III. fig. 8. 8.

* FLUVIATILES. FLUVIATILES.

XXVII.

Pl. III. fig. 8. 8.

NERITA parvus fluviatilis, eleganter maculatus, fasciatus, aut reticulatus. Fluviatilis. Tab. 3. fig. 8. 8.

Nerita fluviatilis, è cæruleo virescens, maculatus, operculo subrufo lunato & aculeato datus. List. H. An. Angl. p. 136. tit. 20. tab. 2. fig. 20.

Nerites fluviatilis è cæruleo virescens, maculatus, operculo suberoceo aculeatoque donatus. —List. H. Conch. tab. 141. fig. 38.

Nerita fluv. exiguus, reticulate variegatus. Small netted Thames nerit.—Mus. Petiv. p. 67. No. 718.—Gaz. tab. 91. fig. 3. 3.

Nerita fluviatilis. River.—Penn. Brit. Zool. No. 142. tab. 87. fig. 142.

Vitta.—Klein Ostrac. p. 20. § 55. spec. 2.

Argenv.

Argenv. Conch. I. p. 372. tab. 31. fig. 3. II. p. 329. tab. 27. fig. 3.

Neritarum fluviatilium varietates, quæ vel ex cæruleo viridi, candido, rosco, fusco, & pallo colore diversimode sunt maculatæ, nebulatæ, punctatæ, undatim vel reticulatim pictæ, &c.—Gualt. I. Conch. tab. 4. fig. LLL.

Fluviatilis. *Nerita testa rugosa labiis edentulis.*—Lin. S. N. p. 1253. No. 723.—F. Suec. I. No. 1318. II. No. 2194.

The *shell* is very thick and strong, of an ovoid *shape*, from half to the full *size* of a cherry kernel, and smooth.

The *base* is flat; the *mouth* semilunar and large, whitish *within*; the *outer lip* even; the *inner* or *pillar lip* even, large, spread, smooth, whitish, often tinged with greenish, and diametrically crosses the mouth.

The *outside* or *body* very convex and glossy, beautifully *variegated* with black, white, bluish, ashen, red, green, &c. sometimes in a pretty *netted* work, sometimes in *zones* or *bands*, sometimes merely *mottled*, and in other very elegant *colourings*.

The *turban* is at the very tip or end of the body, *flat*, and consists of two *whirls*.

The *animal*, says Dr. *Lister*, is whitish, with two capillary horns, near to which lie its eyes, like two black specks; and its *operculum* is semilunar, reddish, thin, shelly, and armed with a kind of spike.

This *species* is frequent in the *Thames*, and other *rivers* of Great Britain; and is not less common in most of the rivers of Europe.

La *coquille* est tres epaisse & forte, d'une *forme* ovoide, de la *grandeur* d'un demi à un pepin entier de cerise, & lisse.

La *base* est applatie; la *bouche* semilunaire & grande, en *dedans* blanchatre; la *levre exterieure* unie; l'*interieure* ou celle de la *columelle* unie, grande, ample, lisse, blanchatre, mais souvent teinte de verdatre, & traverse la bouche diametralement.

L'*exterieur* ou le *corps* fort bombé & lustré, elegamment *varié* de noir, blanc, bleuatre, cendre, rouge, verd, &c. quelquefois dans un joli ouvrage *reticulé*, quelquefois en *zones* ou *bandes*, quelquefois simplement *moucheté*, & en d'autres *bigarrures* tres elegantes.

La *clavicule* est au bout même du corps, *applatie*, & consiste en deux *revolutions*.

L'*animal*, dit le Dr. *Lister*, est blanchatre, avec deux cornes comme de cheveux, proche desquelles se trouve les yeux, comme deux points noirs; son *opercule* est semilunaire, rougeatre, mince, testacé, & armé d'une epine ou pointe.

Cette *espece* est frequente dans la *Tamise*, & autres *rivieres* de la Grande Bretagne; & non pas moins commune dans la pluspart des rivieres de l'Europe.

** MARINÆ. Sea.	** MARINÆ. De Mer.
XXVIII.	XXVIII.
Littoralis. Strand. Tab. III. fig. 7. & Tab. IV. fig. 2. 3.	Pl. III. fig. 7. & Pl. IV. fig. 2. 3.

Nerita vulgaris, unicolor, flavus, aurantiacus, vel fuscus; aut fasciatus, aut reticulatim variegatus. Littoralis. Tab. 3. fig. 7 7. 7. 7. & tab. 4. fig. 2. 3.

Nerita, ex fusco viridescens, aut ex toto flavescens, modo pallide, modo intense ad colorem mali aurantii maturi. Lift. H. An. Angl. p. 164. tit. 11. tab. 3. fig. 11.—*N. fasciatus, unica lata fascia insignitus, caeterum subfuscus ex viridi.* Id. p. 165. tit. 12. tab. 3. fig. 12.—*N. reticulatus.* Id. p. 165. tit. 13. tab. 3. fig. 13.

Nerita, vel citrinus, vel castanei coloris. Lift. H. Conch. tab. 697. fig. 39.—*N. similis fasciatus.* Id. fig. 40.—*N. reticulatus minor, clavicula admodum compressa.* Id. fig. 44.—*N. subrufus, reticulatus, clavicula minus compressa.* Id. fig. 42.—*N. fasciatus.* Id. fig. 43.—*N. subrufus, lineis undatis rarioribus per obliquum depictus.* Id. fig. 41.

Nerita Anglicus maritimus, flavescens, vulgatissimus: & fuscus vulgaris. Muf. Petiv. p. 67. No. 716. 717.—*N. Orcadensis lutea, bifasciis nigris.* Petiv Gaz. tab. 34. fig. 4.—*N. Orcad. fusca fascia unica flavescente.* Id. fig. 5.—*N. Orcad. flava, striis capillaceis.* Id. fig. 6.

Phil. Tranf. at Large, No. 222. p. 321.—*Small sea snail.* Dale Harw. p. 380. 284. No. 1 & 2.—Wallace Orkn. p. 40.—Smith Cork, p. 318.—Wallis Northumb. p. 400. No. 31.

Nerita littoralis. Strand.—Penn. Brit. Zool. No. 143. tab. 87. fig. 143.

Vitta fasciata subrufa. Klein Oftrac. p. 20. § 55. fpec. 6. No. 1.—*Nerita fasciata.* Id. p. 8. § 18. fpec. 2. No. a.—*Cidaris subrufa.* Id. p. 21. § 57. fpec. 2. tab. 2. fig. 33. 34.—*Cricostoma fasciata.* Id. p. 12. § 34. a. b. tab. 1. fig. 25. 26.

Cochlea marina terrestriformis, laevis, citrina.—Gualt. I. Conch. tab 64. fig. N.

Littoralis. Nerita testa laevi, vertice carioso, labiis edentulis.—Lin. S. N. p. 1253. No. 724.—F. Suec. I. p. 377. No. 1320. II. No. 2195.

The *shell* is very thick and ftrong, of a globofe *shape*, and *size* of a hazle nut.	La *coquille* eft extremement epaiffe et forte, de *figure* ronde, & de la *grandeur* d'une noifette.
The *mouth* is circular, whitifh and fmooth *within*; the *outer lip* very thick and even; the *inner* or *pillar lip* fmooth and thick, fpread and oblique.	La *bouche* eft circulaire, blanchatre et liffe en *dedans*; la *levre exterieure* fort epaiffe et unie; l'*interieure* ou celle de la *columelle* liffe & epaiffe, ample & oblique.
The	Le

The *body* is very convex; the *turban* flat or level with it, and consists of three *whirls*.

This species varies greatly in its *colours*, numbers being of an uniform pale yellow, orange colour, or deep brown; others have one or two broad bands or girdles, of a deep brown, running circular, or according to the whirls; and others are dark greenish or olive colour, prettily mottled with a chequer or net work of a light colour. The two last-mentioned varieties are the most scarce.

It is a very common shell on all our British coasts, especially where the shores are rocky.

Le *corps* est fort convexe; la *clavicule* de niveau avec le corps, ou point allongée, & consiste de trois *revolutions*.

Cette espece varie extremement en *couleurs*; un grand nombre etant pale jaune uniforme, ou couleur d'orange, ou brun foncé; d'autres ont une ou deux bandes ou fascies larges, de brun foncé, qui courent circulairement, ou selon le tour des revolutions; et d'autres sont à fond verdatre ou couleur d'olive, tres joliment variées d'un ouvrage en forme d'echiquier ou reticulée, d'une couleur claire. Les deux dernieres varietés sont les plus rares.

C'est une coquille tres commune sur toutes nos côtes Britanniques, especialement ou les rivages sont pleins de rochers.

XXIX.
PALLIDULUS. PALE.
Tab. IV. fig. 4. 5.

XXIX.
Pl. IV. fig. 4. 5.

Nerita Corneus, spira paululum exserta. Pallidulus. Tab. 4. fig. 4. 5.

The *shell* is thin and brittle, semitransparent, of a pale brown *colour*, from the *size* of a cherry kernel to double that bigness, quite *smooth*, except a few slight longitudinal wrinkles, very *convex*, yet the *turban* is somewhat produced.

The *mouth* semilunar, and very patulous for the size of the shell; the *outer lip* thin and even; the *inner* and *pillar lip* thick, white, spread, and concave, with an *umbilicus* at the bottom, large enough to admit a pin's head.

La *coquille* est mince & fragile, demitransparente, d'une *couleur* brune pale, de la *grandeur* d'un pepin de cerise jusques a deux fois cet volume, tout a fait *lisse*, excepté quelques rides longitudinales legeres, fort *convexe*, cependant la *clavicule* est quelque peu allongé.

La *bouche* en forme de croissant, & fort evasée pour la grandeur de la coquille; la *levre exterieure* mince & unie; l'*interieure* & celle de la *columelle* epaisse, blanche, etendue, & concave, avec un *umbilic* au fond, asses grand a admettre une tété d'epingle.

The *body* convex; the *turban* large, not flat, but made up of two well diſtinguiſh'd and ſomewhat prominent *wreaths*.

This *ſpecies* is rare, for I have only received ſome few ſhells from the coaſts of *Kent* and *Dorſet*.

I do not find it *propoſed* or *deſcribed* by any author.

Le *corps* convexe ; la *clavicule* grande, point applatie, et conſiſte en deux *revolutions* bien diſtinguées, & quelque choſe ſaillantes.

Cette *eſpece* eſt rare, car j'ai ſeulement reçu quelqu'unes des côtes de *Kent* & *Dorſet*.

Je ne la trouve *propoſée* ou *decrite* par aucun auteur.

GENUS

GENRE

GENUS IX.

HELIX.

Snails flattish, and whose wreaths or whirls turn horizontal on a plane or level.

GENRE IX.

HELIX.

Limaçons applatis, dont les revolutions tournent horizontalement sur un même plan ou niveau.

* TERRESTRES. LAND.

XXX.

ERICA. HEATH.

Tab. IV. fig. 8. 8.

* TERRESTRES. TERRESTRES.

XXX.

Pl. IV. fig. 8. 8.

HELIX *cinerea albidave, fasciata, ericetorum. Erica.* Tab. 4. fig. 8. 8.

Cochlea cinerea albidave, fasciata, ericetorum.—List. H. An. Angl. p. 126. tit. 13. tab. 2. fig. 13.

Cochlea compressa, umbilicata, fasciata, campestris. List. H. Conch. tab. 78. fig. 78.—*Cochlea subfusca, umbilicata, clavicula modice producta.* Id. fig. 79; and the *two upper figures* of the *tabula muta*, 71.

Cochlea pellucida subflava quinque spirarum magis compressa, umbilicata.—Morton Northampt. p. 416.

Helix albella, Grey.—Penn. Brit. Zool. No. 122. tab. 85. fig. 122.

Nerita integra, striata.—Klein Ostrac. p. 6. § 17. No. 11.

Argenv. Conch. I. p. 372. tab. 31. fig. 2. No. 3. II. p. 329. tab. 27. fig. 2. No. 3.

Cochleæ terrestres depressæ, & umbilicatæ, &c. of Gualt. I. Conch. tab. 2. fig. M. N. O. P. Q. & tab. 3. fig. P. seem all only *varieties* of this *species*.

Itala. Helix testa umbilicata convexa obtusa, anfractibus quinque teretibus, umbilico patulo, apertura suborbiculata.—Lin. S. N. p. 1245. No. 683.

The *shell* is thin and brittle, flat and circular, *size* of a hazle nut, the largest about three quarters of an inch across; the *upper side* somewhat convex, by the swelling of the outermost wreath; the *mouth* circular, with an even edge, and a very large and deep central *umbilicus*.

La *coquille* est mince & fragile, applatie & circulaire, de la *grandeur* d'une noisette, les plus grandes ayant environ trois quarts d'un pouce de diamatre; le *dessus* quelque peu convexe, parceque la revolution exterieure est bombée; la *bouche* circulaire, avec un bord uni, & un *umbilic* central, grand & profond.

The

La

The *bottom* or *under side* near flat, and consists of five cylindrick *wreaths*, gradually diminishing to the center; they are cast or overlaid on one another, and are separated or distinguished by a strong furrow.

The whole *shell* is finely or slightly *striated* across the wreaths. When *young*, they are quite plain, of a horny colour or whitish, and semitransparent; when full *grown*, are opake, of a dull or soil'd white ground, some quite plain, but most generally *fasciated* with pale brown or yellowish circular bands, or according to the turn of the wreaths; but the *order*, *size* or *number* of these brown bands *vary extremely*, tho' *commonly* there is one band in the middle or near the bottom of each wreath, and often other fainter and narrower bands accompany it,

There are great numbers of *dead shells* always found with the others; these are bleached or whitened, and the bands are wash'd out by the rains and weather.

This *species* is very *common all over England*, on heaths and sandy soils, open mountainous fields, and such-like places.

It is a common land snail in most parts of Europe.

La *clavicule*, ou le *dessous*, presque applati, et composé de cincq *revolutions* cylindriques, diminuant proportionellement au centre; elles sont comme jettées ou mises l'une sur l'autre, et sont separées ou distinguées par une strie forte.

Toute la *coquille* est finement *striée* a travers les revolutions. Quand *jeunes*, elles sont tout a fait unies, de couleur de corne ou blanchatre, et demitransparente; quand *adultes*, elles sont opaques, & à fond blanc terne ou sale, quelqu'unes tout a fait unies, mais plus generalement *fasciées* a bandes pale brunes ou jaunatres circulaires, ou selon le tour des revolutions; mais l'*ordre*, la *grandeur* & le *nombre* de ces bandes brunes *varient extremement*, quoique pour l'*ordinaire* il y a une bande au milieu ou près du bas de chaque revolution, & souvent d'autres bandes plus etroites & plus pales l'accompagnent.

Il se trouve toujours parmi ces coquilles un grand nombre des *coquilles mortes*: elles sont blanchies, & leurs bandes sont effacées par les pluies & l'air.

Cette *espece* est tres *commune par tout l'Angleterre*, sur les bruyeres & les terroirs sablonneux, les champs etendus & montagneux, & autres lieux semblables.

C'est un limaçon terrestre commune dans la pluspart des pais de l'Europe.

XXXI.
ACUTA. SHARP.
Tab. IV. fig. 9. 9.

XXXI.
Pl. IV. fig. 9. 9.

Cochlea umbilicata, margine in acie acuta depresso. Acuta. Tab. 4. fig. 9. 9.

Cochlea pulla, sylvatica, spiris in aciem depressis.—List. H. An. Angl. p. 126. tit. 14. tab. 2. fig. 14.—Phil. Transf. No. 105. fig. 13.

Cochlea nostras, umbilicata, pulla.—Lift. H. Conch. tab. 69. fig. 68.

Planorbis terrestris Anglicus, umbilico minore, margine acuto.—Muf. Petiv. p. 69. No. 734.

Cochlea terrestris media acie acuta: Land cheese shell with a sharp edge.—Petiv. Gaz. tab. 92. fig. 11.

Morton Northampt. p. 415—Wallis Northumb. p. 369.

Helix lapicida, Rock.—Penn. Brit. Zool. No. 121. tab. 83. fig. 121.

Serpentulus levis.—Klein Ostrac. p. 8. § 20. No. 4.

Cochlea terrestris depressa & umbilicata, ore ovali, umbilico majore, in quo anfractus spirarum in extima superficie acuminatarum observantur, mucrone tantillum elevato.—Gualt. I. Conch. tab. 3. fig. N.

Cochlea testa utrinque convexa, subtus perforata, spira acuta, apertura ovata transversali.—Lin. F. Suec. I. p. 371. No. 1298. II. No. 2174.

Lapicida. Helix testa carinata umbilicata utrinque convexa, apertura marginata transversali ovata.—Lin. S. N. p. 1241. No. 656.—Muf. Reg. p. 663. No. 362.

The *shell* is gloffy, thin, brittle and semipellucid, flat and circular, of the *size* of a small grape, meafures from a half to three quarters of an inch in diameter, and is very finely *striated* acrofs the wreaths.

The *upper side* somewhat convex or fwell'd; the *mouth* oval, oblique, and bordered with a fine *white edge*, which turns or *spreads outwards*, and has a remarkable *dent* at the place where it joins the edge of the outermoft wreath; *within* it is fmooth, and of a paler colour than the fhell. The *umbilicus* is large, open or fpacious, cylindrick, and very deep, almoft reaching to the bottom of the fhell.

La *coquille* eft luftrée, mince, fragile & demitranfparente, applatie & circulaire, de la *grandeur* d'un petit raifin, fur un demi a trois quarts d'un pouce de diametre, et eft fort finement *striée* a travers les revolutions.

Le *dessus* eft quelque peu convexe ou renflé; la *bouche* ovale, oblique, & entourée par un beau *bord blanc*, qui fe replie ou *s'etend en dehors*, & a une *echancrure* remarquable a l'endroit ou elle s'unit au dos de la revolution la plus exterieure; en *dedans* elle eft liffe, & d'une [illegible] pale que la coquille. L'[illegible] ouverte ou [illegible] profond, pr[illegible] coquille.

The

The *under side* or *turban* rises pretty convex, by the wreaths gradually lying over one another; they are *five* in number, set close together, and distinguish'd only by a fine furrow; their *back* is *carinated*, or runs out into a *very sharp edge*. This *sharp back* of the wreaths, and the *mouth*, are *two specific characters* of this *species*.

The *colours* of the shell are a fine deep brown, variegated with waves, streaks and spaces of a pale yellow, and, when quite fine or *alive*, perfectly resembles *tortoiseshell*.

This elegant *species*, tho' found in many parts of England, is not met with in any plenty, but is *scarce*. I have found them on the rocks at and near *Matlock*, in *Derbyshire*; about *Bath*, in *Somersetshire*, also on rocks; in *Surrey*, *Wiltshire*, and *Hampshire*, in the moss on the bodies of large trees, and in woods. Dr. Lister found them in like places, and on the grass in *Lincolnshire*; Mr. Petiver in hedges, between *Charlton* and *Woolwich*, in *Kent*; Mr. Morton in hedge-bottoms, in *Oakly Parva*, in *Northamptonshire*; and Mr. Wallis on the rocks in *Northumberland*: but they are not *common* or *frequent* any where.

It is also found in other countries of Europe, as, according to Linné, in Sweden, and in Germany and Italy.

Le *dessous* ou la *clavicule* s'eleve asses convexe, parceque les revolutions sont couchées peu à peu l'une sur l'autre; elles sont *cincq* en nombre, serrées ensemble, & distinguées seulement par une strie fine; leur *dos* est en *carenne*, ou s'etend en un *bord fort aigu*. Cet *dos aigu* des revolutions, & la *bouche*, sont *deux caracteres specifiques* de cette *espece*.

Les *couleurs* de la coquille sont un beau brun chargé, marbré des ondes, raies & nuages de jaune pale, &, quand tout a fait bien conservée ou *vivante*, elle ressemble parfaitement à *l'ecaille de tortue*.

Cette *espece* elegante, quoique trouvée en plusieurs lieux de l'Angleterre, ne se rencontre pas en aucun quantité, mais est *rare*. Je l'ai trouvé sur les rochers près de, et a *Matlock*, au comté de *Derby*; autour de *Bath*, au comté de *Somerset*, pareillement sur les rochers; aux comtés de *Surrey*, *Wilts*, & *Hants*, dans la mousse sur les troncs de grands arbres, & dans les bois. Le Dr. Lister les a trouvé en pareils lieux, & sur la verdure, au comté de *Lincoln*; M. Petiver dans les hayes, entre *Charlton* & *Woolwich*, en *Kent*; M. Morton aussi dans les hayes, a *Oakly Parva*, au comté de *Northampton*; & M. Wallis sur les rochers en *Northumberland*: mais elles ne sont pas *communes* ou *frequentes* en aucun lieu.

Elle se trouve aussi en autres pais de l'Europe, comme en Suede, selon Linné, & en Allemagne & l'Italie.

XXXII.

XXXII.

RADIATA. RAYED.

Tab. IV fig. 15. 16.

XXXII.

Pl. IV. fig. 15. 16.

Helix minima umbilicata, pulchre radiata. Radiata. Tab. 4. fig. 15. 16.

Cochlea terrestris compressa, maculata, et leviter striata, D. Dale.—Lift. H. Conch. Appendix, tab. 4. seu tab. 1058. fig. 11. A.

Planorbis hortensis minima, pulchre striata. Dale's spangle shell.—Petiv. Gaz. tab. 31. fig. 5.

Cochlea parva magis compressa umbilicata quinque spirarum, pullo & subflavo colore eleganter tessellata, striis capillaribus transverse depicta.—Morton Northampt. c. 17. p. 416.

Cochlea terrestris depressa & umbilicata, albida, fascia punctata rufa per medium anfractuum, & maculis concoloribus eleganter depicta.—Gualt. I. Conch. tab. 3. fig. Q.

The *shell* very small, *size* of a spangle, or about one quarter of an inch *diameter*, circular, flattish, thin, brittle, not glossy, and with very fine almost imperceptible transverse *striæ*.

The *upper side* is very convex or swell'd; the *mouth* circular and oblique; the *umbilicus* very large, spacious and deep, so as to show all the inner wreaths very distinctly.

The *under side*, or *turban*, flattish, or with a slight convexity, consisting of five *wreaths* closely joined, rising very little above one another, and separated or distinguished by a fine furrow; they are cylindric and subcarinated, or somewhat edged at the lower part of each.

The *shell*, when *fresh* or *alive*, is sullied whitish or pale horn *colour*, thickly set with transverse deep brown rays or waves quite across the shell, so as to exhibit a very pretty radiated colouring.

This small land snail was first discovered by Mr. Sam. Dale in *Kent* and *Middlesex*, and

La *coquille* est fort petite, de la *grandeur* d'une paillette, ou environ le quart d'un pouce de *diametre*, ronde, un peu applatie, mince, fragile, pas lustrée, et avec des *stries* transversales a peine visibles.

Le *dessus* est fort convexe ou bombé; la *bouche* ronde & oblique; l'*umbilic* fort grand, spacieux & profond, ainsi que de montrer les revolutions interieures fort distinctement.

Le *dessous*, ou la *clavicule*, applati, ou peu convexe, & consiste de cincq *revolutions* contigues, s'elevant fort peu l'une sur l'autre, et separées ou distinguées par une strie fine; elles sont cylindriques, & quelque chose aigue ou bordée a la partie inferieure de chacune.

La *coquille*, quand *recente* ou *vivante*, est d'une *couleur* blanchatre sale ou de corne pale, avec des rayes ou ondes transversales brunes chargées, fort serrées, & qui traversent tout a fait la coquille, ainsi que de montrer un fort joli coloris rayonné.

Cet petit limaçon terrestre fut premierement decouvert par M. Sam. Dale en *Kent*

and ſince has been found by Mr. Petiver and myſelf about *London*. Mr. Morton found it in a little wood called *Loteland*, nigh *Harrington*, and in many other *woods* in *Northamptonſhire*, in the moſs at the roots of the aſhes and oaks; and I have received them from my worthy friend Richard Hill Waring, Eſq. from about his ſeat at *Leeſwood* in *Flintſhire*, where it it is pretty frequent between the bark and wood of trees decayed and thrown down, eſpecially of alders, and likewiſe on walls.

Kent & en *Middleſex*, & a été trouvé depuis aux environs de *Londres*, par Monſ. Petiver & moi. M. Morton l'a trouvé dans un petit bois nommé *Loteland*, près de *Harrington*, et dans pluſieurs autres *bois* au comté de *Northampton*, dans la mouſſe aux racines des frenes & chenes; & je les ai reçu de mon tres eſtimé ami Richard Hill Waring, Eſq. des environs de ſa maiſon de campagne a *Leeſwood* au comté de *Flint*, ou il ſe trouve aſſes frequent entre l'ecorce & le bois des arbres pourris & renverſés, particulierement des aunes, et auſſi ſur les murailles.

XXIII.

HISPIDA. BRISTLEY.

Tab. V. fig. 10.

XXIII.

Pl. V. fig. 10.

Helix ſubgloboſa umbilicata, cornea, diaphana, hiſpida. Hiſpida. Tab. 5. fig. 10.

Cochlea terreſtris hirſuta; Small hairy ſtreak'd land cheeſe ſhell: and *Cochlea terreſtris minor lucens.*—Petiv. Gaz. tab. 93. fig. 13 & 14.

Cochlea e compreſſis, coloris ſubfuſci, clavicula productiore, quinque ſpirarum, ex altera parte umbilicata, et ſubtiliter echinata.—Morton Northampt. c. 7. p. 416.

Cochlea teſta utrinque convexa hiſpida, ſpiris quinque rotundatis, ſubtus perforata.—Lin. F. Suec. I. p. 371. No. 1296. II. No. 2182.

Hiſpida. Helix teſta umbilicata convexa hiſpida diaphana, anfractibus quinis, apertura ſubrotundo lunata.—Lin. S. N. p. 1244. No. 675.

The *ſhell* is gloſſy, very thin, brittle, and intirely of a brown or horny *colour*, and ſemitranſparent, ſlightly or finely *ſtriated* acroſs the wreaths; in *ſize* generally ſomewhat ſmaller than a currant; the *ſhape* is near globoſe, or ſwell'd, being convex on both ſides, ſo as to approach the *cochleæ* more than the *helices*, and is a *link* or *medium* between the *genera*.

La *coquille* eſt luſtrée, fort mince, & entierement d'une *couleur* brune ou de corne, & demitranſparente, légérement ou finement *ſtriée* a travers des revolutions; en *grandeur* communement quelque peu moindre que un raiſin de Corinthe; la *forme* eſt preſque ronde, ou bombée, etant convexe des deux cotéz, ainſi qu'elle approche aux *cochleæ* pluſque aux *helices*, & eſt un *lien* ou *milieu* entre ces *genres*.

The *upper side* very convex or ſwell'd; the *mouth* large, ſemilunar, and laid tranſverſe; the *outer lip* even; the *inner* or *piller lip* thickiſh, and ſlightly ſpread; the *umbilicus* central, cylindrick and deep.

The *under ſide*, or *turban*, pretty convex; the *wreaths* five, riſing ſlightly on one another, ſeparated or diſtinguiſhed by a furrow, and are cylindrick or rounded.

This *ſhell*, when *freſh* or *alive*, is thick ſet all over with minute, ſhort, white, brittle, ſilk-like fine *hairs* or *briſtles*, that eaſily *break* and *rub off*; which is the *peculiar character* of this *ſpecies*.

It is found at the bottom of trees, in the moſs, in woods and wet ſhady places in *Lincolnſhire* and *Wiltſhire*; in plenty in *Hampſhire*; and, according to Morton, is not unfrequent in the woods of *Northamptonſhire*, particularly in thoſe about *King's Cliff*.

Le *deſſus* fort convexe ou renflée; la *bouche* grande, en croiſſant, & tranſverſale; la *levre exterieure* unie; l'*interieure* ou de la *columelle* epaiſſe, & légérement etendue; l'*umbilic* au centre, cylindrique & profond.

Le *deſſous*, ou la *clavicule*, aſſes convexe; les *revolutions* cincq, s'elevant un peu l'une ſur l'autre, ſeparées ou diſtinguées par une ſtrie, & ſont arrondies ou cylindriques.

Cette *coquille*, quand *recente* ou *vivante*, eſt parſemée par tout & fort ſerrée avec des *poils* fins ou *ſoies*, courts, blancs, fragiles, & ſoyeux, qui facilement ſe *caſſent* & ſe *frayent*; ce qui eſt le *caractere particulier* de cette *eſpece*.

Elle ſe trouve au bas des arbres, dans la mouſſe, dans les bois & les lieux humides & ombragés aux comtés de *Lincoln* & *Wilts*; en abondance au comté de *Hants*; &, ſelon Morton, elles ne ſont pas rares dans les bois du comté de *Northampton*, & eſpecialement dans ceux aux environs de *King's Cliff*.

D

PALUDOSA. **B**OG.

D

Cochlea compreſſa umbilicata minime trium ſpirarum, ore rotundo reflexo.—Morton Northampt. c. 7. p. 417.

The *ſmalleſt* of all, has only three *ſpires*, with a round *aperture*, whoſe margin is bended a little backward.

It was a *new* or *non-deſcript ſpecies* to Mr. Morton, and is ſo to me, as I cannot find it propoſed by any author.

He found it in boggy places, in *Oxendon* and *Arthingworth*, in *Northamptonſhire*, in great plenty.

La plus *petite* de toutes, elle a ſeulement trois *revolutions*, avec une *bouche* ronde, dont le bord eſt tourné un peu en dehors.

Elle paroiſſoit une *eſpece nouvelle* ou *non-decrite* a M. Morton, & me paroit de même, car je ne la trouve propoſée par aucun auteur.

Il l'a trouvée dans des lieux marecageux, a *Oxendon* & *Arthingworth*, au comté de *Northampton*, en grande abondance.

** FLU-

** FLU-

**FLUVIATILES. River.

XXXIV.

Cornu Arietis. Ram's Horn.

Tab. IV. fig. 13.

Helix fluviatilis depressa major, anfractibus quatuor, ex utraque parte circa umbilicum cava. Cornu arietis. Tab. 4. fig. 13.

Cochlea pulla, ex utraque parte circa umbilicum cava.—Lift. H. An. Angl. p. 143. tit. 26. tab. 2. fig. 26.—Phil. Tranf. No. 105. fig. 23.—App. ad H. An. Angl. p. 7. —App. ad H. An. Angl. in Goed. p. 8.

Cochlea maxima, compressa, fasciata. Lift. H. Conch. tab. 136. fig, 40.—*Cochlea pulla quatuor orbium coccum fundens, purpura lacustris.* Id. tab. 137. fig. 41.

Purpura. S. Cochlea fluviatilis compressa major.—Lift. Exerc. Anat. 2. p. 59.

The flat whirl.—Grew Muf. p. 136.

Planorbis fluviatilis major vulgaris.—Petiv. Gaz. tab. 92. fig. 5.

Helix cornea. Horny.—Penn. Brit. Zool. No. 126. tab. 83. fig. 126.

Cornu ammonis spurium, maximum fluviatile, corrugatum transversim.—Klein Oftrac. p. 5. § 12. No. 1. tab. 1. fig. 7.

Argenv. Conch. I. p. 373. tab. 31. fig. 8. No. 1 & 4. II. p. 330. tab. 27. fig. 8. No. 1 & 4. & p. 75. tab. 8. fig. 7.

Cochlea fluviatilis depressa, pulla, ex utraque parte umbilicata.—Gualt. I. Conch. tab. 4. fig. D. D.

Cornea. Helix testa supra umbilicata plana nigricante, anfractibus quatuor teretibus.—Lin. S. N. p. 1243. No. 671.—F. Suec. I. p. 373. No. 1304. II. No. 2179.—Muf. Reg. p. 665. No. 366.

The *shell* is rather thin and brittle, gloffy, and meafures from three quarters of an inch to an inch and a quarter or a half in diameter, which is the largeft *fize*, and from one quarter to three-eighths in *height* or *depth*, circular and flat; the *adults* of a light brownifh or afhen *colour*; the *young* ones rather afhen or whitifh, and more tranfparent; and the fhell is finely and thickly wrinkled acrofs the wreaths.

The

**FLUVIATILES. Fluviatiles.

XXXIV.

Pl. IV. fig. 13.

La *coquille* eft plutot mince & fragile, luftrée, & portant trois quarts d'un pouce a un pouce & un quart a un demi de diametre, qui eft le plus grand *volume*, & de un quart a trois huitiemes d'un pouce en *hauteur* ou *groffeur*, ronde & applatie; les *adultes* font d'une *couleur* brunatre claire ou cendrée; les *jeunes* plutot cendrées & blanchatres, & ont plus de tranfparence; & la coquille eft finement et ferrement ridée a travers des revolutions.

Les

The *wreaths* are four, feparated or diftinguifhed by a deep furrow; they lie horizontal, or turn on a level; they are all cylindric, and not much inferted or join'd into one another: the *outer wreath* is very large and fwell'd, and protuberates or rifes above the reft, which all gradually decreafe, fo as to run depreffed or concave to the center.

Les *revolutions* font quatre, feparées ou diftinguées par une ftrie profonde; elles font couchées horizontalement, ou tournent fur un même plan; toutes cylindriques, & pas beaucoup enfermées ou jointes l'une dans l'autre: la *revolution exterieure* eft fort grande & s'eleve ou eft enflée, au deffus des autres, qui toutes diminuent proportionellement, ainfi que d'être abbaiffées ou courent concave au centre.

The *upper fide*. The wreaths on this fide do not decreafe fo much as to form a great depreffion, but lie more on a level. The *mouth* is large and oval, in a perpendicular pofition, or lying from furface to furface, or to the depth of the fhell, thick and fpread, and white at the two extremes, where it joins to the wreaths; and *within* is fmooth.

Le *deffus*. Les revolutions de cet coté ne diminuent point tant à former un fi grand abbaiffement, mais font couchées plus de niveau. La *bouche* eft grande & ovale, placée perpendiculairement, ou pofée de furface en furface, ou felon la groffeur ou hauteur de la coquille, epaiffe, etendue, & blanche aux deux extremités, ou elle fe joint aux revolutions; en *dedans* elle eft liffe.

The *under fide*, by the gradual decreafe of the wreaths, runs very depreffed or concave to the center, where it forms a pretty deep *umbilicus*.

Le *deffous*, car l'abbaiffement graduel des revolutions, devient fort abbaiffé ou concave au centre, ou il forme un *umbilic* affès profond.

The *inhabitant* animal is blackifh brown, and has two red capillary horns.

L'animal *habitant* eft noiratre brun, avec deux cornes capillaires rouges.

This fifh emits a fine *fcarlet humour*, if a grain of falt of any kind, or a little pepper or ginger, be put into the mouth of the fhell. It emits this fine fcarlet humour all the year, efpecially in April and September. Dr. Lifter gives a full account of it. *He fays*, This *fcarlet humour* may be readily got, and in great quantity, if a large parcel of thefe fhells be wrapped up in a cloth bag, fprinkling over it a little falt; then the fcarlet liquor will oofe plentifully. This *humour*, if fprinkled with powder'd allum, the *colouring part* immediately fubfides, and the reft of it remains like clear water. The *colouring part*

Cet poiffon jette une belle *humeur ecarlate*, fi un grain de fel d'aucune efpece, un peu de poivre ou de gingembre, font mis dans la bouche de la coquille. Elle jette cette belle humeur ecarlate toute l'année, particulierement en Avril & Septembre. Le Dr. Lifter nous en donne une defcription ample. *Il dit*, Cette *humeur ecarlate* peut être acquife, & en grande quantité, fi un grand tas de ces coquilles font enveloppées dans un fac de linge, en le faupoudrant avec un peu de fel, alors la liqueur ecarlate coulera en abondance. Cette *humeur*, fi faupoudrez avec alun en fpoudre, la *partie colorante* im-

part may be ſtrain'd through a filtring paper, but the elegance of its colour is loſt, and it changes into a dull or unpleaſant ruſty brown. Moreover, if mixed with vinegar, ſpirits of wine, deliquated vegetable ſalts, or common ſalt diſſolved, this elegant ſcarlet colour periſhes in the ſame manner as when mix'd with allum. Neither can this *liquor* be kept by itſelf, pure or unmixed, for in vain did the Doctor ſtrive to preſerve it in narrow-mouthed bottles or phials, perfectly well cloſed, and with oil or honey thrown over it. Thus this *colour* is of ſo *fugitive* a nature, that no *acid* or *aſtringent* has hitherto been found ſufficient to preſerve the elegancy of its tint.

immediatement ſe precipite, & le reſidu reſte comme de l'eau claire. La *partie colorante* peut être paſſé à travers du papier filtrant, mais l'elegance de la couleur eſt perdue, et elle ſe change à une couleur rouille brune, morne & deſagreable. De plus, ſi melée avec du vinaigre, les eſprits du vin, les ſels vegetaux, ou le ſel commun diſſous, cette couleur ecarlate elegante perit de la même maniere que quand melée avec l'alun. Ni peut cette *liqueur* être gardé pure ou ſans melange; car en vain le Docteur a-t'il eſſayé de la conſerver dans des bouteilles ou phioles, à bouches etroites, parfaitement bien fermées, même en la couvrant, avec de l'huile ou du miel. Ainſi cette *couleur* eſt d'une nature ſi *fugitive*, qu'aucune *acide* ou *aſtringente* a été trouvée juſques à preſent, ſuffiſante à conſerver l'elegance de la couleur.

Dr. Liſter further recites ſome obſervations and experiments he made on this *ſcarlet fluid*, to diſcover whether it was a *humour* of the body, to be got by laceration or inciſion, as blood; a *ſaliva* from the throat or ſtomach; or a particular *humour* contained in certain veſſels or parts; but the nicety and difficulty of the experiments render'd it impoſſible for him to determine it preciſely.

Dr. Liſter de plus nous donne un recit de quelques obſervations & experiences qu'il fit ſur cette *humeur ecarlate*, à decouvrir ſi c'eſt une *humeur* du corps qu'on peut tirer par la laceration ou par une inciſion, comme le ſang; une *ſalive* de la gorge ou de l'eſtomac; ou une *humeur* particuliere contenue dans certains vaiſſeaux ou parties; mais la delicateſſe & la difficulté des experiences l'a rendu impoſſible pour lui de la determiner avec preciſion.

This *ſhell* is very common in all ponds, rivers and lakes, throughout *England*, and is as common in other parts of Europe.

Cette *coquille* eſt fort commune dans touts les etangs, rivieres & lacs, par tout l'*Angleterre*, et eſt de mêmê commune dans les autres pais de l'Europe.

Obs. 1.—The *H. Nana*, or *Dwarf*, of Brit. Zool. No. 125. tab. 83. fig. 125. as Mr. Pennant himſelf *ſurmiſes*, is only a *young* ſhell of this *ſpecies*.

Obs. 1.—Le *H. Nana* de la Zool. Brit. No. 125. pl. 83. fig. 125. comme M. Pennant lui même *s'imagine*, eſt ſeulement une *jeune* coquille de cette *eſpece*.

Obs.

Obs.

Obs. 2.—This *species* sometimes has one or two regular circular brownish belts or bands, running to the turn of the wreaths; and Klein figures such a one, which has missed Linné (tho' in his Mus. Reg. he mentions these brownish belts) to propose *two species*, viz. this which he calls *cornea* and the *cornu arietis*; but in reality they are only *varieties* of this *very species*.

Obs. 2. — Cette *espece* quelquefois se trouve avec une ou deux bandes ou zones brunatres, regulieres & circulaires, ou courant selon le tour des revolutions; & Klein donne la figure d'une telle coquille, ce qui à egaré Linné (quoique dans son Mus. Reg. il fait mention de ces bandes brunatres) à proposer *deux especes*, sçavoir, celle ci qu'il nomme *cornea* & le *cornu arietis*; mais en verité elles sont seulement des *varietés* de cette *même espece*.

XXXV.

Limbata. Border'd.

Tab. IV. fig. 10. et Tab. VIII. fig. 8.

XXXV.

Pl. IV. fig. 10. et Pl. VIII. fig. 8.

Cochlea depressa fusca quatuor spirarum altera parte limbata. Limbata. Tab. 4. fig. 10. et tab. 8. fig. 8.

Cochlea fusca altera parte planior, & limbo insignita, quatuor spirarum.—List. H. An. Angl. p. 145. tit. 27. tab. 2. fig. 27.—Phil. Trans. No. 105. fig. 24.

Cochlea fusca limbo circumscripta.—List. H. Conch. tab. 138. fig. 42.

Planorbis minor fluviatilis acie acuta.—Petiv. Gaz. tab. 10. fig. 11.

Depress'd orbicular fresh-water cocklea, with four wreaths.—Wallis Northumb. p. 371.

II. Planorbis. Flat.—Penn. Brit. Zool. No. 123. tab. 83. fig. 123.

Cornu hammonis spurium limbo circumscripta.—Klein Ostrac. p. 5. § 12. No. 2. tab. 1. fig. 8.

Cochlea fluviatilis depressa, altera parte complanata, & limbo insignita quatuor spirarum.—Gualt. I. Conch. tab. 4. fig. F E.

Cochlea testa plana fusca, supra concava, anfractibus quatuor, margine prominulo.—Lin. F. Suec. I. p. 373. No. 1306. II. No. 2176.

Planorbis. Helix testa subcarinata umbilicata plana: supra concava, apertura oblique ovata, utrinque acuta.—Lin. S. N. p. 1242. No. 662.

This *shell* is flat and circular, the most general *size* about that of a silver penny, but of double that size occur, above half an inch in *diameter*, and above one-eighth of an inch *thick*; when *fresh* or *alive*, of a dark

Cette *coquille* est applatie & circulaire, la *grandeur* la plus generale est environ celle d'un sous d'argent, mais de double cette grandeur se trouvent, au delà d'un demi pouce en *diametre*, & plus d'un huitieme

dark brown or chesnut *colour*, with a slight gloss, is pretty thick and strong for its size, and is slightly *striated* in a transverse manner.

tieme de pouce en *epaisseur*; quand *recente*, ou *vivante*, d'une *couleur* brune chargée ou de chataigne, peu lustrée, est assez epaisse & forte pour sa grandeur, & légérement *striée* à travers.

The *upper side* is convex, for the wreaths swell on this side; the *mouth* of a pointed oval shape, large, and placed perpendicular, or to the thickness of the shell.

Le *dessus* est convexe, car les revolutions sont renflées de cet coté; la *bouche* est d'une forme ovale, pointue, grande, & posée perpendiculairement, ou selon l'epaisseur de la coquille.

The *under side* is quite flat, for the wreaths are so, and *border'd* all round on the outermost wreath with a regular and delicate sharp prominent *border*, well defined. This border is not conspicuous on the other wreaths, nor is it seen on the upper side, but only bounds it.

Le *dessous* est tout a fait applati, car les revolutions sont plattes, et *bordé* tout autour sur la revolution exterieure par une *bordure* reguliere, belle, aigue et saillante, tres bien marquée. Cette *bordure* n'est pas visible sur les autres revolutions, ni est elle vue en dessus, mais seulement borne cet coté.

The *wreaths* are four, and lie horizontal, or on a level; they are close, or contiguous, and distinguish'd or separated by a strong furrow; they gradually diminish to the center, which on both sides has a small round shallow *hole* or *umbilicus*.

Les *revolutions* sont quatre, couchées horizontalement, ou de niveau; elles sont serrées, ou contigues, et distinguées ou separées par une strie forte; elles diminuent successivement au centre, qui de deux cotés à un petit *umbilic* ou *cavité*, rond & peu profond.

The *inhabitant animal* is, according to Dr. Lister, black, with two red capillary horns. He observed them to couple about the middle of May; and it *yields* the same *scarlet humour* as the *cornu arietis* last described.

L'*animal habitant*, selon Dr. Lister, est noir, avec deux cornes capillaires rouges. Il les a observé s'accoupler vers la mi Mi-Mai, et elle *produit* la même *humeur ecarlate* que la *cornu arietis* dernierement decrite.

It is a *species* frequent in most stagnant waters and rivers of *England*.

C'est une *espece* frequente dans la plupart des eaux croupissantes & rivieres de l'*Angleterre*.

Tab. 4. *fig*. 10. represents the *upper side*; and the *under side*, with its *border*, is figured *tab*. 8. *fig*. 8.

Pl. 4. *fig*. 10. represente le dessus; & le *dessous*, avec sa *bordure*, est figurée *pl*. 8. *fig*. 8.

XXXVI. XXXVI.

XXXVI.

PLANORBIS. WHIRL.

Tab. IV. fig. 12.

XXXVI.

Pl. IV. fig. 12.

Cochlea exigua plana fine limbo. Planorbis. Tab. 4. fig. 12.

Cochlea exigua fubfufca, altera parte planior, fine limbo, quinque fpirarum.—Lift. H. An. Angl. p. 145. tit. 28. tab. 2. fig. 28.—Phil. Tranf. No. 105. fig. 25.

Cochlea exigua quinque orbium.—Lift. Conch. tab. 138. fig. 43.

Planorbis polygirata minor.—Petiv. Gaz. tab. 92. fig. 6. 7.—Morton Northampt. p. 417.

H. vortex. Whirl.—Penn. Brit. Zool. No. 124. tab. 83. fig. 124.

Cornu hammonis fpurium.—Klein Oftrac. p. 5. § 12. tab. 1. fig. 9.

Cochlea fluviatilis depreffa exigua, altera parte planius, fubflava, fine limbo, quinque fpirarum.—Gualt. I. Conch. tab. 4. fig. G. G.

Cochlea tefta plana fufca : fupra concava, anfractibus quinque, margine acuto.—Lin. F. Suec. I. p. 374. No. 1307. II. No. 2172.

Vortex. Helix tefta carinata ; fupra concava, apertura ovali plana.—Lin. S. N. p. 1243. No. 667.

This *fhell* is about half the *fize* of the foregoing, or about one-third of an inch in diameter, very *flat* and *thin*, fo as fcarcely to exceed card paper in thicknefs. The *under fide* is quite flat, with a *fharp edge* round it, not a *prominent border* as in the preceding. The fhell is of a horny or brownifh *colour*, brittle and fmooth.

The *upper fide* fomewhat convex, by the fwelling of the wreaths; the *mouth* roundifh and tranfverfe; the *wreaths* five, horizontal, or on a level, contiguous, and diftinguifh'd by a line; and it has no depreffion or umbilicus at the center.

This is another *fpecies* of aquatic helix, *frequent* in the ftagnant waters and rivers of England.

Cette *coquille* eft environ la *mi-grandeur* de la precedente, ou environ un tiers d'un pouce de diametre, fort *applatie & mince*, de forte qu'a peine elle furpaffe l'epaiffeur du carton. Le *deffous* eft totalement platte, environné d'un *bord tranchant*, pas une *bordure* qui fait *faillie*, comme dans la precedente. La coquille eft *couleur* de corne ou brunatre, fragile et liffe.

Le *deffus* eft quelque peu convexe, parceque les revolutions font enflées ; la *bouche* arrondie, & pofée à travers ; les *revolutions* font cinco, horizontales, ou fur un même plan, contigues, & diftinguées par une ftrie ; et elle n'a aucun enfoncement ou *umbilic* au centre.

C'eft une autre *efpece* de helix aquatique, *frequente* dans les eaux croupiffantes & les rivieres d'Angleterre.

XXXVII.

Crassa. Thick.

Tab. IV. fig. 11.

XXXVII.

Pl. IV. fig. 11.

Planorbis minima crassa, utrinque umbilicata, anfractibus subdepressis. Crassa. Tab. 4. fig. 11.

Planorbis minima crassa. Tiney many-circled thick river cheese shell.—Petiv. Gaz. tab. 92. fig. 8.

An Lin. F. Suec. I. p. 374. No. 1309? II. N°2177?—An *Helix complanata*. Lin. S. N. p. 1242. No. 663.

The *shell* is about the *size* of a hemp seed, flatted at top and bottom, but very *thick* or *deep*, and therefore seems somewhat globose, or like a small *bead*, of a brownish or horny *colour*, thin, brittle, dull and *smooth*, for even by a magnifier I cannot discover any striæ, &c. on it.

The *upper side* is convex, for the wreaths are swell'd, and they run hollow or are depress'd at the *center*, which has a large and deep *umbilicus*. The *mouth* runs the thickness of the shell, or *transversely*, and is *compressed*, or of a crescent *shape*. The *under side* is pretty strongly *umbilicated*, tho' the wreaths run level, or have no depression. The *wreaths* are five, and lie as it were flatted or compress'd on the fall of the back, or down it; for the *depth* or *globosity* of the *back*, in *regard* to its *size*, causes this apparent compression in the *wreaths* and *mouth*, as at the top and bottom they appear as usual, and are separated by lines or striæ.

This *species* is likewise found in our stagnant waters and rivers, but it is *rarely met with*. I had them from the *Thames*, near this *city*; and Petiver notes his, as sparingly, in the rivulets about *Peterborough-house*, *Westminster*.

Obs.—The *thickness* or *globosity* of the shell is not well express'd in the *figure*.

GENUS.

La *coquille* est environ la *grandeur* d'une semence de chanvre, applatie au sommet et au bas, mais fort *epaisse* ou *grosse*, c'est pourquoi elle paroît quasi ronde, ou comme un *grain* de *collier*, d'une *couleur* brunatre ou de corne, mince, fragile, terne & *lisse*, car même par la loupe je ne peus decouvrir là dessus aucunes stries, &c

Le *dessus* est convexe, parceque les orbes sont renflés, & ils abbaissent au *centre*, qui a un grand & profond *umbilic*. La *bouche* est *transversale*, ou court selon l'epaisseur de la coquille, est *comprimée* & en *forme* de croissant. Le *dessous* est asses fortement *umbiliqué*, quoique les orbes tournent sur un niveau, & ne sont point abbaissés. Les *orbes* sont cincq, & comme applatis ou comprimès sur la descente du dos; car la *profondeur* ou la *convexité* du *dos*, à *l'egard* de la *grandeur* de la coquille, est la cause de cette compression apparente dans les *orbes* & la *bouche*, pusque sur le haut & le bas ils sont comme á l'ordinaire, & sont separès par des lignes ou stries.

Cette *espece* se trouve aussi dans nos rivieres & eaux croupissantes, mais *rarement*. Je les ai trouvé dans la *Tamise*, près de cette *ville*; & Petiver remarque d'avoir trouvé les siens, dans les ruisseaux aux environs de *Peterborough-house* a *Westminster*.

Obs.—L'*epaisseur* ou la *rondeur* de la coquille n'est pas bien exprimée dans la *figure*.

GENRE.

GENUS X.

COCHLEÆ. SNAILS.

Turbinated ſhells, ſwell'd or ventricoſe, with a ſhort turban, or but little produced.

GENRE X.

COCHLEÆ, ou LIMAÇONS.

Coquilles contournées en ſpirale, renflées ou bombées, avec une clavicule courte, ou peu allongée.

* TERRESTRES. LAND.

XXXVIII.

POMATIA. ITALIAN.

Tab. IV. fig. 14. 14.

* TERRESTRES. TERRESTRES.

XXXVIII.

Pl. IV. fig. 14. 14.

COCHLEA magna cinereo rufeſcens, faſciata, leviter umbilicata. Pomatia. Tab. 4. fig. 14. 14.

Cochlea cinerea, maxima, edulis, cujus os operculo craſſo velut Gypſeo per hyemem clauditur. Pomatia.—Geſn. de Aquat. pp. 644. 255.—Liſt. H An Angl. p. 111. tab. 2. fig. 1. —Phil. Tranſ. No. 105. fig. 1.

Cochlea cinereo rufeſcens, faſciata, leviter umbilicata.—Pomatia Geſneri.—Liſt. H. Conch. tab. 48. fig. 46.

Cochlea pomatia edulis Geſneri.—Liſt. Exercit. Anat. I. p. 162. tab. 1.

Cochlea alba major cum ſuo operculo.—Merret Pin. p. 207.—Aubrey's Surrey, p. —Ray's Travels, p. 346.——*Cochlea alba major.* Muſ. Petiv. p. 4. No. 12.—Morton Northampt. p. 415.—Addiſon's Travels, p. 272.—Rutty Dublin, p. 379.

Helix pomatia, Exotic.—Penn. Brit. Zool. No. 128. tab. 84. fig. 128.

Varro de Re Ruſt. l. 3. c. 14.—Plin. Hiſt. Nat. l. 9. c. 56.—Macrob. Saturn. l. 3. c. 13.

Pomatia.—Argenv. Conch. I. tab. 32. fig. 1. p. 383. II. p. 338. tab. 28. fig. 1. p. 81. tab. 9. fig. 4.

Cochlea terreſtris vulgaris, maxima, albicans, pomatia.—Gualt. I. Conch. tab. 1. fig. A.

Cochlea teſta ovata quinque ſpirarum, pomatia dicta.—Lin. F. Suec. I. p. 369. No. 1293. II. No. 2183.

Pomatia. Helix teſta umbilicata ſubovata obtuſa decolore, apertura ſubrotunda lunata.—Lin. S. N. p. 1244. No. 677.

The *ſhell* is about the *ſize* of a ſmall egg, or a very large walnut, of a ſwell'd or globoſe *ſhape*, brittle and thin, and of a dull ſurface, or not gloſſy.

La *coquille* eſt environ la *grandeur* d'un petit oeuf, ou d'une fort grande noix, d'une *forme* ronde ou renflee, fragile & mince, & d'une ſurface terne, ou point luſtrée.

The *mouth* is roundiſh, *within* whitiſh and ſmooth, the edge or contour ſomewhat thickiſh, and turns outwards; likewiſe near and at the *pillar* it ſpreads, is thick, and very much edged outwards, and aſide of it, it has a pretty deep but narrow *umbilicus*.

The *body* is very ſwell'd, or ventricoſe; the *turban* is ſhort, or not much produced. They conſiſt of five convex or ſwell'd *wreaths*, ſeparated by a furrow or line.

The *outſide* is thickly wrought with longitudinal wrinkles. The ground *colour* is ſullied whitiſh, and has broad regular bands of pale brown or ruſſet, that run tranſverſe, or according to the turn of the wreaths; they are generally three: beſides, round the mouth, and about the umbilicus, it is alſo much tinged of the ſame brown or ruſſety colour, and ſome of the longitudinal wrinkles run in like ruſſety ſtreaks.

It is the *largeſt ſpecies* of land ſnail in England, and is found in hedges and woods. It cloſes its ſhell carefully againſt Winter, with a white thick cover or *operculum*, dull and like plaiſter, and in this cloſed ſtate it remains till the beginning of April, or warm weather, at which time it looſens the borders of the cover, and the animal creeps out of the ſhell for its neceſſary occaſions. Dr. Liſter informs us he kept one in his boſom about the beginning of March, when the animal, feeling the warmth, in a few hours diſengaged its cover, and crept out.

The animal being *large, fleſhy*, and not of an *unpleaſant taſte*, has been uſed for *food* in antient times: it was a favourite

diſh

La *bouche* eſt quaſi ronde, en *dedans* blanchatre & liſſe, le bord ou contour quelque choſe epaiſſi, et tourne en dehors; auſſi près de, et à la *columelle* il s'etend, s'epaiſſit & tourne beaucoup en dehors, & à coté il ſe trouve un *umbilic* aſſes profond, mais fort etroit.

Le *corps* eſt fort renflé ou ventru; la *clavicule* eſt courte, ou peu allongée. Ils conſiſtent de cincq *orbes* ou *revolutions*, convexes ou bombés, ſeparés par une ligne ou ſtrie.

L'*exterieur* eſt garni des rides longitudinales fort ſerrées. Le *fond* eſt blanchatre ſale, avec des bandes larges & regulieres d'une couleur brune pale, qui courent tranſverſalement, ou ſelon le tour des revolutions; elles ſont generalement trois: outre, autour de la bouche, & autour de l'umbilic, elle eſt auſſi beaucoup teinte de la même couleur brune pale, & quelques unes des rides longitudinales courent en des pareilles rayes brunatres.

C'eſt la plus *grande eſpece* de limaçon terreſtre en Angleterre, & ſe trouve dans les hayes & les bois. Il ferme ſa coquille ſoigneuſement à l'approche de l'Hyver, avec un couvercle ou un *opercule*, blanc, epais, terne & comme du platre, & en cet etat fermé il reſte juſques au commencement d'Avril, ou du tems temperè, auquel tems il detache les bords de l'opercule, et l'animal ſort de ſa coquille pour ſes occaſions neceſſaires. Le Dr. Liſter nous rapporte, qu'il a gardé un dans ſon ſein environ le commencement de Mars, quand l'animal, ſentant la chaleur, dans peu des heures a detachè ſon opercule, & ſortit.

L'animal etant *grand*, *charnu*, & pas d'un *gout deſagreable*, à eté *uſé en nourriture* anciennement: c'etoit une friandiſe parmi

les

dish with the *Romans*, who had their *cochlearia*, or *snail stews*, wherein they bred and fatten'd them. *Pliny* tells us, that the first inventor of this *luxury* was a *Fulvius Hirpinus*, a little before the civil wars between Cæsar and Pompey. *Varro* has handed down to us a description of the *stews*, and manner of *making* them: *He says*, open places were chose, surrounded by water, that the snails might not abandon them, and care was taken that the places were not much exposed to the sun or to the dews. The *artificial stews* were generally made under rocks or eminences, whose bottoms were water'd by lakes or rivers; and if a natural dew or moisture was not found, they form'd an artificial one, by bringing a pipe to it bored full of holes, like a watering pot, by which the place was continually sprinkled or moistened. The *snails* required *little attendance* or *food*, for as they crawl'd they found it on the floor or area, and on the walls or sides, if not hinder'd by the surrounding water. They were fed with *bran* and *sodden lees of wines*, or like substances, and a few *laurel* leaves were thrown on it.

les *Romains*, qui avoient leur *cochlearia*, ou *escargotoires*, dans lesquels ils nourissoient et engraissoient les limaçons. *Pline* nous dit, que le premier inventeur de cet *luxe* fut un *Fulvius Hirpinus*, un peu avant les guerres civiles entre Cæsar & Pompée. *Varron* nous a transmis la description des *escargotoires*, & la maniere de les *construire*: *Il dit*, des lieux ouverts furent choisis, environnés par de l'eau, afin que les limaçons ne l'abandonneroient pas, & l'on prenoit garde que les lieux n'etoient point beaucoup exposés au soleil, ou a la rosée. Les *escargotoires artificiels* furent construits pour la pluspart au dessous des rochers ou des eminences, dont les bas etoient arrosés par des lacs ou des rivieres; & si il ne se trouvoit une rosée ou une humidité naturelle, ils formoient une artificielle, en portant un tuyau percé de trous, comme un arrosoir, a la place, par lequel elle fut continuellement arrosée ou humectée. Les *limaçons* exigoient peu de *soin* ou de *nourriture*, car comme ils rampoient ils trouvoient leur nourriture sur le plancher ou l'aire, & sur les cotés ou murailles, si l'eau environnante ne les empechoient pas. On les nourrissoit avec du *son* & de *la lie de vin bouillie*, ou telles substances, & quelques feuilles de *laurier* y etoient mises.

Pliny tells us there were many sorts, as the whitish from *Umbria*, the large sort from *Dalmatia*, and the *African*, &c. This *particular kind* seems to be that he mentions, l. 8. c. 39. They propagate very much, and their spawn is very minute.

Pline nous raconte qu'il y avoit plusieurs sortes, comme les blanchatres de l'*Umbrie*, le grand limaçon de la *Dalmatie*, & le limaçon d'*Afrique*. Cette *sorte particuliere* paroit etre celle qu'il propose, l. 8. c. 39. Ils multiplient beaucoup, & leur fray est fort petit.

Varro is scarcely to be credited, when he says some would grow so *large* that their *shells held ten quarts*.

Varron est a peine croyable, quand il dit que quelques unes devenoient si *grands* que leur coquilles etoient capables de *contenir dix quartes*, ou à peu près *dix pintes de Paris*.

They

Ou

They were alſo fed and fattened in large *pots* or *pans*, ſtuck full of holes to let in the air, and lined with bran and ſodden lees, or vegetables.

They are yet uſed as *food* in ſeveral parts of *Europe*, more eſpecially during *Lent*, and are preſerved in *ſtews* or *eſcargotoires*, now a large place boarded in, and the floor covered with herbs, wherein they neſtle and feed.

In *Italy*, in many places, they are ſold in the markets, and are called *Bavoli*, *Martinacci* and *Gallinelle*; in many provinces of *France*, as *Narbonne*, *Franche Comté*, &c. and even in *Paris*. They boil them, ſays Dr. Liſter, in river water, and ſeaſoning them with ſalt, pepper and oil, make a hearty repaſt.

This *ſnail* is not *indigenous*, or originally a *native* of theſe *kingdoms*, but a *naturalized ſpecies*, that has throve ſo well as now to be found in very great quantities. It was firſt imported to us from *Italy* about the middle of laſt century, by a *ſcavoir vivre*, or *Epicure*, as an an article of food. Mr. *Aubrey* informs us, it was a *Charles Howard*, *Eſq*; of the *Arundel family*, who, on that account, ſcattered and diſperſed thoſe ſnails all over the downs, and in the woods, &c. at *Albury*, *an antient ſeat* of that *noble family*, near *Aſhted*, *Boxhill*, *Darking*, and *Ebbiſham* or *Epſom*, in *Surry*, where they have thriven ſo much that all that part of the *county*, even to the confines of *Suſſex*, abounds with them; inſomuch that that they are a nuſance, and far ſurpaſs in number the *common*, or any *other ſpecies* of *Engliſh ſnails*.

The

On les engraiſſoient auſſi dans des grands *pots* ou *terrines*, percès de trous pour admettre l'air, couvertes en dedans de ſon, & des lies bouillies, ou des vegeteaux.

Ils ſont encore en uſage comme un *aliment* en pluſieurs païs de l'*Europe*, plus ſpecialement pendant le *Careme*, & ſont gardés dans des *reſervoirs* ou *eſcargotoires*, qui ſont à preſent, des grandes places entrourées de planches, & l'aire eſt couverte des herbes, dans leſquelles ils ſe niehent & ſe nourriſſent.

En pluſieurs lieux de *l'Italie*, ils ſont vendus dans les marchés, & ſont appellés *Bavoli*, *Martinacci* & *Gallinelle*; en pluſieurs provinces de la *France*, comme *Narbonne*, la *Franche Comté*, &c. & meme a *Paris*. Il les bouillent, dit le Dr. Liſter, dans l'eau de riviere, & les aſſaiſſonnant avec du ſel, du poivre, & de l'huile, ſont un bon repas.

Cet *limaçon* n'eſt point *naturel*, ou primitivement un *natif* de ces *royaumes*, mais une *eſpece naturaliſée*, qui à tant reuſſi qu'il ſe trouve en tres grandes quantités. Il fut nous premierement porté de *l'Italie*, vers le milieu du ſiecle paſſé, par un *ſcavoir vivre*, ou un *Epicurien*, comme un *article* de *aliment*. Mr. *Aubrey* nous enſeigne qu'il etoit un *Charles Howard*, *Eſq*; de la *famille* de *Arundel*, qui, à cet effet, diſperſa ces limaçons ſur toutes les dunes, & dans les bois, &c. à *Albury*, *un ancienne maiſon de campagne* de cette *illuſtre famille*, prés de *Aſhted*, *Boxhill*, *Darking*, & *Ebbiſham* ou *Epſom*, au *comté* de *Surry*, ou ils ont tant reuſſis que toute cette partie du *comté*, meme juſques aux confins de *Suſſex*, en abonde, de ſorte que ils ſont une incommodité, & ſurpaſſent en nombre le *limaçon commun*, ou aucune *autre eſpece* de *limaçon Anglois*.

Les

The *Epicures*, or *sçavoir vivre*, of those days, followed this luxurious folly, and the snails were scattered or dispersed throughout the kingdom, but not with equal success; neither have records transmitted to posterity the *fame* of those *worthies* equal to the *Roman Fulvius Hirpinus*, except of *two*, the *one* Sir *Kenelm Digby*, who dispersed them about *Gothurst*, the seat of that family (now of the *Wrights*) near *Newport Pagnel*, in *Buckinghamshire*, where probably they did not thrive much, as they are not very frequent thereabout: the other *worthy* was a *Lord Hatton*, recorded by Mr. *Morton*, who scattered them in the coppices at his seat at *Kirby*, in *Northamptonshire*, where they did not succeed.

Dr. *Lister* found them about *Puckeridge* and *Ware*, in *Hertfordshire*; and observes they are abundant in the *southern parts*, but are not found in the *northern parts* of *this island*.

In *Surrey*, as before mentioned, they abound; in several other counties they are not uncommon, as in *Oxfordshire*, especially about *Woodstock* and *Bladen*; in *Gloucestershire*, in *Chedworth parish*, and about *Frog Mill*; in *Dorsetshire*, &c. but I have never heard that they are yet met with in any of the *northern counties*.

Les *Epicuriens*, ou *sçavoir vivre*, de ces jours, suivirent cette folie luxurieuse, & les limaçons furent repandus ou dispersés par tout le royaume, mais non pas avec egal succes; ni les archives ont elles transmis a la posterité la *renommée* de ces *grands hommes* egalement au *Romain Fulvius Hirpinus*, hormis de *deux*; un le *Chevalier Kenelm Digby*, qui les à dispersés aux environs de *Gothurst*, sa maison de campagne (a present de la famille de *Wrights*) près de *Newport Pagnel*, au comté de *Buckingham*, ou probablement ils ne reussirent point, car ils ne sont pas frequents la; l'autre *grand homme* etoit un *Lord Hatton*, celebrè par Mr. *Morton*, qui les a repandus dans les taillis a sa maison de campagne a *Kirby*, au comté de *Northampton*, ou ils n'ont point reussi.

Dr. *Lister* les a trouvé autour de *Puckeridge* & *Ware*, au comté de *Hertford*; & il observe qu'ils sont en abondance dans les *parties meridionales*, mais ne se trouvent dans les *parties septentrionales* de *cette isle*.

Au comté de *Surrey*, comme mentionné ci dessus, ils abondent; & en plusieurs autres comtés ils ne sont pas rares, comme au comté de *Oxford*, specialement aux environs de *Woodstock* & *Bladen*; au comté de *Gloucester*, dans la *paroisse* de *Chedworth*, & aux environs de *Frog Mill*; au comté de *Dorset*, &c. mais je n'ay jamais appris qu'ils ont ete trouvés jusques a present dans aucuns des *comtés septentrionaux*.

XXXIX.

XXXIX.

XXXIX.
VULGARIS. COMMON.
Tab. IV. fig. 1. 1.

Cochlea vulgaris fusca, maculata & fasciata. Vulgaris. Tab. 4. fig. 1. 1.

Cochlea vulgaris major, pulla, maculata & fasciata, hortensis.—Lift. H. An. Angl. p. 113. tit. 2. tab. 2. fig. 2.—Phil. Tranf. No. 105. fig. 5.

Cochlea hortensis nostra fusca, maculata & fasciata.—Lift. H. Conch. tab. 49. fig. 47. & tab. Anat. 53. & 102 a 105.—Merret Pin. p. 207.—Sibbald Scotiæ, 34.—Hooke Microgr. obf. 40.—Muf. Petiv. p. 4. No. 13.—Rutty Dublin, p. 379.

H. hortensis, Garden.—Penn. Brit. Zool. No. 129. tab. 84. fig. 129.

Argenv. Conch. I. p. 383. tab. 32. fig. 3. II. p. 338. tab. 28. fig. 3.

Cochleæ terrestris vulgaris, cinerea, aliquando pulla, fasciis quatuor fulvis distincta.—Gualt. I. Conch. tab. 1. fig. C. C. & Variety, fig. D. D.

Lucorum. Helix testa imperforata subrotunda levi fasciata, apertura oblonga fusca.—Lin. S. N. p. 1247. No. 962.

The *shell* in *shape* like the laft, but not quite half the *size*, thin and brittle, with a fomewhat fmooth but dull furface.

The *mouth* is roundifh; the *lip* thick, very much turned outwards when full grown, and forms a broad white border within fide. The *inner* or *pillar lip* is alfo thick, very fpread, and milk white. It has no *umbilicus*.

The *body wreath* very ventricofe; the *turban* confifts of four other convex *wreaths*, feparated by a furrow or line, but is little produced.

The *outfide* of the fhell has fine longitudinal wrinkles, fet clofe; the *ground* is of a dufky yellowifh brown, mottled and variegated by dark brown, or chocolate coloured fpots, which run in belts or rows, generally two narrow ones near the upper part of each wreath, and a very broad and narrow one at the bottom; but thefe two latter belts are much confufed together in moft fhells. The interftices between the dark fpots

XXXIX.
LIMAÇON COMMUN.
Pl. IV fig. 1. 1.

La *coquille* en *forme* reffemble à la derniere, mais n'eft pas tout à fait la moitie de fa *grandeur*, mince & fragile, avec une furface quelque peu liffe mais terne.

La *bouche* eft quafi ronde; la *levre* epaiffé, fort retrouffée en dehors quand adulte, & forme une bordure large & blanche en dedans. La *levre interieure* ou de la *columelle* eft auffi epaiffe, fort etendue, & blanc de lait. Elle n'a point d'*umbilic*.

L'*orbe exterieur*, ou le *corps*, eft fort renflé; la *clavicule* confifte de quatre autres *orbes* convexes, feparés par une ftrie ou ligne, mais elle eft peu allongée.

L'*exterieur*, ou la *robe*, à rides longitudinales fines, très ferreés; le *fond* eft jaunatre brun terne, variolé & bigarré des taches brunes obfcures, ou couleur de chocolat, qui font rangées en bandes, ou zones, généralement deux etroites près de la partie fuperieure de chaque orbe, & deux autres dont une eft large, & l'autre etroite au bas; mais ces dernieres zones font beaucoup brouillées enfemble dans la plufpart des

ſpots being ſtreaked and mottled by the ground colour, form a very agreeable marbled variegation in ſome ſhells; but in moſt the ground colour alſo forms regular intermediate belts to the brown ones.

The *animal* has four *horns*, the *two upper* and *larger* ones have each a *black ſpeck* which ſeem like *eyes*; but the *lower horns* have none.

It *feeds* on all kinds of vegetables and their fruits. In *Winter time*, it hides itſelf in hollow trees, crevices in the barks of old trees, holes, &c. It then *cloſes up* its mouth with a thin membranous *cover*, or *operculum*. They couple about the end of May, or beginning of June, lay plenty of eggs in July, which are large, whitiſh, ſlightly cluſtered together like grapes, and covered with a ſoft membrane. The ſnails have a quick growth.

Dr. *Liſter* made experiments on the *ſaliva*, or *ſpittle*, which they copiouſly emit from their mouths. It ſeems to be a very different *humour* from that got from them by *laceration* or *inciſion*. The *latter* is bluiſh, and more fluid; the *former*, or the *ſaliva*, which they freely emit, is thick, more ſlimy, and a little more coloured, or approaching to yellowiſh. *He* firſt tried plenty of both *liquors mixed together*, in a glaſs bottle, and, being well ſhaken, they united into a yellow jelly-like ſubſtance. In another bottle, he put ſome of the *humour* only, or by itſelf, which, though often and ſtrongly ſhaken, did not *congulate*, or *form* a *ſubſtance*.—Another *experiment* was, *he* put a number of ſnails in a cloth,

des coquilles. Les intervalles entre les taches obſcures etant rayés & variolés par le fond, forment une bigarrure marbrée tres agreable dans quelques coquilles; mais dans la plupart la couleur du fond forme auſſi des bandes ou faſcies intermediates aux faſcies brunes.

L'*animal* a quatre *cornes*, dont les *deux ſuperieures* & *plus longues* ont chacune une petite *tache noire* qui reſemble *des yeux*; mais les *cornes inferieures* ſont depourvues.

Il ſe *nourrit* de toutes ſortes des vegetaux & leur fruits. En *Hyver* il ſe cache dans les arbres creux, les crevaſſes dans l'ecorce des vieux arbres, & dans les trous, &c. Alors il *ferme* ſa bouche avec un *couvercle* ou *opercule* mince & membraneux. Ils accouplent vers la fin de May, ou le commencement de Juin, pondent leur oeufs en abondance en Juillet, qui ſont grands, blanchatres, & en pelotons, comme une grape de raiſins, & couverts d'une membrane tendre. Les limaçons croiſſent vitement.

Le Dr. *Liſter* à fait des experiences ſur la *ſalive* qu'ils jettent en abondance de leur bouche. Elle paroit etre une *humeur* fort differente de celle qu'ils donnent par la *laceration*, ou par une *inciſion*. La *derniere* eſt bleuatre, & plus fluide; la *premiere*, ou la *ſalive*, quils jettent librement, eſt epaiſſe, plus gluante, & une peu plus colorée, ou approchante au jaunatre. *Il* a premierement fait l'experience ſur une grande quantité des deux *liquers melées enſemble*, dans une bouteille de verre, & qui, etant bien branlées, elles ſe ſont unies dans une ſubſtance jaune comme une gelée. Dans une autre bouteille il mit de l'*humeur* ſeule, ou par ſoi meme, qui, quoique ſouvent, & fortement ebranlée,

cloth, and ſprinkled them with a little ſalt; but they did not emit any other *humour* than a *yellowiſh, ſlimy ſaliva. He* then took them out of the cloth, and wounding them by inciſions, they emitted plenty of a *blueiſh humour*, which put into a ſpoon, and ſuffered to evaporate gently over a fire, became inſtantly a *white jelly*, like the *ſerum* of the *blood*, with which he thinks it has great affinity. This *blueiſh humour* tried with *ſalt* or *allum*, does not *precipitate* any *colouring parts*, as the *ſcarlet humour* of the *cornu arietis*, deſcribed *ſpecies XXXIV. ſupra*; but *hardens* into a kind of *blueiſh, gluey matter*.

lée, ne s'eſt pas *caillée* ou produit une *ſubſtance*. Une autre *experience* fut, *il* mit un nombre de limaçons dans une toile, & les ſaupoudra avec un peu de ſel, mais ils n'ont donné d'autre *humeur* que une *ſalive jaunatre gluante*. Alors *il* les ota hors de la toile, & les bleſſant par des inciſions, ils jetterent quantité d'une *humeur bleuatre*, qui miſe dans une cuillere, & la laiſſant evaporer doucement ſur le feu, elle devint immediatement une *gelée blanche*, comme le *ſerum* du *ſang*, avec lequel il penſe qu'elle a beaucoup de rapport. Cette *humeur bleuatre* miſe à l'*epreuve* avec du *ſel* ou *alun*, ne *precipite* pas des *parties colorantes*, comme l'*humeur ecarlate* du *cornu arietis*, decrite *eſpece* XXXIV. *ſupra*; mais ſe *durcit* dans une ſorte de *matiere bleuatre*, & *comme la colle*.

The *ſnails* are often uſed with ſucceſs in *hectic* or *conſumptive* caſes. The *ſaliva* or *ſpittle* is uſed in the *manufactories* for *bleaching wax* at *Montpelier*, where, *ſays* the *Doctor*, he ſaw them bruiſe the ſnails, and line the inſides of the moulds for the melted wax, that the wax might not ſtick in them, but come out freely, or with eaſe. Dr. *Liſter* alſo gives a receipt for a ſort of *cement* to be made of the *humour* from *ſnails*. Take, *ſays he, equal parts* of *quick lime, white* of *eggs*, and *this humour* got from the *ſnails* by *inciſion*, mix them together, and grind them on a marble; it makes an excellent *cement* to join broken *ſtones, china*, &c. ſo as to faſten the fractured pieces together immediately.

Les *limaçons* ſont ſouvent donnés avec ſucces dans les *fiévres* étiques ou les *conſomptions*. La *ſalive* eſt uſée dans la *manufacture* à *blanchir la cire* a *Montpelier*, ou, *dit* le *Docteur*, il les a vu concaſſer les limaçons, & induire le dedans des moules pour la cire fondue, de ſorte que la cire ne s'attache aux moules, mais ſe tire librement, & avec facilité. Dr. *Liſter* nous donne auſſi la recette à faire une ſorte de *ciment* de cette *humeur* de *limaçons*. Prenés, *dit il*, des *parties egales* de *chaux vive*, de *blanc des oeufs*, & cette *humeur* des *limaçons* tirée par l'*inciſion*, melées les enſemble, & broyées les ſur un marbre; il fait un *ciment* excellent à cimenter des *pierres* ou de la *porcelaine caſſée*, &c. de ſorte que de les affermir a l'inſtant.

This *common ſnail* is found in woods, hedges, gardens, &c. in very great *abundance* throughout theſe *kingdoms*.

Cet *limaçon commun* ſe trouve en grande *abondance* dans les bois, les haies, les jardins, &c. partout ces *royaumes*.

XL.

XL.

XL.

UNIFASCIATA. SINGLE STREAK.
Tab. XVII. fig. 6.

XL.

Pl. XVII. fig. 6.

Cochlea subumbilicata, maculata, unica fascia angusta, per medium anfractus insignita. Unifasciata. Tab. 17. fig. 6.

Cochlea maculata, unica fascia pulla, angustioreque, per medium anfractus, insignita.—List. H. An. Angl. p. 119. tit. 4. tab. 2. fig. 4.

Cochlea subflava, maculata, atque unica fascia castanei coloris per medium anfractus, insignita. Phil. Trans. No. 105. fig. 4.—*Dr. Lister's marbled meadow snail.* Mus. Petiv. p. 5. No. 15.

Cochlea maculata, unica fascia fusca, per medium orbem insignita.—List. H. Conch. tab. 56. No. 53.

Helix arbustorum. Shrub.—Penn. Brit. Zool. No. 130. tab. 85. fig. 130.

Cochlea testa utrinque convexa subcinerea: fascia solitaria grisea, labro reflexo.—Lin. F. Suec. I. p. 370. No. 1295. II. No. 2184.

Arbustorum. Helix testa umbilicata convexa acuminata, apertura suborbiculari bimarginata: antice elongata.—Lin. S. N. p. 1245. No. 680.—Mus. Reg. p. 668. No. 371.

The *shell* is of the *size* of a grape, rather thick, *round*, smooth, or with but a few fine or scarcely visible longitudinal wrinkles, and glossy.

The *mouth* semilunar; the *lip* thick, and much turned outward, and brown, but *within* it forms a thick, broad, regular milk white border all round, and at the *pillar* is a little spread. The *rest* of the *inside* of the *mouth* is brownish. Aside of the *pillar* it is *subumbilicated*, or has a narrow shallow *cavity*.

It consists of five *wreaths*, only laid on each other, and separated by a furrow; the *turban* is therefore depress'd, or not produced.

The *colours* vary a little in the tints; but when a good *live shell*, the *ground* is of a light chesnut colour, finely and thickly marbled with small short transverse streaks and

La *coquille* est de la *grandeur* d'un raisin, plûtôt epaisse, *ronde*, lisse, ou avec peu de rides longitudinales fines ou à peine visibles, & lustrée.

La *bouche* est en croissant; la *levre* epaisse, beaucoup retroussée en dehors, & brune, mais en *dedans* elle forme une bordure reguliere, epaisse, & blanche de lait, tout autour, & à la *columelle* est un peu etendue. Le *reste* du *dedans* de la *bouche* est brunatre. A coté de la *columelle* elle est un peu *umbiliquée*, ou à une *cavite* peu profonde & etroite.

Elle a cinq *orbes*, surmontant l'un l'autre, & separés par une strie; c'est pourquoi la *clavicule* est applatie, ou point allongée.

Les *couleurs* varient un peu dans leur coloris; mais quand c'est une bonne *coquille* & *vivante*, le *fond* est couleur de chataigne clair, finement & grandement marbrée

and dots, of a dark chocolate colour, in a very *beautiful* manner, and has always a *single* regular narrow *line* or *belt*, also of a dark chocolate colour, that runs along the *middle* of each *wreath*, *spirally*, or according to the *turn* of them.

marbrée par des petites lignes courtes, placées à travers, & des points, d'une couleur de chocolat foncée, dans une maniere tres *elegante*, & a toujours une *seule fascie* ou *bande*, reguliere & etroite, aussi de couleur de chocolat foncée, qui court le long du *milieu* de chaque orbe, en *spirale*, ou selon leur *tour*.

This *species*, according to Dr. *Lister* and Mr. *Petiver*, is found in shady hedges, and is frequent in rushy, moist or marshy meadows. It is also found in coppices and shrubberies; *but*, by my *observations*, is not *frequent*, or a *common kind*. I have seen them from *Surrey*, *Hampshire*, *Wiltshire*, and some other counties.

Cette *espece*, selon le Dr. *Lister* & M. *Petiver*, se trouve dans les hayes ombragées, & est frequente dans les lieux humides, ou marecageaux & plein de joncs. Elle se trouve aussi dans les taillis & arbrisseaux; *mais*, par mes *observations*, elle n'est pas *frequente*, ou *commune*. Je les ai vu des comtés de *Surrey*, *Hants*, & *Wilts*, & quelques autres comtés.

XLI.

FASCIATA. GIRDLED.

Tab. V. fig. 1. 2. 3. 4. 5. 8. 14 & 19.

XLI.

LA LIVRÉE.

Pl. V. fig. 1. 2. 3. 4. 5. 8. 14. & 19.

Cochlea imperforata, interdum unicolor, interdum variis fasciis depicta. Fasciata. Tab. 5. fig. 1. 2. 3. 4. 5. 8. 14 & 19.

Cochlea citrina aut leucophæa, non raro unicolor, interdum tamen unica, interdum etiam duobus, aut tribus, aut quatuor, plerumque vero quinis fasciis pullis distincta.—List. H. An. Angl. p. 116. tit. 3. tab. 2. fig. 3.

Cochlea vulgaris, & colore & fasciis multa varietate ludens.—Phil. Transf. No. 105. fig. 3. Muf. Petiv. p. 5. No. 14.—*Cochlea vulgaris testa variegata.* Merret Pin. p. 207. Wallace Ork. p. 39.

Cochlea interdum unicolor, interdum variegata, item variis fasciis depicta.—List. H. Conch. tab. 57. fig. 54.

Cochlea terrestris vulgatissima variegata. Common girdled hedge snail.—Petiv. Gaz. tab. 91. fig. 9. 10. 11. 12. tab. 92. fig. 9. 10.

Helix nemoralis. Variegated.—Penn. Brit. Zool. No. 131.

Argenv. Conch. I. p. 383. tab. 32. fig. 8. II. p. 338. & p. 82. tab. 28. fig. 8. & tab. 9. fig. 5. *La livrée.*

Gualt. I. Conch. tab. 1. fig. H. I. L. M. P. & Q. tab. 2. fig. D. F.

Cochlea

Cochlea testa utrinque convexa flava: fascia subsolitaria fusca, labro reflexo.—Lin. F. Suec. I. p. 370. No. 1294. II. No. 2186.

Nemoralis. Helix testa imperforata subrotunda, lævi diaphana fasciata, apertura subrotunda lunata.—Lin. S. N. p. 1247. No. 691.—Muf. Reg. p. 670. No. 376.

The *shell* is very *thin*, *pellucid* and *brittle*, *glossy*, and somewhat *larger* than a cherry, *globose*, and thick set with fine longitudinal *striæ*.

The *mouth* is semilunar, the *exterior lip* thin and turned outward, the *interior* or *pillar lip* very much spread, and it has no *umbilicus*.

It has five convex *wreaths*, but the *turban* is very little produced.

This *species* varies exceedingly, not only in its *colours*, but in the *situation* and *number* of its *bands* or *girdles*. All the *sorts* are very *beautiful shells*. It might be *imagined* that this *variety* of colourings is owing to their being different *species*, of different *growths*, or to the difference of *sexes*; but, as all the *sorts* indiscriminately *couple together*, that cannot be the *cause*.

The *principal varieties* are,

1. *Uniform*, of a *pale citron colour*, or *yellow* of *different shades*: the *mouth* finely border'd, within and without, with a dark brown, and with a brownish shade or cloud on so much of the body wreath as lies within the mouth, or from the outer lip quite across to the edge of the pillar. Pretty *frequent*.

2. *Uniform*, of a *flesh colour* of *different shades*, with the *mouth* in like manner border'd with dark brown, and the body wreath also shaded exactly the same as the last. Not very *frequent*.

3. *Uniform*,

La *coquille* est fort *mince*, *transparente* & *fragile*, *lustrée*, & au delà en *grandeur* d'une cerise, *ronde*, & à *stries* fines & longitudinales fort serrées.

La *bouche* est en croissant, la *levre exterieure* mince & retroussée en dehors, l'*interieure* ou celle de la *columelle* fort etendue, & elle n'a point d'*umbilic*.

Elle a cinq *orbes* convexes, mais la *clavicule* est fort peu allongée.

Cette *espece* varie extremement, non seulement dans ses *couleurs*, mais dans la *situation* & le *nombre* de ses *bandes* ou *fascies*. Toutes les *sortes* sont de tres *belles coquilles*. On se pouvoit *imaginer* que cette *varieté* de coloris est due à cause qu'elles sont de differentes *especes*, de differents *ages*, ou à la difference des *sexes*; mais, comme toutes les *sortes s'accouplent ensemble* sans distinction, cela ne peut etre la *cause*.

Les *varietés principales* sont,

1. D'une *couleur uniforme* ou *simple*, *pale citron*, ou *jaune* de *differentes nuances*: la *bouche* fort bien bordée, en dedans & au dehors, d'une couleur brune foncée, avec une ombre ou nuage obscur brunatre sur autant de l'orbe du corps qui est au dedans de la bouche, ou à travers de la levre exterieure jusques au bord de la columelle. Asses *frequentes*.

2. D'une *couleur uniforme* ou *simple*, de *chair* ou *incarnate*, de *differentes nuances*, avec la *bouche* bordée de la même maniere de brun foncé, & l'orbe du corps aussi ombragé exactement de même que la derniere. Point fort *frequente*.

3. D'une

3. *Uniform* of *different degrees* of *brown*, with the ſame circumſtances. *Common.*

4. The *ground yellow*, or *greeniſh yellow* of *different ſhades*, with a regular *ſingle* ſpiral *girdle*, or according to the turn of the wreaths, in the very *middle* of each wreath, with the brown border round the *mouth*, and the ſhade or cloud on the *body*. Pretty *frequent.*

5. The *ground fleſh colour* of *different ſhades*, variegated in like manner with a *ſingle girdle*, the *border* round the *mouth*, and on the *body*. *Not very frequent.*

6. *Many* dark-brown ſpiral *girdles* on the *yellow*, *fleſh*, or *browniſh grounds*, ſometimes to *five girdles* at leaſt on the body wreath; ſometimes only *four*. Theſe *girdles* are of different *breadths*, ſome being very *narrow*, like ſtreaks, others *broader*, like belts; and others ſo *extremely broad* as to cover the parts, and make the *ground colour* only appear in *girdles*. They are alſo not *equidiſtant* or regularly ſet; but the *very broad girdles* lie moſt generally on the upper part of the ſhells. Theſe *girdled ſorts* are the *moſt frequent* or *common.*

This *beautiful ſpecies* is extremely *common every where* in woods, hedges, gardens, &c. throughout theſe *kingdoms.*

XLII.

3. D'une *couleur uniforme* de *differentes nuances* de *brun*, avec les memes circonſtances. *Communes.*

4. Le *fond jaune*, ou *verdatre jaune* de *differentes nuances*, avec une *ſeule faſcie* reguliere & ſpirale, ou ſelon le tour des orbes, dans le *milieu* meme de chaque orbe, avec la bordure brune autour de la *bouche*, & l'ombre ou le nuage ſur le *corps.* Aſſez *frequente.*

5. Le *fond couleur de chair* ou *incarnate* de *differentes nuances*, varié de la meme maniere avec une *ſeule faſcie*, la bordure autour de la *bouche*, & le nuage ſur le *corps.* *Point fort frequentes.*

6. Avec *pluſieurs faſcies* ſpirales & brunés foncées ſur les fonds *jaunes incarnates* & *brunatres*, quelquesfois juſques à *cinq faſcies* pour le moins ſur l'orbe du corps; quelquefois ſeulement *quatre*; ces *faſcies* ſont de differentes *largeurs*, quelques unes etant fort *etroites*, comme des raies, d'autres *plus larges*, comme des baudriers; & d'autres ſi *extremement larges* que de couvrir ces parties, & faire le *fond* paroitre ſeulement comme en *faſcies*. Elles ne ſont auſſi regulieres ni poſées à *meme diſtances*; mais les *faſcies fort larges* ſe trouvent le plus ſouvent ſur la partie ſuperieure des coquilles. Ces *ſortes à faſcies* ſont les plus *frequentes* ou *communes.*

Cette *eſpece elegante* eſt extremement *commune par tout* dans les bois, les haies, les jardins, &c. dans ces *royaumes.*

XLII.

XLII.
VIRGATA. STRIPED.
Tab. IV. fig. 7.

XLII.
Pl. IV. fig. 7.

Cochlea umbilicata alba Virgata. *Virgata* tab. 4. fig. 7.

Cochlea alba leviter umbilicata pluribus fasciis circumdata, clavicula productiore. Lift. H. Conch. tab. 59. fig. 56.

Cochleola alba fasciata Cantabrigiensis, umbilico parvo. *Newmarket heath shell.*—Petiv. Gaz. tab. 17. fig. 6.

Helix Zonaria. Zoned.—Penn. Brit. Zool. No. 133. tab. 85. fig. 133.

Zonaria. H. testa umbilicata convexa depressiuscula, apertura oblongiuscula marginata.—Lin. S. N. p. 1245. No. 681.—Gualt. I. Conch. tab. 3. fig. L. LL.

The *shell* is thin, semipellucid and brittle; *size* of a currant, or small grape, *round*, the *surface* smooth, but not glossy.

The *mouth* roundish, the *outer lip* thin, and turns backward; the *inner* or *pillar lip* alfo turns backward, and is very much spread. It has a round, deep, central *umbilicus*.

The *upper part* very convex or swelled, the *bottom* or *turban* is somewhat depressed; the whole shell has five *wreaths*, separated from each other by a furrow.

The *ground* is most generally white, with a glance of a pale rust colour. On the *top*, almost to the edge of it, or round the umbilicus, it is white; from near the edge to the middle of the body wreath, it is adorned with numerous *circular streaks*, fine as hairs, some whereof are transversly *mottled*: They are all dark brown. In the *middle* of the *wreath* succeeds a broad white girdle, then a broad one of dark brown, and a very broad white one follows to the *bottom* of the *body wreath*; but all the *wreaths* of the *turban* have only *two broad girdles*, the upper one dark brown, the under one white.

This

La *coquille* est mince, demitransparente & fragile; de la *grandeur* d'un raisin de corinthe, ou un petit raisin, *rond*; la *surface* lisse, mais point lustrée.

La *bouche* est quasi ronde, la *levre exterieure* mince, & retroussée en dehors; l'*interieure* ou celle *de la columelle*, aussi retroussée & beaucoup etendue. Elle a un *umbilic* rond, central & profond.

La *partie superieure* est fort convexe ou bombée, le *bas* ou *le clavicule* est quelque peu applati; la coquille entiere a cinq *orbes*, separés par une strie.

Le *fond* est pour l'ordinaire blanc, avec une nuance couleur de rouille pale. Sur le *sommet*, presque à son bord, ou autour de l'umbilic, elle est blanche; de près du bord jusques au milieu de l'orbe du corps, elle est ornée de *rayes circulaires* nombreuses fines comme des cheveux, quelques unes desquelles sont *tacbetées* transversalement: Elles sont toutes brune-obscure. Dans le *milieu* de l'*orbe* suive une bande large blanche, apres une autre large brune obscure, & une bande blanche tres large, suit jusques au *bas* de l'*orbe du corps*; mais touts les *orbes* de la *clavicule* ont seulement *deux bandes larges*, la superieure brune-obscure, l'inferieure blanche.

Cette

This *pretty species* inhabits dry sandy soils and banks. It is *common* only in *some parts*, as in the grass on *Heddington heath*, in *Oxfordshire*, in plenty; also in plenty in *Hampshire*. It is likewise found in *Cornwall*; and Mr. Petiver found it on *Newmarket heath*, in *Cambridgeshire*.

Cette *jolie espece* se trouvé dans les bancs & les terreins secs & sablonneux. Elle est *commune* seulement dans *quelques parts*, comme sur l'herbe en la *Bruyere de Heddington*, au comté de *Oxford*, en abondance; aussi en abondance au comté de *Hants*. Elle se trouvé pareillement en *Cornwall*; & M. Petiver les a trouvé dans la *Bruyere de Newmarket*, au comté de *Cambridge*.

XLIII.

RUFESCENS. REDDISH.

Tab. IV. fig. 6.

XLIII.

Pl. IV. fig. 6.

Cochlea umbilicata & striata dilute rufescens. Rufescens. Tab. 4. fig. 6.

Cochlea dilute rufescens, aut subalbida, sinu ad umbilicum exiguo, circinata.—Lift. H. An. Angl. p. 125. tit. 12. tab. 2. fig. 12. & H. Conch. tab. 71. *muta*; the *two lower figures* mark'd *A*.—Morton Northampt. p. 415.

Helix rufescens. Mottled.—Penn. Brit. Zool. No. 127. tab. 85. fig. 127.

Cochlea terrestris depressa, & umbilicata mellei coloris, labio candido repando, sinu ad umbilicum exiguo circinato.—Gualt. I. Conch. tab. 3. fig. M.

The *shell* is thin and semipellucid, *roundish*, and of the *size* of a currant; the *surface* somewhat *glossy*, and thickly and finely *striated* lengthways.

The *mouth* oblong; the *outer lip* turn'd a little backward; the *inner* or *pillar lip* also, and very spread or broad: the *umbilicus* central, large, and very deep.

The *top* is convex or swell'd. The *wreaths* are five: at the *middle* of the *body wreath* it edges very slightly out, or is somewhat *subcarinated*. The *turban* is much depress'd, for the four *wreaths* are merely laid on one another, and are separated by furrows.

The

La *coquille* est mince & demi-transparente, *quasi ronde*, & de la *grandeur* d'un raisin de Corinthe; la *surface* quelque peu *lustrée*, & *striée* à stries fines & serrées selon sa longeur.

La *bouche* est oblongue; la *levre exterieure* retroussée un peu en dehors; l'*interieure*, ou celle de la *columelle*, aussi retroussée, & fort etendue ou large: l'*umbilic* central, grand, & fort profond.

Le *sommet* est convexe ou bombé. Les *orbes* cincq en nombre: sur le *milieu* de l'*orbe du corps* elle deborde fort légerement, ou est quelque peu *en careune*. La *clavicule* est fort applatie, car les quatre *orbes* sont simplement posés l'un sur l'autre, & separès par des stries.

La

The *shell* is *pale brownish* (for the *whitish* ones, or *subalbidæ* of Dr. *Lister*, are *discoloured* by the weather) and has a few longitudinal rays of a darker shade, especially round the mouth, and a *circular* light *streak* appears on the body whirl, where it *subcarinates*.

This *species* is not *very common*. I have received it from *Cornwall* and *Hampshire*. It is pretty frequent about *Leeswood*, the *seat* of my honoured friend Richard Hill Waring, Esq; in *Flintshire*, between the bark and wood of trees thrown down and decayed, especially alders. Dr. *Lister* found it in plenty about *Tadcaster*, in the woods, and hedges of marshy and shady meadows, and in like places throughout *Craven* in *Yorkshire*; and *observes* that there is a *variety* (if not a different *species*) in *Kent*, somewhat larger, lighter coloured, and with a smaller umbilicus: and Mr. Morton found it at *Morsley*, and the other *Northamptonshire* woods.

La *coquille* est *pale brunatre* (car les *blanchatres*, ou *subalbidæ* du Dr. *Lister*, sont *decolorées* par le tems) et à quelques rayes longitudinales plus foncées, especialement autour de la bouche, & une *raie circulaire* plus claire paroit sur l'orbe du corps, ou il est *un peu à carenne*.

Cette *espece* n'est point *fort commune*. Je l'ai reçu de *Cornwall* & du comté de *Hants*. Elle est asses frequente aux environs de *Leeswood*, la *maison de campagne* de mon tres honoré ami Richard Hill Waring, Esq; au comté de *Flint*, entre l'ecorce & le bois des arbres renversés & pourris, particulierement des aunes. Le Dr. *Lister* l'a trouvé en abondance aux environs de *Tadcaster*, dans les bois, & les haies des prairies marecageuses & ombragées, & en tels lieux, par tout *Craven*, au comté de *York*; & il *observe*, qu'il y à une *varieté* (si non une *espece* differente) en *Kent*, un peu plus grande, de couleur plus claire, & avec un umbilic plus petit: & M. Morton l'a trouvé a *Morsley*, & les autres bois du comté de *Northampton*.

** FLUVIATILES. River.

XLIV.

Vivipara. Viviparous.

Tab. VI. fig. 2. 2.

** FLUVIATILES. Fluviatiles.

XLIV.

Pl. VI. fig. 2. 2.

Cochlea fusco viridescens trifasciata. Vivipara. Tab. 6. fig. 2. 2.

Cochlea maxima fusca sive nigricans, fasciata. List. H. An. Angl. p. 133. tit. 18. tab. 2. fig. 18.—*Cochlea fasciata ore ad amussim rotundo.*. Phil. Trans. No. 105. fig. 17. —*Cochlea maxima viridescens fasciata vivipara.* List. Exercit. Anat. 2. p. 17. tab. 2.— *C. vivipara fasciata fluviatilis.* List. H. Conch. tab. 126. fig. 26.—*C. vivipara altera nostras testa tenuiori fluvii Cham.* Ib. Mantissa, tab. 1055. fig. 6.

Plot Oxfordsh. c. 7. p. 186.—Morton Northampt. p. 417.—Mus. Petiv. p. 84. No. 814 & 815.

Helix vivipara. Viviparous.—Penn. Brit. Zool. No. 132. tab. 84. fig. 132.

Saccus ore integro.—Klein Oſtrac. p. 43. § 121. ſpec. 2. No. 3.

Argenv. Conch. II. Zoom, p. 73. tab. 8. fig. 2.

Buccinum fluviatile fuſcum, ſive nigricans, faſciatum, quinque orbibus prœditum.—Gualt. I. Conch. tab. 5. fig. A.

Cochlea teſta oblongiuſcula obtuſa, anfractibus teretibus, lineis tribus lividis.—Lin. F. Suec. I. p. 375. No. 1312. II. No. 2185.

Vivipara. Helix teſta imperforata, ſubovata obtuſa cornea: cingulis fuſcatis, apertura ſuborbiculari.—Lin. S. N. p. 1247. No. 690.

The *ſhell* is thin and tranſparent, *ſize* of a walnut, of a produced or taper *ſhape*, *ſmooth*, or at moſt ſet with fine longitudinal *ſtriæ* or wrinkles, and ſomewhat gloſſy.

La *coquille* eſt mince & tranſparente, de la *grandeur* d'une noix, d'une *forme* allongée, *liſſe*, ou tout au plus à *ſtries* ou rides longitudinales fines, & quelque peu luſtrée.

The *mouth* is oval, the *outer lip* thin, the *inner* or *piller lip* very much ſpread, and aſide of it is a narrow and pretty deep *umbilicus.*

La *bouche* eſt ovale, la *levre exterieure* mince, l'*interieure* ou celle de la *columelle* beaucoup etendue, & à coté ſe trouve un *umbilic* etroite & aſſes profond.

The *wreaths* are five, cylindric and ſwell'd; they run taper to the bottom, which ends very ſharp pointed, and are ſeparated by a great depreſſion.

Les *orbes* ſont cincq, cylindriques & bombés; ils vont en appetiſſant au bas, qui finit dans une pointe fort aigue; ils ſont ſeparès par un grand abbaiſſement.

The *ground colour* is yellowiſh olive green, with three narrow regular ſpiral *girdles*, or according to the run of the wreaths, on each wreath, of a *duſky brown colour.*

Le *fond* eſt de *couleur* jaunatre verd d'olive, avec trois *bandes*, etroites, regulieres & ſpirales, ou ſelon le tour des orbes, ſur chaque orbe, d'une *couleur brune obſcure.*

The *inhabitant animal* has a head ſomewhat like that of a bull. The *Swedes*, from that *circumſtance*, *vulgarly call* it (according to *Linné Faun. Suec. I.*) *bull head*, and ſome *French authors*, *limaçon à tete de bœuf.* The *horns* are thick and pointed, and it is all over elegantly *mottled* with numerous ſmall brown ſpots, on a *yellow ground.* It's *lid*, or *operculum*, is teſtaceous, thinniſh, reddiſh, tranſparent, and ſet with concentric ſtriæ. It *feeds* on duck weed and other aquatic plants.

L'*animal habitant* a la tete quelque choſe reſſemblante à celle d'un bœuf. De cette *circonſtance* les *Suedois la nomment communement* (ſelon *Linné Faun. Suec. I.*) *tete de bœuf*, & quelques *auteurs François*, *limaçon à tete de bœuf.* Les *cornes* ſont groſſes & pointues, & il eſt par tout elegamment *bigarré* par des petites taches brunes, en grand nombre, ſur un *fond jaune.* Son *couvercle*, ou *opercule* eſt teſtacé, un peu mince, rougeatre, tranſparent, & à ſtries concentriques. Il ſe *nourrit* du lentille ſauvage & autres plantes aquatiques.

This *kind* is found, in great abundance, in all our *rivers* and *ſtagnant waters.*

Cette *eſpece* ſe trouve, en abondance, dans toutes nos *rivieres* & *eaux croupiſſantes.*

XLV.

*** MARINÆ. Sea.

XLV.

Catena. Chain.

Tab. V. fig. 7.

Cochlea umbilicata albo rufescens fasciis maculatis, maxime ad imos orbes distincta. Catena. Tab. 5. fig. 7.

Cochlea rufescens, fasciis maculatis, maxime ad imos orbes distincta.—List. H. An. Angl. p. 163. tit. 10. tab. 3. fig. 10.

Cochlea sublivida, ore fusco, ad basin cujusq; orbis velut funiculus depingitur.—List. H. Conch. tab. 568. fig. 19. & fig. 20? & tab *muta* 561. fig. 8.

Cochlea marina orbibus catenatis. Petiv. Gaz. tab. 93. fig. 7.—*Sea snail,* & *Cochlea parva.* Dale Harw. p. 379. No. 1. 3. Phil. Transf. No. 249.—*Cochlea alba, lunaris, rufescens, fasciis maculatis distincta.* Borlase Cornw. p. 276. Wallace Orkn. p. 40.

Nerita glaucina. Livid.—Penn. Brit. Zool. No. 141. tab. 87. fig. 141.

Platystoma & cassis.—Klein Ostrac. p. 15. § 40. No. 6. & p. 92. § 234, 235. No. 7.

Cochlea marina umbilicata, lævis, punctis subrufis, vel ex livido citrino depicta, vel aliquando candida.—Gualt. I. Conch. tab. 67. fig. E. F. G.

Cochlea subrotunda obtusa umbilicata: fasciis quinque maculis ferrugineis sagittatis, secunda lineis undulatis.—Lin. F. Suec. I. p. 378. No. 1324. II. No. 2197.

Glaucina. Nerita testa umbilicata lævi, spira obtusiuscula, umbilico semiclauso: labio gibbo dicolore.—Lin. S. N. p. 1251. No. 716.—Muf. Reg. p. 674. No. 384.

The *shell* is thick and strong, *smooth,* except a few longitudinal wrinkles, very *glossy, size* of a large walnut, and *round.*

The *mouth* is oblong oval; the *inside* dingy purplish, or brownish; the *outer lip* thick and even; the *inner* or *pillar lip* greatly spread on the body wreath: the *umbilicus* very large and deep.

The *wreaths* are five; the *body one* extremely ventricose; the *four* others or the *turban,* which is but little produced, are also swell'd and cylindric, and separated by a strong furrow.

 The

*** MARINÆ. De Mer,

XLV.

Pl. V. fig. 7.

La *coquille* est epaisse & forte, *lisse,* excepté quelques rides longitudinales, tres *lustrée,* de la *grandeur* d'une noix, & *ronde.*

La *bouche* est ovale oblong; le *dedans* de couleur terne pourprée, ou brunatre; la *levre exterieure* epaisse & unie; l'*interieure* ou celle de la *columelle* beaucoup etendue sur l'orbe du corps: l'*umbilic* fort grand & profond.

Les *orbes* sont cincq; *celui* du *corps* extremement enflé; les *quatre* autres ou la *clavicule,* qui est peu allongée, sont aussi enflés & cylindriques, & separès par une strie forte.

Les

The *shells* vary in *colour*, but most generally are white, shaded with pale and strong chesnut colour; others are livid and white, and others almost white. At the very *bottom* of each *wreath* it has a remarkable *girdle*, made up of distant parallel rays or streaks, curved or oblique, of a deep brown colour, which appears like a very pretty *chain work:* it runs circular, or to the turn of the wreaths. The *adult shells*, and many others, are *thus coloured*; but numbers of *young ones* are also *fish'd* on our *coasts*, about the *size* of small grapes, which are elegantly adorned with three or four other *chain-like girdles*, especially on the *body wreath*, the spots whereof are triangular, or like arrow heads, and emulate the fine West-India shells of this kind.

Les *coquilles* varient en *couleur*, mais plus généralement elles sont blanches nuancées de couleur de chataigne, pale & foncée; d'autres sont livides, & d'autres presque blanches. Au *bas* même de chaque *orbe* se trouve une *bande*, formée de raies ou lignes distantes & paralleles, courbées ou obliques, d'une couleur brune foncée, qui paroit comme un joli *ouvrage en chaine:* elle court circulairement, ou selon le tour des orbes. Les *coquilles adultes*, & plusieurs autres, sont *ainsi colorées*; mais quantité de *jeunes coquilles*, qui sont *prises* sur nos *cotes*, environ la *grandeur* de petits raisins, sont elegamment ornées de trois ou quatre autres *bandes en chaine*, specialement sur l'*orbe du corps*, dont les taches sont triangulaires, ou comme les bouts des fleches, & egalent les plus belles coquilles de cette espece qui se trouvent aux Indes Occidentales.

This *species* is not *uncommon* on *most* of our *coasts*, about *Harwich* and other *places* of the *Essex shores*, says *Dale*; on the *Kentish* coast, about *Margate*, &c. on the *Dorsetshire*, *Devonshire*, and *Cornish* shores; on the sandy shores of *Lincolnshire*, and the *mouth* of the *Humber*, but *not common*, says *Lister*. *In short*, on *most* of our *English coasts*, and in the *Orkneys*, according to *Wallace*.

It is also a native of the *Mediterranean*, the coast of *Senegal* in *Africa*, and the *West Indies*, where they are with *many girdles*, and very *beautiful*.

Cette *espece* n'est *pas rare* sur la *plupart* de nos *cotes*, autour de *Harwich*, & autres *lieux* des *rivages* de *Essex*, selon *Dale*; sur les cotes de *Kent*, autour de *Margate*, &c. sur les rivages des comtés de *Dorset*, *Devon*, & *Cornwall*; aux rivages sablonneux du comté de *Lincoln*, & à l'*embouchure* du *Humber*, mais *point frequemment*, dit *Lister*. *Enfin*, sur la *plupart* de nos *cotes Angloises*, & dans les *Isles Orcades*, selon *Wallace*.

Elle est aussi native de la *Mediterranée*, de la cote de *Senegal* en *Afrique*, & des *Indes Occidentales*, ou elles se trouvent avec *plusieurs bandes*, & tres *belles*.

XLVI.

PARVA. SMALL.

Tab. VIII. fig. 12.

XLVI.

Pl. VIII. fig. 12.

Cochlea parva umbilicata interdum tota alba apice violaceo, interdum fasciata. Parva. Tab. VIII. fig. 12.

The *shell* is thick and *small*, not so *large* as a tare, *round*, *smooth*, and very *glossy*.

The *mouth* is round; the *outer lip* even; the *inner* and *pillar lip* very thick, and greatly spread; on it, at the bottom, is a small *umbilicus*, like a pin-hole.

The *body wreath* very ventricose. The *turban* consists of two other cylindric wreaths, and is slightly produced or lengthened.

This *species* varies in its *colours*: some *shells* are *milk white*, with the *end* of the *turban pale violet*; others are *livid* or *whitish*, and have two or three regular *brown belts* on the *body wreath*, also with the end of the turban violet.

I do not find this *species described* by any *author*. Mine were from the coast of *Devonshire*, by the hands of William Watson, jun. M. D. of Bath.

This *species* seems to be a *link* between the *nerits* and *cochleæ*; for the having only *three wreaths* and the *spread pillar* brings them to the *former*, while the *form* and *produced turban* demands this latter *genus*.

La *coquille* est epaisse & *petite*, pas si *grande* que une vesse, *ronde*, *lisse*, & fort *lustrée*.

La *bouche* est ronde; la *levre exterieure* unie; l'*interieure* ou celle de la *columelle* fort epaisse & beaucoup etendue; sur elle, à son bas, se trouve un petit *umbilic*, comme la piqueure d'une epingle.

L'*orbe* du *corps* est fort enflé. La *clavicule* continent deux autres orbes cylindriques, & est un peu allongée.

Cette *espece* varie en ses *couleurs*; quelques *coquilles* sont *blanches de lait*, avec le *bout* de la *clavicule couleur de violette pale*; d'autres sont *livides* ou *blanchatres*, & ont deux ou trois *bandes brunes*, regulieres, sur l'*orbe* du *corps*, aussi avec le bout de la clavicule violet.

Je ne trouve cette *espece decrite* par aucun *auteur*. Les miens me furent envoyé de la cote du comté de *Devon*, par Guillaume Watson, jun. M. D. de Bath.

Cette *espece* paroit etre un *lien* entre les *nerites* & les *limaçons*; car ayant seulement *trois orbes* & la *columelle etendue* l'a raproche au *premier*, pendant que la *forme* & la *clavicule allongée* demande qu'on la met dans le dernier *genre*.

GENUS

GENRE

GENUS XI.

TURBO.

Snails not round, but of a taper form, the turban being produced or lengthened, and the upper wreaths or ſpires generally ventricoſe.

GENRE XI.

TURBO.

Limaçons d'une forme point ronde, mais allongée, la clavicule etant effilée ou allongée, & les orbes ou revolutions ſuperieurs généralement renflés.

* TERRESTRES. LAND.

XLVII.

STRIATUS. STRIATED.
Tab. V. fig. 9.

* TERRESTRES. TERRESTRES.

XLVII.

Pl. V. fig. 9.

TURBO *albeſcens rufo variegatus, eleganter ſtriatus. Striatus.* Tab. 5 fig. 9.

Cochlea cinerea, interdum leviter rufeſcens, ſtriata, operculo teſtaceo cochleato donata. Cochlea terreſtris turbinata & ſtriata Columnæ de purpura, c. 9. p. 18. *ubi etiam delineatur ſub hoc titulo, Cochlea turbinata.*—Liſt. H. An. Angl. p. 119. tit. 5. tab. 2. fig. 5.

Cochlea terreſtris turbinata & ſtriata. Fab. Col.—Liſt. H. Conch. tab. 27. fig. 5.

Phil. Tranſ. No. 105. fig. 2.—Morton Northampt. p. 415.—Petiv. Muſ. p. 5. No. 16.

Turbo tumidus. Tumid.—Penn. Brit. Zool. No. 110. tab. 82. fig. 110.

Turbo lunaris teſſellatus & ſtriatus.—Klein Oſtrac. p. 55. § 161. ſpec. 3.—Argenv. Conch. I. p. 384. tab. 32. fig. 12. II. p. 339. tab. 28. fig. 12.

Turbo terreſtris tenuiſſime ſtriatus, ipſo ore circinato, cui etiam limbus latus, & ſtriatus, albidus.—Gualt. I. Conch. tab. 4. fig. B.

An *Turbo reflexus.*—Lin. S. N. p. 1238. No. 638?

The *ſhell* is *ſpiral* and *produced,* thinniſh and tranſparent, of the *ſize* of a horſe bean or filberd kernel, and ſomewhat gloſſy. It is moſt elegantly *ſtriated* all over with very fine hair-like ſtriæ, thickly ſet, and *circular,* or according to the run of the ſpires.

The *mouth* quite round; the *lips* are thick, edged, and turn outward, the *inner* or *pillar* one being thicker and more turn'd. Aſide of it lies the *umbilicus,* a narrow ſhallow cavity.

It

La *coquille* eſt contournée en *ſpirale & allongée,* un peu mince & tranſparente, de la *grandeur* d'une feve de cheval, ou le noyau d'une noiſette, & quelque choſe luſtrée. Elle eſt elegamment *ſtriée* par tout à ſtries tres fines comme des cheveux, ſerrées, & *circulaires,* ou ſelon le tour des orbes.

La *bouche* tout à fait ronde; les *levres* ſont epaiſſes, bordées, & retrouſſées en dehors, l'*interieure* ou celle de la *columelle* etant plus epaiſſe & plus retrouſſée. A ſon coté ſe trouve l'*umbilic,* une cavité etroite & peu profonde.

Elle

It has five *spires* or *wreaths*, all cylindric, swelled, and separated by a deep furrow. The *upper spires* are very *ventricose*, the others *taper*, but not much, to an obtuse tip.

The *ground* is whitish, *clouded, rayed, and streaked* in a very pretty manner with *pale brown* and *chesnut colour*; the rays and streaks are all longitudinal.

This *delicate* and *pretty species (the most elegant of all our snails, says Dr. Lister)* is *not* very *common* in *England*. In *Surrey*, about *Darking*, and its neighbourhood. Mr. Petiver found it about *Charlton* in *Kent*; and Dr. Lister in that *county*, but not *frequent*. Mr. Morton in *Northamptonshire*, as in *Wakerly Lordship*; but notes it is not *very common* in that *county*. Mr. Pennant in the woods of *Cambridgeshire*; and, lastly, Dr. Lister near *Oglethorpe* and *Burwell* woods in *Lincolnshire*, and also in *Yorkshire*.

In France, says Dr. Lister, it is frequent every where, and it is also found in Italy.

Elle a cinq *orbes* ou *revolutions* touts cylindriques, bombés, & separés par une strie profonde. Les *orbes superieurs* sont fort *enflés*, les autres *appetissent*, mais pas beaucoup, en un bout obtus.

Le *fond* est blanchatre, *nuagé*, *rayonné*, & *rayé* d'une maniere tres agreeable de *couleur brune pale & de chataigne*. Les rayons & les raies etant touts longitudinales.

Cette *jolie* & *agreable espece (le plus elegant de touts nos limaçons, dit le Dr. Lister)* n'est pas tres *commune* en *Angleterre*. En *Surrey* autour de *Darking*, & ses environs. M. Petiver la trouvé autour de *Charlton* en *Kent*, & le Dr. Lister dans cet meme *comté*, mais point *frequemment*. M. Morton dans le comté de *Northampton*, comme dans la *Seigneurie de Wakerly*, mais il rapporte qu'elle n'est point *fort frequente* dans cet *comté*. M. Pennant dans les bois au comté de *Cambridge*; & dernierement, le Dr. Lister près des bois de *Oglethorpe* & *Burwell* au comté de *Lincoln*, & aussi dans le comté de *York*.

En France, dit le Dr. Lister, elle se trouve par tout frequemment, & elle se trouve aussi en Italie.

XLVIII.
GLABER. SMOOTH.
Tab. V. fig. 18.

XLVIII.
Pl. V. fig. 18.

Turbo imperforatus glaber interdum totus albus & opacus, interdum corneus & pellucidus. Glaber. tab. 5. fig. 18.

Buccinum exiguum, quinque anfractuum, mucrone acuto.—Lift. H. An. Angl. p. 122. tit. 7. tab. 2. fig. 7.—Morton Northampt. p. 415.

Buccinum alterum exiguum in musco degens quinque anfractuum mucrone acuto.—Phil. Transf. No. 105. fig. 7.

Buccinulum oblongum avenaceum.—Petiv. Gaz. tab. 30. fig. 7.

Stagnalis.

Stagnalis. Helix testa subperforata subturrita, anfractibus quinque, apertura ovata.—Lin. S. N. p. 1248. No. 697.

The *shell* is of a very taper or produced *shape*, for the spires do not jut out, or advance beyond one another; about the *size* of an oat, extremely *glossy* and *smooth*, except a few longitudinal wrinkles.

The *mouth* is round, and bordered with a thick white edge; and it has no *umbilicus*.

The spires are *six*, all *gradually tapering* to a *point*, and are only to be distinguished by a strong *furrow*.

There are *two varieties* or *growths* of this *species*, *one* white and opake, the *other* very thin, transparent, and of a brown horn colour.

It is not an *uncommon kind* in many parts of *England*, and inhabits woods, and the moss on old trees and their roots, and on walls. Very *plenty*, says Lister, at *Estrope* in *Lincolnshire*. According to Morton, it is one of the commonest sorts in *Northamptonshire*, not only in the mosses, but also in the sedge on the boggy sides of several standing springs, as at *Oxendon* and *Arthingworth*. I have observed it in *Surrey* and *Middlesex*; and Mr. Petiver found it in Lord Wooton's grove at *Hampstead*.

Obs.—*Bester*, and after him *Linné*, are erroneous in proposing it as an *aquatic snail*.

La *coquille* est d'une *forme* fort effilée ou alongée, car les revolutions ne dejettent ou faillent au delà l'une de l'autre; environ la *grandeur* d'une avoine, extremement *lustrée* & *lisse*, excepté quelques rides longitudinales.

La *bouche* est ronde, bordée d'une belle bordure blanche & epaisse; elle n'a point de *umbilic*.

Les *revolutions* sont *six*, toutes *diminuant par degrés* en une *pointe*, & ne peuvent etre distingués que par une *strie* forte.

Il se trouve *deux varietés* ou *ages* de cette *espece*, *un* blanc & opaque, *l'autre* mince, transparent, & d'une couleur brune de corne.

Elle est une *sorte asses commune* dans plusieurs lieux de l'*Angleterre*, & habite les bois, & la mousse sur les vieux arbres, & leur racines, & sur les murailles. En *grande abondance*, selon *Lister*, à *Estrope* au comté de *Lincoln*. Selon Morton, elle est une sorte de plus commune au comté de *Northampton*, non seulement dans les mousses, mais aussi dans l'herbe sur le bords marecageux de plusieurs eaux croupissantes, comme à *Oxendon* & *Arthingworth*. Il les a observé en *Surrey* & *Middlesex*; & M. Petiver les a trouvé dans le boccage de Lord Wooton à *Hampstead*.

Obs.—*Bester*, & apres lui *Linné*, erronnement la proposent comme un *limaçon aquatique*.

XLIX.

Cydindraceous. Cylindric.

Tab. V. fig. 16.

XLIX.

Pl. V. fig. 16.

Turbo minimus mucrone obtuso, five vere cylindraceus.—*Cylindraceus*, tab. 5. fig. 16.

Buccinum exiguum subflavum, mucrone obtuso, five cylindraceum.—Lift. H. An. Angl. p. 121. tit. 6. tab. 2. fig. 6.—Phil. Tranf. No. 105. fig. 6.

Buccinulum minimum ovale. Petiv. Gaz. tab. 35. fig. 6.—Morton Northampt. p. 415.

Turbo mufcorum. Mofs.—Penn. Brit. Zool. No. 118. tab. 82. fig. 118.

Cochlea tefta fubpellucida, fpiris fex dextrorfis, fubcylindracea obtufa.—Lin. F. Suec. I. p. 372. No. 1301. II. No. 2173.

Mufcorum. Turbo tefta ovata obtufa pellucida: anfractibus fenis fecundis, apertura edentula. Lin. S. N. p. 1240. No. 651.

This *fhell* is *very fmall*, not above one quarter part of a barley-corn in *fize*; *tranfparent*, very *thin* and *brittle*, *fmooth*, *gloffy*, and of a brown or horny *colour*.

In *fhape* it is *thick* and *cylindric*; for the fpires are all equal, except the laft, which ends fuddenly in an obtufe, or blunt tip. The *mouth* is oval, and the *fpires* are fix, feparated by a ftrong furrow.

It is not a *very rare kind* in *England*; it inhabits the moffes on old walls, thatches, trees, &c. I have found it in *Middlefex* and *Surrey*. Mr. Petiver on the fandy banks of the Thames at *Kingfton*, in the latter *county*. Dr. Lifter in *plenty* at *Eftrope* in *Lincolnfhire*. Mr. Morton, in *great plenty*, in a ground nigh *Morfley Wood* in *Northamptonfhire*; and I have received it from old walls at *Leefwood* in *Flintfhire*.

E.

La *coquille* eft *fort petite*, pas plus en *grandeur* que la quatrieme partie d'un grain d'orge; *tranfparente*, fort *mince* & *fragile*, *liffe*, *luftrée*, & d'une *couleur* brune ou de corne.

Sa *forme* eft *groffe* & *cylindrique*; car les revolutions font toutes egales, excepté la derniere, qui finit fubitement dans un bout obtus ou emouffé. La *bouche* eft ovale, & les *revolutions* font fix, feparées par une ftrie forte.

Elle n'eft pas une *forte fort rare* en *Angleterre*; & habite dans les mouffes fur les vielles murailles, chaumieres, & vieux arbres, &c. Je les ai trouvé en *Middlefex* & *Surrey*. M. Petiver fur les rivages fablonneux de la Tamife à *Kingfton*, dans le *dernier comté*. Le Dr. Lifter en *abondance* à *Eftrope* au comté de *Lincoln*. M. Morton, en *grande abondance*, dans un terrein près du *bois de Morfley* au comté de *Northampton*; & je les ai reçu des vielles murailles à *Leefwood* au comté de *Flint*.

E.

E. E.

RUPIUM. ROCK.

Buccinum rupium, majusculum, circiter senis orbibus circumvolutum.—List. H. An. Angl. p. 122. tit. 8. tab. 2. fig. 8.—Phil. Trans. No. 105. fig. 8.—Morton Northampt. p. 415.—Wallace Orkn. p. 39.

Size somewhat larger than a grain of wheat, transparent, of a chesnut *colour*, with about eight *spires* gradually tapering. —*Lister*.

The Doctor found this *kind* on the rocks over the torrents in the mountainous part of *Yorkshire*, called *Craven*, and elsewhere he also found this snail *plentifully*. Mr. Morton in the moss at the roots of old trees, and in the sedge upon the boggy sides of several standing springs, in *Northamptonshire*, but not very *common*; and Dr. Wallace observed it in the *Orkney Isles*.

Grandeur quelque chose au delà d'un grain de froment, transparente, *couleur* de chataigne, avec huit *revolutions* diminuant par degrès.—*Lister*.

Le Docteur a trouvé cette *espece* sur les rochers au dessus des torrents dans la partie montagneuse du comté de *York*, nommée *Craven*, & ailleurs il les trouva aussi en *abondance*. M. Morton dans la mousse aux racines des vieux arbres, & dans les herbes sur les bords marecageaux de plusieurs eaux croupissantes, au comté de *Northampton*, mais non pas *frequemment*; & le Dr. Wallace l'a observé dans les *Isles Orcades*.

F. F.

FASCIATUS. FASCIATED.

Buccinum exiguum fasciatum & radiatum. List. H. Conch. tab. 19. fig. 14.—*Turbo fasciatus. Fasciated.* Penn. Brit. Zool. No. 119. tab. 82. fig. 119.

The *shell* has six *spires*; *white, marbled* or *fasciated* with *black*. *Length* half an inch. Very *frequent* in *Anglesea*, in sandy soils near the coasts.—*Pennant*.

Dr. Lister notes that Mr. Lhwyd found it in *Wales*, and in the *Isle of Alderney*.

La *coquille* a six *revolutions*; *blanche, marbrée* ou *bandée* de *noir*. Un demi pouce en *longeur*. Fort *commune* en *Anglesea*, dans les terreins sablonneux près des cotes.—*Pennant*.

Le Dr. Lister rapporte que M. Lhwyd les trouva en *Galles*, & dans l'*isle de Alderney*.

**FLUVIATILES. River. L.

Nucleus. Kernel.

Tab. V. fig. 12.

**FLUVIATILES. Fluviatiles. L.

Pl. V. fig. 12.

Turbo imperforatus parvus subrufus, Levis, quinque spirarum. Nucleus. Tab. 5. fig. 12.

Cochlea parva, subflava, intra quinque spiras finita.—List. H. An. Angl. p. 135. tit. 19, tab. 2. fig. 19.

Buccinum subflavum alterum, quinque spirarum, atque operculo tenui & pellucido, testacco tamen cochleatoque donatum.—Phil. Transf. No. 105. fig. 21.

Cochlea parva pellucida, operculo testaceo cochleatoque clausa.—List. H. Conch. tab. 132. fig. 32.

Cochleola oblonga fluviatilis. Common small river snail. Petiv. Gaz. tab. 18. fig. 8.—*Small fresh-water turbo, with five wreaths.* Wallis Northumb. p. 370.

Helix tentaculata. Olive.—Penn. Brit. Zool. No. 140. tab. 86. fig. 140.

Cochlea testa oblonga obtusa; anfractibus quatuor laxis cinereis opacis; apertura subovata.—Lin. F. Suec. I. p. 376. No. 1313. II. No. 2191.

Tentaculata. Helix testa imperforata ovata obtusa impura, apertura subovata.—Lin. S. N. p. 1249. No. 707.

The *shell* is *thinnish*, semitransparent, and about the *size* of a cherry-stone kernel, of a uniform yellowish chesnut *colour* (but when exposed to the weather, white, as most shells are) very *smooth* and *glossy*, and of a neat, graceful, produced or subconic *shape*.

The *mouth* is oval; the *lips* even and bordered. It has no *umbilicus*. The *spires* are five; the body one very large and swell'd; the others decrease gradually to a point, and they are separated by a strong furrow.

This pretty shell is found, in *plenty*, in most *rivers* and *stagnant waters* throughout these *kingdoms*.

La *coquille* est *quasi mince*, demitransparente, & environ la *grandeur* d'un noyau de cerise, d'une *couleur* jaunatre chataigne uniforme (mais quand exposée aux injures du tems, elle blanchit, comme font toutes les coquillages) fort *lisse* & *lustrée*, & d'une *forme* quasi conique, allongée, bien tournée & agreable.

La *bouche* est ovale; les *levres* unies & bordées. Elle n'a point d'*umbilic*. Les *orbes* sont cinq; celui du corps fort grand & enflé; les autres diminuent par degrès à une pointe, & sont separès par une strie forte.

Cette jolie coquille se trouve, en *abondance*, dans la plupart des *rivieres* & *eaux croupissantes* partout ces *royaumes*.

LI.

TRIANFRACTUS. THREE SPIRED.

Tab. V. fig. 13.

LI.

Pl. V. fig. 13.

Turbo subflavus pellucidus imperforatus, testa prætenui fragili, trium spirarum. Trianfractus. Tab. 5. fig. 13.

Buccinum subflavum pellucidum trium spirarum.—Lift. H. An. Angl. p. 140. tit. 24. tab. 2. fig. 24.—Phil. Tranf. No. 105. fig. 18.

Buccinum subflavum pellucidum trium orbium.—Lift. H. Conch. tab. 123. fig. 23.

Bucc. fluviatile nostras, testa prætenui fragili. Muf. Petiv. p. 83. No. 808. Morton Northampt. p. 417.—*Small turbo with three wreaths.* Wallis Northumb. p. 370.

Helix putris. Mud.—Penn. Brit. Zool. No. 137. tab. 86. fig. 137.

Neritostoma vetula. Klein Ostrac. p. 55. § 159. tab. 3. fig. 70.

Argenv. Conch. I. p. 385. tab. 32. fig. 23. II. p. 340. tab. 28. fig. 23.

Buccinum fluviatile, testa fragili, pellucida, albida, prima spira admodum elongata & ventricosa. Gualt. I. Conch. tab. 5. fig. C. C.—Et *Bucc. fluv. subflavum pellucidum, ore ad plausum aperto, trium spirarum.* Id. tab. 5. fig. F.

Cochlea testa membranacea subflava oblonga, mucrone obtuso, anfractibus tribus.—Lin. F. Suec. I. p. 377. No. 1317. II. No. 2189.

Putris. Helix testa imperforata ovata obtusa flava, apertura ovata.—Lin. S. N. p. 1249. No. 705.

The *shell* so extremely *thin* as to be *membranaceous*, and is therefore very *transparent* and brittle, of an uniform brownish yellow or horn *colour*; about the *size* of a cherry-stone; the *surface* near smooth, being only set with very fine almost imperceptible longitudinal wrinkles, very *glossy*, and of a produced or subconic *shape*.

The *mouth* is oblong oval, very large, and patulous. It has no *umbilicus*.

The *spires* are three; the body one vastly large, tumid, and takes up three quarters of the shell; the last hardly equals a pin's head in size. The spires are separated by a furrow.

This *species* is found in plenty in *rivers* of clear waters, among reeds and bullrushes; as also in the *stagnant waters* throughout these *kingdoms*.

It

La *coquille* est si extremement *mince* à etre *membraneuse*, & par la extremement *transparente* & fragile, d'une *couleur* brunatre jaune ou de corne; environ la *grandeur* d'un noyau de cerise; la *surface* presque lisse, ayant seulement quelques rides longitudinales à peine visibles, fort *lustrée*, & d'une *forme* quasi conique.

La *bouche* est ovale oblongue, fort grande, & evasée. Il n'a point d'*umbilic*.

Les *revolutions* sont trois; celle du corps extremement grande, enflée, & fait trois quarts du volume de la coquille; la derniere à peine egale la tete d'une epingle en grandeur. Les revolutions sont separées par une strie.

Cette *espece* se trouve en abondance dans les *rivieres* d'eau claire, entre les jones; comme aussi dans les *eaux croupissantes*, partout ces *royaumes*.

C'est

It is an *amphibious animal*, says Dr. Lister, insomuch that it will freely *get out* of the rivers in Summer, *lie* on the *grass*, or *creep* up to the tops of the slender twigs of *willows*.

C'est un *animal amphibie*, dit le Dr. Lister, de telle maniere que librement il *sortira* des rivieres dans l'Eté, se *coucher* sur l'*herbe*, ou *ramper* jusques aux bouts des rejettons deliés des *saules*.

LII.

STAGNALIS. LAKE.

Tab. V. fig. 11.

LII.

Pl. V. fig. 11.

Turbo longus et gracilis in tenue acumen mucronatus, imperforatus & pellucidus. Stagnalis. Tab. 5. fig. 11.

Buccinum longum sex spirarum, omnium & maximum & productius, subflavum, pellucidum, in tenue acumen ex amplissima basi mucronatum. Turbo lævis in stagnis degens. Aldrov. de Testaceis, l. 3. p. 359. No. 3. *Ubi ejus figura habetur.*—List. H. An. Angl. p. 137. tit. 21. tab. 2. fig. 21.—Phil. Trans. No. 105. fig. 22.

Buccinum subflavum pellucidum, sex orbium, clavicula admodum tenui, productiore.—List. H. Conch. tab. 123. fig. 21.

Bucc. minus fuscum, sex spirarum, ore angustiore. List. H. An. Angl. p. 139. tit. 22. tab. 2. fig. 22.—*Bucc. quinque spirarum plenarum, mucrone sæpius mutilato obtusoque.* Phil. Trans. No. 105. fig. 20.

Bucc. fluviatile nostras oblongum majus. Mus. Petiv. p. 82. No. 805.—Et *B. fluv. nostras oblongum minus.* Id. No. 806.

Morton Northamp. p. 417.—Et *Bucc. quinque spirarum pellucidum subflavum.* Id. p. 418.

Helix stagnalis. Lake. Penn. Brit. Zool. No. 136. tab. 86. fig. 136.—*Fresh-water turbo with six wreaths.* Wallis Northumb. p. 369.

Auricula stagnorum subflava. Klein Ostrac. p. 54. § 157. spec. 1. tab. 3. fig. 69.—Argenv. Conch. I. p. 372. tab. 31. fig. 6. No. 1. II. p. 329. tab. 27. fig. 6. No. 1.

Buccinum fluviatile, oblongum, ore angusto, fuscum, sex spirarum. Gualt. I. Conch. tab. 5. fig. E.—Et *B. fluv. testa tenuissima & fragilissima, prima spira notabiliter ventricosa & elongata, in mucronem aculeatum desinens, subflavum pellucidum.* Id. tab. 5. fig. I. L.

Cochlea testa producta acuminata opaca, anfractibus senis subangulatis, apertura ovata.—Lin. F. Suec. I. p. 374. No. 1310. II. No. 2188.

Stagnalis. Helix testa imperforata ovato-subulata, subangulata, apertura ovata.—Lin. S. N. p. 1249. No. 703.

The *shell* is very *brittle*, *thin*, and *transparent*, from one and three quarters to two inches

La *coquille* est fort *fragile*, *mince*, & *transparente*, de un & trois quarts à deux pouces en

inches *long*, and about three quarters of an inch *over*, where broadeſt; of a ſlender or graceful *ſhape*; *colour* whitiſh or ſullied light brown, and thick ſet with longitudinal wrinkles. Theſe *wrinkles*, in the larger or full grown ſhells, for ſome ſpace around the mouth or fore part, *riſe thick and prominent*, and are often *croſs'd* or *latticed* by like *tranſverſe* ones, which, as *Linné* juſtly obſerves, ſomewhat *ſubangulates* the ſhell.

en *longeur*, & environ trois quarts d'un pouce en *largeur*, ou elle eſt plus large; d'une *forme* deliée & bien tournée, de *couleur* blanchatre ou brune claire ſale, & à rides longitudinales fort ſerrées. Ces *rides*, dans les plus grandes ou coquilles adultes, pour quelque eſpace autour de la bouche, ou ſur le devant, *s'elevent groſſes & en vive arrete*, & ſont ſouvent *miſes en travers* ou en *barreux* par des autres ſemblables *tranſverſales*, qui, comme *Linné* à juſtement *obſervé*, fait la coquille en quelque ſorte *angulaire*.

The *mouth* is oblong-oval, and large; the *outer lip* thin; the *inner* or *pillar lip* forms a very thick edge, which twirls obliquely, and is greatly ſpread on the body whirl. It has no *umbilicus*.

La *bouche* eſt ovale oblongue, & grande; la *levre exterieure* mince; la *levre interieure* ou de la *columelle* forme une bordure fort epaiſſe, qui tourne obliquement, & eſt fort etendue ſur la revolution du corps. Elle n'a point d'*umbilic*.

The *ſpires* are ſix; the body one very large and ventricoſe; the *others*, or the *turban*, gradually taper quite to a fine ſharp point. They are ſeparated by a ſtrong declivity and furrow.

Les *revolutions* ſont ſix; celle du corps eſt fort grande & enflée; les *autres*, ou la *clavicule*, vont en appetiſſant juſques à une pointe tres aigue. Elles ſont ſeparées par un grand penchant & une ſtrie.

This is the *largeſt* and *moſt produced* of all the *Britiſh river ſnails*, and is found in *plenty* in all rivers, lakes, ponds, and other waters, throughout *theſe kingdoms*.

C'eſt la *plus grande* & la *plus allongée* de toutes les *limaçons fluviatiles Britanniques*, & ſe trouve en *abondance* dans toutes les rivieres, etangs, & autres eaux, partout *ces royaumes*.

Obs.—Dr. *Liſter* has made *two ſpecies* of this *kind*, by his *maximum* and *minus*; Mr. *Petiver* has followed him; and Mr. *Morton*'s laſt quoted *kind*, p. 418, is probably only a *variety* of this *ſpecies*.

Obs.—Le Dr. *Liſter* a fait *deux eſpeces* de *celle ci*, par ſon *maximum* & *minus*; M. *Petiver* l'a ſuivi; & la derniere *ſorte* de M. *Morton* citée, p. 418, eſt vrai ſemblablement que une *variété* de cette *eſpece*.

LIII.

Patulus. Wide Mouth.

Tab. V. fig. 17.

LIII.

Pl. V. fig. 17.

Turbo subflavus pellucidus quatuor spirarum ore patulo. Patulus. Tab. 5. fig. 17.

Buccinum pellucidum subflavum, quatuor spirarum, mucrone acutissimo, testæ apertura omnium maxima.—Lift. H. An. Angl. p. 139. tit. 23. tab. 2. fig. 23.

Buccinum subflavum pellucidum, quatuor orbium, ore amplissimo, mucrone acuto.—Lift. H. Conch. tab. 123. fig. 22.

Buccinum fluviatile pellucidum, subflavum, quatuor spirarum, mucrone acuto, testæ apertura patentissima.—Lift. Exerc. Anat. 2. p. 54.

Buccinum fluviatile nostras breve, ore patulo. Muf. Petiv. p. 83. No. 807.—*Turbo with four wreaths, a remarkable large mouth, and a short acute apex.* Wallis Northumb. p. 370.

Helix auricularia. Ear.—Penn. Brit. Zool. No. 138. tab. 86. fig. 138.

Auricula pellucida.—Klein Oftrac. § 157. p. 55. fpec. 2.

Argenv. Conch. I. p. 373. tab. 31. fig. 7. II. p. 330. tab. 27. fig. 7.

Buccinum fluviatile pellucidum, subflavum, mucrone acutissimo & brevi; prima spira insigniter ventricosa, testæ apertura omnium maxima.—Gualt. I. Conch. tab. 5. fig. G.

Cochlea testa diaphana anfractibus quatuor, mucrone acuto brevissimo, apertura acutissima.—Lin. F. Suec. I. p. 376. No. 1315. II. No. 2192.

Auricularia. Helix testa imperforata ovata obtusa, spira acuta brevissima, apertura ampliata.—Lin. S. N. p. 1250. No. 708.

The *shell* is *thin, brittle,* and *transparent,* of a light brown or horny *colour,* of the *size* of a filberd or fmall olive, but in *shape* very round or fwell'd, the *turban* forming only a fhort point.

It is *glossy,* and near *smooth,* having only very fine longitudinal ftriæ, hardly perceptible. *Umbilicus* it has none, only a narrow depreffion lies afide the edge of the pillar lip, which is not perforated.

The *mouth* is oval, extremely large and wide; the *lips* or edges of it are not even, but finuated or waved, turn fomewhat out-

La *coquille* eft *mince, fragile,* & *transparente,* d'une *couleur* de corne ou brune claire, de la *grandeur* d'une noifette ou d'une olive, mais en *forme* fort ronde ou enflée, la *clavicule* formant feulement une pointe courte.

Elle eft *lustrée,* & prefque *lisse,* ayant feulement des ftries longitudinales fort fines à peine vifibles. Elle n'a point d'*umbilic,* mais feulement un enfoncement etroit fe trouve à coté du bord de la levre de la columelle, qui n'eft point percé en trou.

La *bouche* eft ovale, extremement grande & evafée, fes *levres* ou bords ne font pas unies, mais finueufes ou ondoyantes, & un

outwards, and alſo often outwardly border'd or wrinkled. The *pillar lip* turns greatly outwards, and forms a high edge; and the *pillar* is very thick, and *ſinuous* or *twirl'd*, not unlike a *human ear*.

The *ſpires* are four; the *body one* makes up almoſt the whole ſhell, it is ſo extremely large and ſwell'd, for the *three* others, or the *turban*, are ſo ſhort, or little produced, that they do not gradually proceed from it; they end in an acute point or tip, are about an eighth of an inch long, the *ſecond ſpire* being only of the ſize of a ſmall bead, and the *two others* like pins heads or tubercles.

This *ſpecies* is alſo found in plenty in our rivers and other waters.

un peu retrouſſées, & ſouvent bordées & ridées exterieurement. La *levre interieure* ou de la *columelle* eſt fort retrouſſée, & forme un bord elevé ou haut; & la *columelle* eſt fort epaiſſe, & *ſinueuſe* ou *tournée*, pas diſſemblable à une *oreille humaine*.

Les *revolutions* ſont quatre; *celle du corps* fait preſque le volume de la coquille, elle eſt ſi extremement grande & enflée, car les *trois* autres, ou la *clavicule*, ſont ſi courtes, ou peu allongées, qu'elles ne procedent par degrés du corps; quoique elles finiſſent en une pointe aigue ou bout, elles ſont environ un huitieme d'un pouce en longueur, la *ſeconde revolution* etant ſeulement de la grandeur d'un petit grain de collier, & les *deux autres* comme des tetes d'epingles ou des tubercules.

Cette *eſpece* ſe trouve auſſi en abondance dans nos rivieres & autres eaux.

LIV.
ADVERSUS. CONTRARY.
Tab. V. fig. 6. 6.

LIV.
Pl. V. fig. 6. 6.

Turbo ſiniſtrorſus ſive contrarius exiguus, bullæ formis, trium ſpirarum. Adverſus. Tab. 5. fig. 6. 6.

Buccinum exiguum, trium ſpirarum à ſiniſtra in dextram convolutarum. Liſt. H. An. Angl. p. 142. tit. 25. tab. 2. fig. 25.—*Buccinum fluviatile à dextra ſiniſtrorſum tortile, triumque orbium, ſive neritæides.* Id. H. Conch. tab. 134. fig. 34. Wallace Orkn. p. 39. Morton Northampt. p. 417.

Argenv. Conch. I. p. 373. tab. 31. fig. 6. No. 7. II. p. 330. tab. 27. fig. 6. No. 7.

Buccinum fluviatile, ſubflavum, pellucidum, trium ſpirarum, tenuiſſimum, ore magno ovali elongato, mucrone vero breviſſimo.—Gualt. I. Conch. tab. 5. fig. H.

Cochlea teſta pellucida flava ovata ventricoſa ſiniſtrorſa: apertura ovato-oblonga longitudinali: ſpira introducta.—Lin. F. Suec. I. p. 372. No. 1302. II. No. 2159.

Hypnorum. Bulla teſta ovata pellucida contraria, ſpira prominente, apertura ovato-lanceolata.—Lin. S. N. p. 1185. No. 387.

The La

The *shell* is *thin*, very *brittle*, and *pellucid*, *size* of a pepper corn, of an oval *shape* like to a *bulla*, very ventricose or *swell'd*, of a pale horny *colour*, quite *smooth*, and *glossy*.

The *mouth* is oblong-oval, very large and patulous, and extends to near the bottom of the shell, *similar* to the *bulla genus*. The *inner* or *pillar lip* is somewhat spread, or forms a narrow flat ledge. The *spires* are three, the body one making almost the whole shell; the other two spires, or the *turban*, being like small knobs, and no ways produced, but flat or blunt, though very conspicuous. All the *spires*, *contrary* to *most shells*, (when held *perpendicular*, with the *mouth upwards*, and *fronting* the person) turn from the *left to the right*, and have the *mouth* on the *right side*, or is an *heterostrophum* shell.

This *curious small species* is not very *common*. It is found in some of our rivers and stagnant waters; as in the *Thames*, where I have got it in different places, from *Wandsworth* to *Windsor*; in several ponds and stagnant waters about *London*, and in *Middlesex*; also in like places in *Surrey*, and in the river *Wandle*. Dr. Lister says it is pretty plenty in a rivulet and in ponds about the village *Heyworth*, near *York*: in some parts of the *Ise*, between *Arthingworth* and *Newbottle bridge*, and in some few fen ditches in *Northamptonshire*, according to Mr. Morton; and Mr. Wallace mentions it to be *common* in the small rivulets in the *Orkneys*.

La *coquille* est *mince*, fort *fragile*, & *transparente*, de la *grandeur* d'un grain de poivre, de *forme* ovale comme une *bulla*, fort ventrue ou *enflée*, d'une *couleur* de corne pale, tout à fait *lisse*, & *lustrée*.

La *bouche* est oblongue-ovale, fort grande & evasée, & s'etend à près du bas de la coquille, *semblable* au *genre* du *bulla*. La *levre interieure*, ou de la *columelle*, est un peu etendue, ou forme un bord etroit & applati. Les *revolutions* sont trois, celle du corps faisant presque tout le volume de la coquille; les autres deux revolutions, ou la *clavicule*, etant comme de petites bosses, en aucune maniere allongée, mais platte ou obtuse, quoique fort distinctes. Toutes les *revolutions*, au *contraire* de la *plupart* de *coquilles* (quand tenues *perpendiculairement*, avec la *bouche en haut*, & *envers* la personne) tournent de *gauche à droite*, & ont la *bouche* à *coté droite*, ou est une coquille *heterostrophe* ou *unique*.

Cette *petite* & *curieuse espece* n'est pas fort *commune*. Elle se trouve dans quelques unes de nos rivieres & eaux croupissantes; comme dans la *Tamise*, ou je l'ai trouvée en differents endroits, de *Wandsworth* à *Windsor*; dans plusieurs etangs & eaux croupissantes aux environs de *Londres*, & dans *Middlesex*; aussi en pareils lieux au comte de *Surrey*, & dans la riviere *Wandle*. Le Dr. Lister dit qu'elle est asses abondante dans un petit ruisseau & dans des etangs aux environs du village *Heyworth*, près de *York*: en quelques endroits de l'*Ise*, entre *Arthingworth* & le *pont* de *Newbottle*, & quelques fossés du pais marecageux, au comté de *Northampton*, selon M. Morton; & M. Wallace rapporte qu'elle est fort *commune* dans les petits ruisseaux aux isles *Orcades*.

Obs.—*Linné* has *ranked* this shell as a species of *bulla*, in his *Syst. Nat.* It certainly *verges* much on that *genus*, but nevertheless has more *affinity* to this of the *turbo*.

Obs.—*Linné* à *rangé* cette coquille comme une espece de *bulla*, dans son *Syst. Nat.* Certainement elle *approche* beaucoup à cet *genre*, mais neantmoins à plus de *rapport* à cellui ci du *turbo*.

*** MARINÆ. Sea.

*** MARINÆ. De Mer.

LV.

Littoreus. Periwinkle.

Tab. VI. fig. 1. 1.

Pl. VI. fig. 1. 1.

Turbo pyramidalis crassus, fuscus, striis crebris præditus. Littoreus. Tab. 6. fig. 1. 1.

Cochlea fusca fasciis crebris angustisque prædita, couvins à piscatoribus Scarborgensibus Anglice dicta.—List. II. An. Angl. p. 162. tit. 9. tab. 3. fig. 9. & App. An. Angl. in Goedart, p. 25.

Cochlea nigricans dense & leviter striata.—List. H. Conch. tab. 585. fig. 43.

Cochlea marina: whilks or periwinkles. Merret Pin. p. 193. Wallace Orkn. p. 40. Martin W. Isles, p. 145.—*Trochus.* Leigh Lancash. tab. 3. fig. 10.—*Cochlea maritima nostras crassa fasciata.* Mus. Petiv. p. 83. No. 813.—*Cochlea Orcadensis maxima, crassa, fasciata.* Ib. No. 813. & *Great Orcade covin.* Id. Gaz. tab. 36. fig. 11.—*Couvins, periwinkles, pinpatches.* Dale Harw. p. 379. No. 4. Rutty Dublin, p. 381. Smith Cork, p. 318.

Turbo littoreus. Periwinkle.—Penn. Brit. Zool. No. 109. tab. 81. fig. 109.

Saccus nigricans & Galea fusca.—Klein Ostrac. p. 43. § 121. spec. 3. No. 1. & p. 57. § 167. spec. 1. No. 4.

Buccinum parvum integrum, ore obliquo, globosum, crassum, lineatum, & obscure striatum, ex albido, fusco, livido depictum.—Gualt. I. Conch. tab. 45. fig. G.

Littoreus. Turbo testa subovata acuta striata, margine columnari plano.—Lin. S. N. p. 1232. No. 607.—F. Suec. II. No. 2169.

The *shell* is extremely *thick* and *strong*, from the *size* of a filberd to that of a small walnut; the *surface* generally dull, and cover'd with filth, but, when cleans'd, *smooth*; the *ground colour* is dark brown or blackish, thickly and circularly lineated, or according to the turn of the spires, with fine light brown and whitish hair-

La *coquille* est *epaisse* & *forte*, de la *grandeur* d'une noisette jusques à celle d'une petite noix; la *surface* est generalement terne, & couverte de saleté, mais, quand nettoyée, *lisse*; & à *fond* brun obscur ou noiratre, rayée à lignes fines capillaires, brunes claires & blanchatres, spirales, ou selon le tour des revolutions, & tres serrées; la *forme*

hair-like ftreaks or lines; the *fhape* is pyramidal, and tapers to a very fharp point. This is the *appearance* of the *larger* or *adult* fhells.

The *young* or *fmall* fhells of the *fize* of peas, and double that bignefs, are often *reddifh*, *yellowifh*, *light brown* or *whitifh*, and fometimes prettily *girdled*. They are thickly and fpirally *ftriated* with fine ftriæ, fo as to be *rough*, inftead of being *fmooth*, or only ftreaked, as the adults are.

The *mouth* is oval; the *lips* even, and from within the mouth they thin outwardly to a fharp edge, fo that the mouth within forms a broad fhelving border all round. The *pillar lip* is much fpread; and it has no *umbilicus*.

The *infide* of the *mouth*, in the *young* fhells, is dark and brownifh; in the *adults*, milk white; but all have a border of dark brown or chefnut colour along the outer lip.

The *fpires* or *wreaths* are five; the *body one* very fwell'd and large, and makes three quarters of the fhell: the *four* others that compofe the *turban* are fmall, and taper to a very fharp point. All the *fpires* are level, that is, they do not jut out beyond one another, and are feparated by a deep furrow.

This *fpecies* is found in great *abundance* on moft of the fhores of thefe *kingdoms*, more efpecially the rocky fhores. It is one of the fhell fifh greatly eaten by the poor, and commonly fold in the markets.

forme eft pyramidale, & appetiffe dans une pointe fort aigue. Ceci eft l'*apparence* des *adultes* ou coquilles *grandes*.

Les *jeunes* ou *petites* coquilles de la *grandeur* de pois, ou de double cet volume, font fouvent *rougeatres*, *jaunatres*, *brunes claires* ou *blanchatres*, & quelquefois joliment *fafciées*. Elles font *ftriées* en fpirale à ftries fines & ferrées, deforte qu'elles font *apre*, au lieu de *liffe*, ou feulement rayée, comme les adultes font.

La *bouche* eft ovale; les *levres* unies, & du dedans de la bouche elles deviennent minces vers l'exterieur, ainfi que d'etre à bord tranchant, deforte que la bouche autour du dedans fonne un bord large, qui va en pente. La *levre* de la *columelle* eft fort etendue; & elle n'a point d'*umbilic*.

Le *dedans* de la *bouche*, dans les *jeunes* coquilles, eft foncé & brunatre; dans les *adultes*, blanc de lait; mais toutes ont une bordure de couleur brune foncée ou de chataigne, le long de la levre exterieure.

Les *orbes* ou *revolutions* font cincq; *celui* du *corps* fort enflé & grand, & fait trois quarts du volume de la coquille: les *quatre* autres qui compofent la *clavicule* font petits, & appetiffent en une pointe fort aigue. Touts les *orbes* font de niveau, c'eft à dire, ils ne debordent l'un l'autre, & font feparès par une ftrie forte.

Cette *efpece* fe trouve en grande *abondance* fur la plupart des cotes de ces *royaumes*, plus fpecialement fur les cotes pleines de rochers. C'eft une de coquillages beaucoup mangée par le menu peuple, & vendues communement aux marchès.

The *periwinkles* of the *Orkneys* are generally quadruple the ſize of thoſe of the *Iriſh* or *Engliſh* coaſts.

The name of *periwinkle* is a corruption of *petty winkle, i. e.* ſmall *winkle* or *whelk.*

Les *coquilles* des *Orcades* ſont generalement quatrefois la grandeur de celles des cotes *Angloiſes* ou *Irlandoiſes.*

Le nom Anglois de *periwinkle* eſt une corruption de *petty winkle,* c'eſt à dire, *petit buccin.*

LVI.

Lineatus. Streaked.

Tab. VI. fig. 7.

LVI.

Pl. VI. fig. 7.

Turbo trochiformis cinereus lineis aut lituris nigris inſignitus, columella ſubdentata. Lineatus. Tab. 6. fig. 7.

This *ſhell* is nearly of the ſame *ſize* as the laſt, or periwinkle, very *thick* and *ſtrong,* of a *ſmooth* ſurface; the *ground colour* dark aſhen, very thickly variegated with fine hair-like ſhort ſtreaks or lines of black, in a pretty manner. The *ſhape* is ſubconic, like a trochus, but depreſſed or not very produced, and ends in a ſomewhat obtuſe tip.

The *mouth* is round. The *outer lip* even, but on the turn to the inner or pillar lip it ſpreads, and forms a flat border; from thence, or along the *pillar lip,* it alſo ſpreads and flattens extremely much, eſpecially on the body ſpire: it is alſo *ſinuous* or *twirl'd,* and at the *top* of the *pillar* ſwells into a thick *bump* or *rude tooth,* and aſide of the *pillar,* at the *bottom* of it, is a long narrow ſhallow *dent,* as if *ſubumbilicated.* The *inſide* of the *mouth* is fine *mother of pearl*; and the *outer lip* has a regular pretty broad brown border along it.

The

Cette *coquille* eſt près de la méme *grandeur* que la derniere, tres *epaiſſe* & *forte,* d'une ſurface *liſſe*; le *fond* eſt de couleur de cendres foncée, bigarrée à lignes ou raies fines capillaires, courtes, noires, & tres ferrées, d'une maniere tres agreable. La *forme* eſt quaſi conique, comme un ſabot, mais applatie ou point fort allongée, & finit dans un bout un peu obtus.

La *bouche* eſt ronde. La *levre exterieure* unie, mais ſur le tour à la levre interieure ou de la columelle elle s'etend, & forme une bordure platte; delà, ou le long de la *levre* de la *columelle,* elle s'etend & s'applatit auſſi extremement, ſpecialement ſur la revolution du corps: elle eſt auſſi *ſinueuſe* ou *tortillée,* & au *haut* de la *columelle* elle s'enfle dans une groſſe *boſſe,* ou une *dent groſſiere*; & à coté de la *columelle,* à ſon *bas,* ſe trouve une *entailleure* longue, etroite & peu profonde, comme ſi elle etoit *quaſi umbiliquée.* Le *dedans* de la *bouche* eſt de tres belle *nacre*; & la *levre exterieure* à une jolie bordure brune, reguliere, & aſſes large, qui court ſon long.

Les

The *spires* are five, all swell'd or prominent, and very distinct: they are separated by a depression and a strong furrow. The *body spire* is large; the *second* is also large; the *others* gradually decrease to an obtuse end.

When the *outer coat* is taken off, the *shell* is of *mother of pearl*.

I do not find this *species described* by any *author*. I have received it from *Plymouth*, and other places on the *Devonshire* coast; also from *Cornwall*, the *Dorset* coasts, and from near *Pwllbely* in *Carnarvonshire*. The coasts of *Hampshire* likewise yield it; and it is found in *plenty* on the coasts of *Norfolk*.

Les *revolutions* sont cincq, toutes enflées ou avancées, & fort distinctes: elles sont separées par un enfoncement & une strie forte. *Celle* du *corps* est grande; la *seconde* est aussi grande; les *autres* appetissent par degrés à un bout obtus.

Quand elle est depouillée de sa *robe*, la *coquille* est de *nacre*.

Je ne trouve cette *espece decrite* par aucun *auteur*. Je l'ai reçu de *Plymouth*, & autres lieux sur la cote du comté de *Devon*; aussi de *Cornwall*, & des cotes du comté de *Dorset*; pareillement de près de *Pwllbely* au comté de *Carnarvon*. Les cotes du comté de *Hants* les fournit aussi; & elles se trouvent en *abondance* sur les cotes de *Norfolk*.

LVII.

Ovalis. Oval.

Tab. VIII. fig. 2. 2.

LVII.

Pl. VIII. fig. 2. 2.

Turbo ovalis striatus rubicundus fasciis albis, columella uniplicata & unidentatá. Ovalis. Tab. 8. fig. 2. 2.

Buccinum parvum, rostro integro, tenuiter striatum, fasciatum, clavicula paulo productiore, unico dente ad columellam.—List. H. Conch. tab. 835. fig. 58.

Voluta tornatilis. Oval.—Penn. Brit. Zool. No. 86. tab. 71. fig. 86.

Auris Midæ fasciata.—Klein Ostrac. p. 37. § 96. spec. 1. No. 1.

Tornatilis. Voluta testa coarctata ovata substriata, spira elevata acutiuscula, columella uniplicata.—Lin. S. N. p. 1187. No. 394.

The *shell* is *thin*, *brittle*, and *transparent*, about the *size* of a small olive, of an oval *shape*, pointed at each end, *spirally* and thickly *striated* with extreme fine hair-like striæ, of a pale red *colour*, with *two* very regular and pretty broad white *girdles* on each *spire*, and is *glossy*.

La *coquille* est *mince*, *fragile*, & *transparente*, environ la *grandeur* d'une petite olive, d'une *forme* ovale, pointue aux deux bouts, *striée spiralement* & tres serrée à stries fines & capillaires, d'une *couleur* rouge pale, avec *deux bendes* blanches, fort regulieres & asses larges, sur chaque *revolution*, & est *lustrée*.

The

La

The *mouth* is oblong-oval and narrow; the *outer lip* even; the *inner* or *pillar lip* thick and edged, finuous and twirl'd on the upper part into a fingle plait or fold, which at its *turn* is *thick* and *bump'd*, fo as to refemble a *tooth*.

It has five *fpires*, which gradually diminifh to a fharp point, and are feparated by a ftrong folding furrow.

This *pretty fpecies* I have received from *Tinmouth* and *Exmouth*, in *Devonfhire*; and Mr. Pennant notes it from *Anglefea*.

La *bouche* eft oblongue-ovale & etroite; la *levre exterieure* unie; l'*interieure* ou de la *columelle* epaiffe & bordée, finueufe & tortillé fur la partie fuperieure dans un feul pli, qui, à fon *tournant*, eft *epais* & en *boffe*, deforteque de reffembler à une *dent*.

Elle a cincq *revolutions*, qui diminuent par degrès à une pointe aigue, & font feparées par une ftrie forte en pliffure.

J'ai reçu cette *jolie efpece* de *Tinmouth* & *Exmouth*, au comté de *Devon*; & M. Pennant dit qu'elle fe trouve en *Anglefea*.

LVIII.

Carinatus. Keel'd.

Tab. VIII. fig. 10.

LVIII.

Pl. VIII. fig. 10.

Turbo minimus albus, porcis elevatis membranaceis circulatim cinctus, anfractibus fubtus angulatis. Carinatus. Tab. 8. fig. 10.

Striatulus. Turbo tefta fubcancellata turrita, anfractibus contiguis, cingulifque varicofis interceptis.—Lin. S. N. p. 1238. No. 635.

A very *fmall fhell*, about the *fize* of a hemp-feed, quite *white*, rather *thin* and *tranfparent*, and of a lengthened *fhape*. It is fet with circular thick and fharp high *ridges*, or according to the turn of the fpires. Thefe *ridges* are generally *three*, *equi-diftant*, and placed on the *bottom* part of each *fpire*, for the upper part has only fine circular capillary ones.

The *mouth* is oval, and *border'd* all round on the outfide by a thick prominent *ledge*. It has no *umbilicus*.

The *fpires* are four, and gradually leffen to a fharp tip; the *bottom* of *each* is *flat* and *broad*, but *fharp* or *angulated* at the edge of

Une *fort petite coquille*, environ la *grandeur* d'un chenevi, totalement *blanche*, plutot *mince* & *tranfparente*, & d'une *forme* allongée. A *fillons* elevés, aigus, gros & circulaires, ou felon le tour des revolutions. Ces *fillons* font generalement *trois*, à *diftances egales*, & fitués fur la partie *inferieure* de chaque *revolution*, car la partie fuperieure a feulement de fillons circulaires fins ou capillaires.

La *bouche* eft ovale, & *bordée* alentour de l'exterieur d'un *rebord* epais & faillant. Elle n'a point d'*umbilic*.

Les *revolutions* font quatre, qui diminuent par degrés à un bout aigu; le *bas* de *chacune* eft *plat* & *large*, mais *aigu* ou *angulaire*

of the body, by means of the *lower ridge* of each *spire*.

I received this *species* from *Cornwall*; and, according to Linné, it is also a Mediterranean shell.

angulaire au bord du corps, à cause du *sillon inferieur* de chaque *revolution*.

J'ai reçu cette *espece* de *Cornwall*; et, selon Linné, elle est aussi une coquille de la Mediterranée.

LIX.

Pictus. Painted.

Tab. VIII. fig. 1. 3.

LIX.

Pl. VIII. fig. 1. 3.

Turbo minimus lævis, albo & rubro perbelle pictus. Pictus. Tab. 8. fig. 1. & 3.

Turbo minimus lævis, variegatus, albo rubicundus: The small red and white variegated whelke.—Borlase Cornw. p. 277.

Pullus. Turbo testa imperforata ovata lævi, apertura antice diducta.—Lin. S. N. p. 1233. No. 610.

This *shell* is of a lengthened *shape*, about the *size* of a cherry-stone kernel, *thin* and *transparent*, quite *smooth* and *glossy*. The *ground colour* is *white*, prettily variegated with *red*; sometimes only *mottled* or *shaded* with *pale red*; sometimes with longitudinal broad *waved stripes* of a fine *deep purple red*; and sometimes the pale red runs in circular *girdles*, very regular, and adorned with short transverse streaks of dark brown. It is a *beautiful shell* in all these *varieties*.

The *mouth* is oval; the *outer lip* even, but its lower end lengthens on the body spire much beyond the opening of the mouth: the *inner* or *pillar lip* is thick and spread. It has no *umbilicus*.

The *spires* are four, gradually diminishing to a sharp point, and are separated by a furrow.

I have received it from the coast of *Cornwall*, and from *Exmouth* in *Devonshire*. According to Linné, it is also an inhabitant of the Mediterranean sea.

LX.

Cette *coquille* est d'une *forme* allongée, environ la *grandeur* d'un noyau, *mince* & *transparente*, tout à fait *lisse* & *lustrée*. Le *fond* est *blanc*, joliment bigarrè de *rouge*; quelquefois seulement *taché* ou [illegible] de *rouge pale*; quelquefois avec des *rayes* longitudinales, larges & *ondées*, d'une belle *couleur pourpre rouge foncée*; & quelquefois le rouge pale est en *fascies* circulaires & regulieres, & ornées à travers de petites rayes courtes, d'un brun foncé. Elle est une *coquille* tres *elegante* dans toutes ces *varietés*.

La *bouche* est ovale; la *levre exterieure* unie, mais vers le bas s'allonge sur la revolution du corps beaucoup au delà de l'ouverture de la bouche: la *levre interieure*, ou de la *columelle* est epaisse & etendue. Elle n'a point de *umbilic*.

Les *revolutions* sont quatre, diminuant par degrès à une pointe aigue, & sont separées par une strie.

Je l'ai receu de la cote de *Cornwall*, & de *Exmouth* au comte de *Devon*. Selon Linné, c'est aussi une habitante de la mer Mediterranée.

LX.

LX.

Cancellatus. Latticed.

Tab. VIII. fig. 6. 9.

LX.

Pl. VIII. fig. 6 & 9.

Turbo minimus albus cancellatim vel decussatim striatus. Cancellatus. Tab. 8. fig. 6 & 9.

Buccinum, parvum, integrum, ore perpendiculari, minimum, striis minutissimis cancellatum, subalbidum. Gualt. I. Conch. tab. 44. fig. X. X.

C. iev. Turbo testa oblongo-ovata, striis decussatis: punctis eminentibus. Lin. S. N. p. 1233. No. 609.

The *shell* is *very small*, hardly double, the *size* of a caraway seed, of a taper *shape*, totally *milk white*, rather *thick*, and with no *gloss*. It is wrought all over with a *deep latticed work*, formed by thick *ridges*, which run *spiral* and *longitudinal*, so as to *cross* or *decussate* each other.

The *mouth* is round, and it has no *umbilicus*.

The *spires* are four. They gradually decrease to a bluntish tip, and are seperated by a strong furrow.

I have received it from the island of *Guernsey*, and from *Cornwall*. It is also a native of the Mediterranean, and of the coast of Senegal in Africa.

La *coquille* est *forte petite*, à peine de double la *grandeur* d'une graine de carvi, d'une *forme* allongée, totalement de *couleur blanche* de *lait*, plutot *epaisse*, & sans *lustre*. Elle est partout travaille en un *ouvrage profond* à *treillis*, formé par des *sillons* epais, qui sont *spirales* & *longitudinales*, desorte qu'ils se *croisent* & s'*entrecoupent*.

La *bouche* est ronde, & il n'a point de *umbilic*.

Les *revolutions* sont quatre. Elles diminuent par degrés à un bout obtus, & sont separées par une strie forte.

Je l'ai receu de l'isle de *Guernsey* & de *Cornwall*. Elle est aussi une native de la Mediterranée, & de la cote de Senegal en Afrique.

LXI.

Parvus. Small.

LXI.

Turbo parvus interdum lacteus, interdum violaceus aut fuscus, costis longitudinalibus confertus. Parvus.

Lacteus. Turbo testa cancellata turrita: striis longitudinalibus elevatis confertis. Lin. S. N. p. 1238. No. 634.

A *very small shell*, about the *size* of a common ant, of a subconic or taper *shape*, ending in a sharp point. The *colours* vary, often

Une *fort petite coquille*, environ la *grandeur* d'une fourmi commune, d'une *forme* quasi conique ou allongée, finissant dans une

often *milk white* and *glossy*, which rather seem worn shells that have lost their colour, as the work on them is also impaired, or not so sharp and fine. Many are of a fine *violet colour*, and glossy, and have the work pretty strong; but the most *perfect shells*, and in greater number, are *brown* and *white*, as they have the work on them very strong and in relief.

une pointe aigue. Ses *couleurs* varient, bien souvent *blanche de lait* & *lustrée*, qui me paroissent de coquilles mortes qui ont perdu leur couleur, comme leur travail est aussi affoibli, ou point si aigu & fin. Plusieurs sont d'une belle *couleur de violette*, & lustrée, avec le travail asses fort ou vif; mais les plus *parfaites coquilles*, & en plus grand nombre, sont *brunes* & *blanches*, comme elles ont le travail tres fort & en relief.

The *mouth* is round, and *surrounded* on the *outside* by a thick prominent *ledge* or *border*. It has no *umbilicus*.

La *bouche* est ronde, & *environnée* à l'*exterieur* d'une *bordure* epaisse & saillante, ou *bourrelet*. Elle n'a point de *umbilic*.

The *spires* are five, all gradually tapering to a sharp tip, and separated by a depression. They are *wrought* with very prominent *longitudinal ribs*, thick set.

Les *revolutions* sont cincq, diminuant par degrés à une pointe aigue, & separées par un abbaissement. Elles sont à *cotes* ou *sillons* fort elevés & *longitudinales*, tres serrès.

I have received these shells from the coast of *Guernsey*, from *Cornwall*, and from *Devonshire*. It is also a shell of the Mediterranean.

J'ai reçeu ces coquilles des cotes de *Guernsey*, de *Cornwall*, & du comté de *Devon*. Elle est aussi une coquille de la Mediterranée.

LXII.
ULVÆ. ULVA.

LXII.

Turbo parvus lævis et fuscus. Parvus.
Turbo ulvæ. Ulva.—Penn. Brit. Conch. No. 120. tab. 86. fig. 120.

About same *size* as the last, *thin*, of a produced subconic *shape*, tapering to a sharp point, quite *smooth*, and of a dirty white or brownish *colour*.

Environ la *grandeur* de la derniere, *mince*, d'une *forme* allongée & quasi conique, appetissante en une pointe aigue, tout à fait *lisse*, & d'une *couleur* blanche sale ou brunatre.

The mouth *oval*, has a capillary shallow *sent* aside of the pillar, but is not pierced, or an *umbilicus*.

La bouche est *ovale*, elle a une *entailleure* fine ou capillaire à coté de la columelle, mais elle n'est point percée, ou en *umbilic*.

The *spires* are five, tapering to a point, do not *jut* out beyond one another, and are separated by a strong furrow.

It is found on *several* of our *coasts*, generally on the *ulva lactucæ*, or *sea lettuce*.

Les *revolutions* sont cincq, appetissantes en une pointe; elles ne *debordent* pas l'une l'autre, & sont separées par une strie forte.

Elle se trouve sur *plusieurs* de nos *cotes*, generalement sur le *ulva lactucæ*, ou la *laitue marine*.

GENUS. GENRE

GENUS XII.

Strombiformis. Needle Snail.

Snails of a very long, ſlender and taper ſhape, like the ſtrombi, or needles, from which they differ only in the round and circumſcribed mouth.

GENRE XII.

Strombiformis, ou Limaçon Vis.

Limaçons d'une forme extremement allongée ou tres effilée, comme la vis, de leſquelles elles different ſeulement par la bouche ronde & entiere.

* TERRESTRES. Land.

LXIII.

Perversus. Reversed, or Oat.

Tab. V. fig. 15. 15.

* TERRESTRES. Terrestres.

LXIII.

Pl. V. fig. 15. 15.

Strombiformis parvus pullus, ore compreſſo, anfractibus contrariis ſtriatis. Perverſus. Tab. 5. fig. 15. 15.

Buccinum pullum, opacum, ore compreſſo, circiter denis ſpiris faſtigiatum.—Liſt. H. An. Angl.. p. 123. tit. 10. tab. 2. fig. 10.—Phil. Tranſ. No. 50. 72. & 105. fig. 10.—App. H. A. A. 4to. p. 23. tab. 1. fig. 7.—App. H. A. A. in Goedart. p. 43. tab. 1. fig. 7.

Buccinum exiguum pullum, duodecim orbium.—Liſt. H. Conch. tab. 41. fig. 39. [illegible] et *Min.*

Buccinum alterum pellucidum ſubflavum, intra ſenas circiter ſpiras mucronatum.—Liſt. H. An. Angl. p. 124. tit. 11. tab. 2. fig. 11.—Phil. Tranſ. No. 105. fig. 11.

The ſmall whirl ſnail with numerous rounds, and winding from the mouth towards the right-hand. Grew. Muſ. p. 132. Morton Northampt. p. 415. Et *Buccinum heteroſtrophum minutum fuſcum ſex ſpirarum ore ſubrotundo.* Id. p. 416. tab. 13. fig. 1.—*Buccinulum Anglicum heteroſtrophon oblongum ſtriis capillaceis.* Petiv. Muſ. p. 65. No. 703.—*Little torcular cochlea.* Wallis Northumb. p. 368.

Turbo perverſus. Reverſed.—Penn. Brit. Zool. No. 116. tab. 82. fig. 116.

Argenv. Conch. I. p. 385. tab. 32. fig. 19. 20. II. p. 340. & p. 83. tab. 28. fig. 19 & 20. & tab. 9. fig. 14.

Turbo terreſtris rufeſcens ore denticulato, à dextra in ſiniſtram convolutus. Gualt. I. Conch. tab. 4. fig. C. D. E. & tab. 4. fig. G. H. I. L.

Cochlea teſta pellucida oblonga, ſpiris decem ſiniſtrorſis, apertura ſubrotunda.—Lin. F. Suec. I. p. 372. No. 1300. II. No. 2172.

Perverſus. Turbo teſta turrita pellucida, anfractibus contrariis apertura edentula..—Lin. S. N. p. 1240. No. 650.

The *shell* is about the *size* and *shape* of an oat, being long and slender, of a dark brown *colour*, with very little gloss, *thin*, somewhat *transparent*, and set all over thickly with very fine hair-like longitudinal *striæ*.

The *mouth* is oblong-oval and *sinuous*, much border'd and turned outwards, *perpendicular*, *compress'd*, and narrows at bottom into a *sinus*, which is generally set with a small *angle* like a *tooth*; it has a *plait* or *folding* between it and the body, aside of the outer lip, and *two others*, very strong, at the *top* of the *head*, over it.

The *spires* are *ten*, *often less*, tapering, tho' not gradually, to a point, for they swell a little towards the middle, and narrow towards the two extremes, on a *level*, or do not *jut* out beyond one another, and are *separated* by a slight folding furrow. It is an *heterostrophe shell*, for the *mouth* lies on the *right side*, and *all the spires turn from the left to the right*.

These smaller ones are the *young* shells, but always with them are found *old* ones of *double* or *treble* the *size*; in every other respect like these, but *proportionably larger* and *stronger* in their several *parts* and *work*. The *plaits* or *foldings* near the *mouth* are *deep* and very *strong*; the *striæ stronger* and *more distinct*; the *border round the mouth* greatly turned outwards, very broad, flat, thick, and milk white, and the *sinuosities*, *jags* or *teeth*, within, are large, white, and very conspicuous; some are *bidentated*, and most of these *old* ones have *eleven*, and and some even *twelve spires*.

From these *circumstances*, *authors* run into *confusion*, by making the *different*

growths

La *coquille* est environ la *grandeur* & la *forme* d'une avoine, etant longue & deliée, d'une *couleur* brune foncée, peu lustrée, *mince*, quelque chose *transparente*, & à *stries* capillaires, fort fines, longitudinales, & tres serrées.

La *bouche* est oblong-ovale & *sinueuse*, tres bordée, & retroussée en dehors, *perpendiculaire*, *comprimée*, & etrecie au bas dans un *sein*, qui est generalement garni d'un petit *angle* comme une *dent*; il y a un *pli* entre elle & le corps, à coté de la levre exterieure, & *deux autres*, tres forts, sur le *sommet* de la *tete*, par dessus la bouche.

Les *revolutions* sont *dix*, *souvent moins*, appetissantes, mais non pas graduel, à une pointe, car elles sont un peu enflées vers le milieu, & etrecies vers les deux bouts, de *niveau*, ou ne *debordant* l'une l'autre, & sont *separées* par une strie en pli fort legere. C'est une *coquille unique*, ou *heterostrophe*, car la *bouche* est *à droite*, & *toutes les revolutions tournent de gauche à droite*.

Ces petites sont les coquilles *jeunes*, mais toujours parmi eux se trouvent des *vielles* de *double* ou *triple* leur *grandeur*; en touts autres egards comme celles la, mais *à proportion plus grandes* & *plus fortes* dans leur differentes *parties* & *ouvrage*. Les *plis* près de la *bouche* sont *profonds* & tres *forts*; les *stries plus fortes* & *plus distinctes*; la *bordure autour de la bouche* extremement retroussée en dehors, tres large, applatie, epaisse, & blanche de lait, & les *sinuosités*, *decoupures* ou *dents*, au dedans, sont grandes, blanches, & fort visibles; quelques unes sont *à deux dents*, & la plupart de ces *vielles coquilles* ont *onze*, & quelques unes même *douze revolutions*.

De ces *circonstances*, les *auteurs* tombent en *confusion*, en faisant les *differents*

ages

growths different species. The *accurate* and *judicious Lister himself* has formed *two species,* in his *tit.* 10 & 11, on the difference of the number of the spires and other slight particulars. The several figures in *Gualtieri* are only *varieties*; and the *bidens* of *Linné, Syst. Nat. p.* 1240. *No.* 649. and of *Mr. Pennant, Brit. Zool. No.* 117. *tab.* 81. *fig.* 117. is apparently no other than an *old shell,* for such *large* and *bidentated* ones I have not *unfrequently found* nestled with these *common smaller shells.*

ages, differentes especes. Le *limé* & *judicieux Lister même* a formé *deux especes,* dans ses *tit.* 10 & 11. sur la difference du nombre des revolutions, & autres particularités legeres. Les differentes figures dans *Gualtieri* sont seulement des *variétés*; et le *bidens* de *Linné, Syst. Nat. p.* 1240. *No.* 649. & de *M. Pennant, Zool. Brit. No.* 117. *tab.* 81. *fig.* 117. n'est apparemment autre que une *vielle coquille,* parceque de telles *grandes* & *à deux dents* j'ay *trouvé pas rarement* nichées parmi ces *petites* & *communes coquilles.*

Though the *number* of *spires* in a *shell* is a *criterion,* yet it is not an *infallible one,* for the number of spires vary in some species, either from the different growths or sexes: in such cases the *young shells* have always a *less number,* and the *males* have their spires *less numerous* than the *females.* This *very species* is, perhaps, as strong an instance of the *difference* in the *number* of the *spires* as *can be,* for *it* is found from *six* to *twelve spires,* as *Linné* has also noted in his *Fauna Suecica.*

Quoique le *nombre* de *revolutions* dans une *coquille* est un *caractere,* il n'est pourtant un qui est *emancable,* car le nombre des revolutions differe en quelques especes, par les differents ages ou sexes: en ces cas, les *jeunes coquilles* ont toujours un *moindre nombre,* & les *males* ont leurs revolutions *moins nombreuses* que les *femelles.* Cette *espece même* est, peutetre, un exemple aussi fort de la *difference* dans le *nombre* des *revolutions* que *peut etre,* car *elle* se trouve de *six* à *douze revolutions,* comme *Linné* a aussi remarqué dans son *Fauna Suecica.*

This *shell* is not *uncommon* in *most counties* of *England,* in the mofs, and in the fents or crevices in the barks of old trees, also in the mofs in the crevices and holes of old walls, and such-like places.

Cette *coquille* n'est *pas rare* dans *la plupart des comtés* d'*Angleterre,* dans la mousse, & dans les fentes ou crevasses en l'ecorce de vieux arbres, comme aussi dans les crevasses & trous de vielles murailles, & autres semblable lieux.

* * MA-

* * MA-

** MARINÆ. Sea.

LXIV.

Bicarinatus s. Torcular.
Two-ridged Screw.
Tab. VI. fig. 3.

** MARINÆ. De Mer.

LXIV.

Pl. VI. fig. 3.

Strombiformis major rubro luteſcens aut pullus: anfractibus duabus carinis ſive ſtriis acutis inſignitis. Bicarinatus s. Torcular. Tab. 6. fig. 3.

Buccinum craſſum, duobus acutis & inæqualiter altis ſtriis in ſingulis duodecim minimum ſpiris donatum. An Buccinum ſtriatum σαλπιγξ Fab. Columnæ?—Liſt. H. An. Angl. p. 160. tit. 7. tab. 3. fig. 7.

Cochlea longiſſima clavicula, medio quoque orbe unica valde acuta & eminente ſtria. Liſt. H. Conch. tab. 591. fig. 59. A?—*Bucc. ſtriatum σαλπιγξ Fab. Colum. cochlea alba, denſe & acute ſtriata.* Id. tab. 590. fig. 54?—Et *Cochlea alba, medio quoque orbe late excavato.* Id. tab. 591. fig. 58. A?

Turbo duplicatus. Doubled.—Penn. Brit. Zool. No. 112. tab. 81. fig. 112.

Strombus ſtriatus, & *Strombus ſagitta.*—Klein. Oſtrac. p. 29. § 74. No. c. & p. 30. No. a.

Torculum.—Argenv. Conch. I. p. 276. tab. 14. fig. C. II. p. 232. tab. 11. fig. C.

Turbo integer, vulgaris, craſſus, ponderoſus, in medio anfractuum coſta duplici acuta diſtinctus, albidus.—Gualt. I. Conch. tab. 58. fig. C.

Duplicatus. Turbo teſta turrita: anfractibus carinis duabus acutis.—Linn. S. N. p. 1239. No. 643.—Muſ. Reg. p. 662. No. 359.

The *ſhell* is very *thick, heavy* and *ſtrong,* of a ſlender *ſhape,* and tapering quite to a ſharp point, from four to five inches *long,* and about an inch *broad* at the mouth, and *gloſſy.*

The *mouth* is quite oval, the *outer lip* even, the *pillar lip* thick and flatted, and has no *umbilicus.* The whole ſhell is of an uniform pale yellowiſh *colour,* with a tint of red.

The *ſpires* are fourteen, gradually tapering to a ſharp point, ſwell or jut out a little beyond one another, and are ſeparated by a ſtrong declivity and a furrow.

Each

La *coquille* eſt tres *epaiſſe, peſante* & *forte,* d'une *forme* effilée, & appetiſſante à une pointe aigue, de quatre a cincq pouces en *longeur,* & environ un pouce en *largeur* à la bouche, & *luſtrée.*

La *bouche* eſt tout à fait ovale, la *levre exterieure* unie, celle de la *columelle* epaiſſe & applatie; elle n'a point d'*umbilic.* Toute la coquille eſt d'une *couleur* uniforme pale jaunatre, avec un teint de rouge.

Les *revolutions* ſont quatorze, appetiſſantes par degrés à une pointe aigue; elles ſont enflées, ou debordent l'une l'autre, & ſeparées par une grande pente & une

ſtrie.

Each ſpire has a *gutter*, bounded by *two* ſtrong, ſharp, and very prominent *ridges*, that runs ſpiral along the middle of it. The ſhell is beſides thickly ſet with fine ſlight *ſtriæ*, and a few *longitudinal wrinkles*.

I ſhould not have inſerted *this ſhell* in my *Britiſh Conchology*, but for the *authority* of Dr. *Liſter*, an *author* who, for *knowledge* and *veracity*, has not his equal. The *Doctor*, after having *deſcribed it*, *acquaints* us he had not ſeen any *live ones*, that it is a *very rare ſhell*, and that he had ſometimes *purchaſed* them of the *Scarborough fiſhermen*; therefore, *ſays he*, it ſeems to be a *pelagian* ſhell, or only got in *open ſea, far from the ſhores*.

However, I ſhall take the liberty to acquaint my readers that I have great reaſon to imagine the Doctor is in an *error*, and was *impoſed on*, as I have made particular enquiries at *Scarborough*, and *on that coaſt*, by *ſeveral perſons* of *knowledge* and *curioſity*, who reſided there through all the ſeaſons of the year, *even on purpoſe* to *ſtudy* and *collect Natural Hiſtory*, who all *confirm* that the *ſhell* is not known to them as a *native* of that *coaſt*, neither is it at all *known* to the *inhabitants*, *fiſhermen*, or other *ſeaſaring people*.

This *ſpecies* is pretty *common* in our *collections*. ſome giving it for a *Weſt-Indian ſhell*, others for an *Eaſt-Indian*, and others for an *European*. *Linné* ſays it is a *native* of the *Mediterranean ſea*.

ſtrie. *Chaque revolution* a une *goutiere*, bornée par *deux ſillons* forts & ſaillants en vive arrete, qui courent ſpirales le long de ſon milieu. La coquille outre eſt à *ſtries* fines & legeres tres ſerrées, & quelques *rides longitudinales*.

Je n'aurois pas inſeré *cette coquille* dans ma *Conchologie Britannique*, que ſur l'*autorité* du Dr. *Liſter*, un *auteur* qui, pour ſa *ſcience* & ſa *verité*, n'a point ſon egal. Le *Docteur*, apres l'avoir *decrit*, nous *inſtruit* qu'il n'a jamais vu des *vivantes*, qu'elle eſt une *coquille tres rare*, & que il les avoit quelquefois *acheté* des *pecheurs de Scarborough*; ainſi, *dit il*, elle paroit etre une *coquille pelagienne*, ou ſeulement à etre priſe en *haute mer*, *tres eloignée des cotes*.

Cependant, je prendrai la liberté de faire ſavoir à mes lecteurs que j'ai lieu à bon droit de penſer que le Docteur eſt en *erreur*, & qu'on *lui à trompè*, comme j'ay fait des recherches particulieres a *Scarborough* & *les cotes aux environs*, par de *perſonnes* de *ſcience* & *curioſité*, qui ont fait leur reſidence dans cet lieu pendant toutes les ſaiſons de l'année, *même à deſſein* de *etudier* & *faire une collection d'Hiſtoire Naturelle*, qui toutes *confirment* que la *coquille* ne leur eſt pas connue comme une *native* de cette *cote*, ni eſt elle *connue de tout* aux *habitants*, *pecheurs*, & autres *gens de mer*.

Cette *eſpece* eſt aſſes *commune* dans nos *collections*, quelques uns la faiſant une *coquille* des *Indes Occidentales*, d'autres des *Indes Orientales*, & enfin d'autres la diſent etre une *Européene*. *Linné* dit que elle eſt une *habitante* de la *mer Mediterranée*.

LXV.

LXV.

LXV.

TEREBRA. AUGER.

Tab. VII. fig. 5. 6.

LXV.

Pl. VII. fig. 5. 6.

Strombiformis medius albus rufo variegatus, anfractibus striatis. Terebra. Tab. 7. fig. 5. 6.

Buccinum tenue, dense striatum, duodecim minimum spiris donatum.—Lift. H. An. Angl. p. 161. tit. 8. tab. 3. fig. 8.

An *Cochlea ex fusco rufescens dense et leviter striata.* Lift. H. Conch. tab. 590. fig. 55?—aut *Cochlea alba, dense striata, orbis primi superiore parte paulo depressiore.* Ejusd. tab. 591. fig. 57?

Wallace Orkn. p. 40.—*Turbo terebra. Auger.* Penn. Brit. Zool. No. 113. tab. 81. fig. 113.

Buon. Ricr. p. 198. fig. 115.—Rumph. Muf. tab. 30. fig. M.—*Strombus cochloides planus,* & *Strombus cochloides striatus.* Klein Oftrac. p. 28. § 74. fpec. 2. No. A. & p. 29. No. g.—*Vis.* Argenv. Conch. I. p. 276. tab. 14. fig. D. II. p. 232. tab. 11. fig. D.

Cochlea testa longa subulata, spiris duodecim striatis.—Lin. F. Suec. I. p. 378. No. 1322. II. No. 2171.

Terebra. Turbo testa turrita anfractibus carinis sex acutis.—Lin. S. N. p. 1239. No. 605. Muf. Reg. p. 662. No. 360.

The *fhell* is rather *thinnifh*, and quite of a flender *fhape*, tapering to a very fharp point; in *length* from one and a half to two and a half inches, and at the top better than half an inch *over*.

The *ground colour* of the *perfect* or *live fhells* is whitifh, thickly and very prettily mottled with pale brownifh red, in fmall ftreaks and dots, and it is fomewhat *gloffy*.

The *mouth* is quite round; the *inner lip* even; the *pillar lip* very thick, fomewhat twirl'd, and greatly fpread on the body fpire. *Within* it is very fmooth and gloffy. It is not *umbilicated*.

The *fpires* are twelve, all gradually tapering to a fharp point; they do not jut out beyond one another, and are feparated by a ftrong furrow. Each *fpire* is thickly fet

La *coquille* eft plutot *mince*, & tout à fait d'une *forme* deliée, appetiffante à une pointe aigue, de un pouce & demi à deux pouces & demi en *longeur*, & la *largeur* à la tete furpaffe un demi pouce.

Le *fond* des *coquilles parfaites* ou *vivantes* eft blanchatre, joliment pointillée, & d'une maniere tres ferrée, de pale brunatre rouge, en petites rayes & points, & elle eft un peu *luftrée*.

La *bouche* eft tout à fait ronde; la *levre interieure* unie; *celle* de la *columelle* fort epaiffe, quelque peu tortillée, & beaucoup etendue fur la revolution du corps. En *dedans* elle eft fort liffe & luftrée. Elle n'a point d'*umbilic*.

Les *revolutions* font douze, toutes appetiffantes à une pointe aigue; elles ne debordent pas au delà l'une de l'autre, & font feparées par une ftrie forte. Chaque *revolution*

set with pretty prominent *ridges*, like threads; but *one* near to the *top* of each *spire* is greatly prominent or raised, and shelves on the other side to the next spire above; and two other ridges, *middle ones*, are also remarkably strong, and *regularly chequer'd* with transverse spots in a beautiful manner, not unlike the girdles on some *volutes* of the *admiral kind*. The other *ridges* are small and very many.

This *species* is not *uncommon* on many of our *coasts*, and in great plenty on some, as at the *Scilly Islands*; at *Liverpool*, where they are called *cockspurs*, and on the other shores of *Lancashire*; at *Scarborough*, after winter storms, according to Lister; at *Exmouth*, and other places on the *western shores*; and I have received very fine and perfect ones from the coasts of *Wales*, as *Flintshire*, *Pwlbely* in *Carnarvonshire*, and *Barmouth* in *Merionethshire*. It is also a shell of the *Orkneys*.

Though this *shell* is *common*, yet a *good specimen* is very *difficult* to be procured, for the *mouth* or *upper part* is always much *broken*, and the *colours* are *lost*. On this *account* I must beg the *indulgence* of my *readers*, for though I have given *two figures*, yet *neither of them represent the object well*.

revolution est à *fillons* asses saillants, comme des fils, & fort serrés; mais *un* près du *haut* de chaque *revolution* est fort saillant ou elevé, & penche de l'autre coté à la revolution prochaine au dessus d'elle; & deux autres fillons, *au milieu*, sont aussi remarquablement forts, & *regulierement marquetés* de taches à travers d'une maniere tres belle, & point dissemblable aux bandes de quelques *volutes* de la *sorte* des *amiraux*. Les autres *fillons* sont petits & nombreux.

Cette *espece* n'est *pas rare* sur plusieurs de nos *cotes*, & en grande abondance sur quelques unes, comme aux *Isles de Scilly*; à *Liverpool*, ou on les appelle *cockspurs*, c'est à dire *des ergots*, & sur les autres cotes du comté de *Lancaster*; à *Scarborough*, apres des tempetes d'hyver, selon Lister; à *Exmouth*, & autres lieux sur les *cotes occidentales*; & j'ai receu de fort belles & parfaites des cotes de *Galles*, comme du comté de *Flint*, *Pwlbely* au comté de *Carnarvon*, & *Barmouth* au comté de *Merioneth*. C'est aussi une coquille des *Isles Orcades*.

Quoique cette *coquille* est *commune*, neantmoins *un bon morceau* ou *exemple* est tres *difficile* à procurer, car la *bouche* ou *partie d'enhaut* est toujours fort *cassée*, & les *couleurs* sont *perdues*. Pour cette *raison* il faut supplier l'*indulgence* de mes *lecteurs*, car quoique j'ay donné *deux figures*, cependant *ni l'une* ou *l'autre represente bien l'objet*.

LXVI.

CINCTUS. GIRDLED.
Tab. VII. fig. 8.

LXVI.

Pl. VII. fig. 8.

Strombiformis medius albus pullo variegatus, anfractibus porcis tumidis latis & spirabilibus cinctis. Cinctus. Tab. 7. fig. 8.

An *Turbo integer, vulgaris, spiris gradatim complanatis, striis minutissimis circumdatus, ex albo & roseo eleganter variegatus.*—Gualt. I. Conch. tab. 58. fig. E?

An *Exoletus. Turbo testa turrita: anfractibus carinis duabus obtusis distantibus.*—Lin. S. N. p. 1239. No. 644?

The *shell* is *thickish* and *strong*, from two to two and a half inches in *length*, and three quarters of an inch *broad* at the upper part, of a taper *shape*, and ending in a point, but is rather swell'd than slender. The *ground colour* milk white, mottled with chesnut coloured longitudinal streaks or waves; but *often* also the *brown colour* is so *prevalent* that it rather appears a *brown* shell, the *white* being *little* and very *sullied*, and the brown longitudinal *waves* or *variegations* extremely *strong* and *deep* coloured.

The *mouth* is quite round, the *lips* not remarkable, and it is not *umbilicated.*

The *spires* are twelve, swell'd or rounded so as to jut out from each other, and are separated by a great declivity and a furrow. Each *spire* is wrought with several spiral rounded *ridges*, *two* of which, *viz.* the *two* in the *middle*, are remarkably large, broad and flat.

It is *a rare shell.* I have received it only from the coasts of *Lincolnshire* and *Lancashire.*

La *coquille* est *quasi epaisse* & *forte*, de deux à deux pouces & demi en *longeur*, & trois quarts d'un pouce en *largeur* à la partie superieure, d'une *forme* effilée, & finissant en une pointe, mais elle est plutot enflée que deliée. Le *fond* est blanc de lait, bigarré des rayes ou ondes longitudinales, couleur de chataigne; mais *souvent* aussi la *couleur brune domine* ainsi que elle paroit une coquille *brune*, y ayant *peu* de *blanc* fort *terne*, & les *ondes* longitudinales ou *bigarrures* brunes etant tres *fortes* & de couleur *foncée.*

La *bouche* est tout à fait rond, les *levres* aucunement remarquables, & elle n'est point *umbiliquée.*

Les *revolutions* sont douze, enflées ou arrondies ainsi qu'elles debordent l'une l'autre, & sont separées par un grand penchant & une strie. Chaque *revolution* est à plusieurs *sillons*, spirales & arrondis, *deux* desquels, *à sçavoir*, les *deux* du *milieu*, sont remarquablement grands, larges & applatis.

C'est une *coquille rare.* Je l'ai receu seulement des cotes des comtés de *Lincoln* & *Lancaster.*

LXVII.

CLATHRATUS. BARRED, or, FALSE WENTLETRAP.

Tab. VII. fig. 11.

LXVII.

FAUSSE SCALATA.

Pl. VII. fig. 11.

Strombiformis minor albus aut pullo variegatus, costis longitudinalibus elatis eleganter distinctus. Clathratus. Tab. 7. fig. 11.

Cochlea variegata, striis raris admodum eminentibus exasperata. — List. H. Conch. tab. 588. fig. 51.

Buccinum album minus costis eleganter elatis. Small winckle trepe.—Mus. Petiv. p. 66. No. 705.

Cochlea alba, striis raris admodum eminentibus exasperata. Striated white cochlea, or bastard wentletrap.—Borlase Cornw. p. 276. tab. 28. fig. 9.

Turbo clathrus. Barred.—Penn. Brit. Zool. No. 111. tab. 81. fig. 111. 111. A.

Rondelet, testac. 89. fig. 5.—Buon. Ricr. p. 197. fig. 111.—*Buccinum scalare.* Rumph. Mus. tab. 29. fig. W.—*Scala spuria.* Klein Ostrac. p. 52. § 150. spec. 2. No. a. tab. 3. fig. 66.

Turbo integer, &c. Gualt. I. Conch. tab. 58. fig. H.—*Fausse scalata.* D'Avila, p. 221. No. 427.

Cochlea oblonga: striis longitudinalibus marginatis. Lin. F. Suec. I. p. 379. No. 1325. II. No. 2170.—*Clathrus. Turbo testa cancellata turrita: anfractibus contiguis lævibus.* Id. Mus. Reg. p. 658. No. 352.—Et *Clathrus. Turbo testa cancellata turrita exumbilicata, anfractibus contiguis lævibus.* Id. S. N. p. 1237. No. 631.

The *shell* is *pretty thick* and *strong*, of a taper *shape*, but somewhat *swell'd*, or not *slender*, from one to two inches or more in *length*, and about half an inch *broad* at the head, chiefly *milk white*, but often with *small brownish* or *livid spots* set in *rows*, and also sometimes of a *livid* or *dusky appearance*.

The *mouth* is perfectly round, and *border'd outwardly* with a very thick prominent rounded *border*, like a *ring*. It has no *umbilicus*.

The *spires* are most generally nine, all *swell'd*, and jut out from one another, but taper to a sharp point; they are *separated* by a large and deep channel. All the *shell* is

La *coquille* est *asses epaisse* & *forte*, d'une *forme* effilée, mais quelque peu *enflée*, ou point *deliée*, de un à deux pouces ou plus en *longeur*, & environ un demi pouce en *largeur* à la tete, principalement *blanches de lait*, mais souvent à *petites taches brunatres* ou *livides*, en *rangées*, & aussi quelquefois d'une *apparence livide* ou *terne*.

La *bouche* est tout à fait ronde, & *bordée en dehors* par une *bordure* fort epaisse, saillante & arrondie, comme un *anneau*. Elle n'a point de *umbilic*.

Les *revolutions* sont pour la plupart neuf, toutes *bombées*, & debordent l'une l'autre, mais appetissent dans une pointe aigue; elles sont *separées* par un grand & profond canal.

is ſet its *length*, or from top to bottom, with regular thick and very prominent *ribs*, moſt generally *eight*; they run a little oblique, but equi-diſtant, and riſe from the *ring* round the mouth as from an *axis*, and are very thick or bump'd at the *commiſſures* of the *ſpires*. This *work* is very *elegant and curious*, and *analogous* to that on the famous ſhell called the *ſcaleris*, or *wentletrap*, from which particular it has obtained its *Engliſh* and *French trivial names*.

It is found on the *Corniſh*, *Devonſhire*, *Lancaſhire*, and ſome other of our coaſts; in *plenty* on the *eaſtern coaſts* of *Wales*; on the *Iriſh* coaſts; and at *Leith*, and other coaſts of *Scotland*.

This ſpecies is in great plenty alſo on the ſhores of Holland and France. It is likewiſe an inhabitant of the North, the Mediterranean, and the American ſeas.

canal. Toute la *coquille* eſt à *cotes* regulieres, epaiſſes & tres ſaillantes, qui courent *ſa longeur*, ou du ſommet au bas, plus communement *huit*; elles ſont un peu obliques, à diſtances egales, & s'elevent de l'*anneau* qui environne la bouche comme d'un *axe*, & ſont fort groſſes & boſſues aux *interſtices* ou *ſeparations* des *revolutions*. Cet *ouvrage* eſt tres *elegant & curieux*, & *analogue* à celui la de la fameuſe coquille nommée *ſcalaris*, ou *wentletrap*, de laquelle particularité elle à obtenue ſes *noms de guerre Anglois* & *François*.

Elle ſe trouve ſur les cotes des comtés de *Cornwall*, *Devon*, & *Lancaſter*, & quelques autres de nos cotes; ſur les *cotes orientales* de *Galles*, en *abondance*; ſur les cotes *Irlandoiſes*; & à *Leith*, & autres cotes de l'*Ecoſſe*.

Cette eſpece abonde auſſi ſur les cotes de la Hollande & de la France. Elle eſt pareillement une habitante des mers du Nord, la Mediterranée, & de l'Amerique.

LXVIII.

Albus. White.

Strombiformis parvus albiſſimus Levis. White.

Turbo minimus lævis albus. Milk-white ſmooth whelke. Borlaſe Cornw. p. 277.—An *Turbo albus. White.* Penn. Brit. Zool. No. 114. tab. 79?

A *ſmall ſpecies*, not half an inch *long*, of a taper *ſhape*, not ſlender, but very ſwell'd or bodied, *milk white*, *ſmooth*, and very *gloſſy*, pretty *thick*, yet *ſemitranſparent*.

The *mouth* is oblong-oval: the *ſpires* are eight, level, or not prominent beyond one another, only ſeparated by a ſlight furrow, and taper to a ſharp point.

This

LXVIII.

Une *petite eſpece*, point un demi pouce en *longeur*, d'une *forme* effilée, pas deliée, mais fort enflée ou groſſe, *blanche de lait*, *liſſe*, & fort *luſtrée*, aſſes *epaiſſe*, neantmoins *demitranſparente*.

La *bouche* eſt oblong-ovale: les *revolutions* ſont huit, unis, ou ne debordant l'une l'autre, ſeulement ſeparées par une ſtrie legere, & appetiſſent en une pointe aigue.

Cette

This *ſpecies* is found on the ſhores of *Cornwall*, about *Fowey*, *Whitſand Bay*, the *Land's End*, &c. alſo in *Devonſhire*; and, according to Mr. Pennant, on the ſhores of *Angleſea*.

Cette *eſpece* ſe trouve ſur les cotes de *Cornwall*, à *Fowey*, la *Baie* de *Whitſand*, le *Land's End*, &c. auſſi au comté de *Devon*; &, ſelon M. Pennant, ſur les rivages de *Angleſea*.

LXIX.

GLABER. SMOOTH.

Strombiformis parvus corneus glaber. Smooth.
An *Turbo Levis. Smooth.*—Penn. Brit. Zool. No. 115. tab. 79. *the upper figure?*

A *ſmall ſpecies*, about half an inch *long*, very *taper* and *ſlender*, *thin* and *tranſparent*, extremely *ſmooth* and *gloſſy*, of a light horn *colour*, with a few ſlight ſpiral *ſtreaks*, ſometimes *opake white*, ſometimes *red*, eſpecially at the commiſſures of the ſpires, and well diſtinguiſh their ſeparation.

The *mouth* is oblong: the *ſpires* about ten, all level, or not prominent, and tapering to a very fine ſharp point; they are well diſtinguiſhed by the ſpiral white lines.

I received them from *Exmouth* in *Devonſhire*; and three were found in the ſtomach of a *five finger* or *common ſtella marina*. Mr. Pennant notes them from the ſhores of *Angleſea*.

LXIX.

Une *petite eſpece*, environ un demi pouce en *longeur*, fort *effilée* & *deliée*, *mince* & *tranſparente*, extremement *liſſe* & *luſtrée*, d'une *couleur* de corne claire, avec quelques *rayes* ſpirales legeres, quelquefois *blanches* & *opaques*, quelquefois *rouges*, ſpecialement aux interſtices des revolutions, & diſtinguent tres bien leur ſeparation.

La *bouche* eſt oblongue: les *revolutions* environ dix, toutes de niveau, ou ne debordant l'une l'autre, & appetiſſent en une pointe fort fine & aigue; elles ſont bien diſtinguées par les lignes ſpirales blanches.

Je les ai receu de *Exmouth* au comté de *Devon*; & trois furent trouvées dans l'eſtomac d'une *etoile marine commune* ou *à cinq rayons*. M. Pennant les notifie des rivages de *Angleſea*.

LXX.

RETICULATUS. LATTICED.
Tab. VIII. fig. 13.

LXX.

Pl. VIII. fig. 13.

Strombiformis parvus ſubfuſcus reticulatus. Reticulatus. Tab. 8. fig. 13.
Turbo minimus, ſubfuſcus, acus inſtar acuminatus. The ſmall needle whelke.—Borlaſe Cornw. p. 277.

Turbo

Turbo integer, acuminatus, ſtriatus, papillis minimis exaſperatus, ſubalbidus.—Gualt. I. Conch. tab. 58. fig. G.

The *ſhell* is very *thick* and *ſtrong* for its *ſize*, which generally is from one quarter to half an inch in *length*, of a perfect taper *ſhape*, but not very *ſlender*, not *gloſſy*, of a light brown *colour*, and wrought all over with a ſtrong *lattice* or *net work*, by *longitudinal* and *tranſverſe* or *ſpiral ridges*, thickly ſet, which croſs or decuſſate one another.

The *mouth* is roundiſh. It has nine *ſpires*, all tapering to a point, level or not prominent, and ſeparated by a ſtrong deep furrow.

This *ſpecies* is found in *plenty* on the ſhores of *Cornwall*, in *Whitſand Bay*, at the *Land's End*, &c. and I found it, in immenſe quantities, on ſand-banks thrown up by the ſea about five miles from *Falmouth*. It is alſo found in ſome *places* on the *Devonſhire* coaſt.

La *coquille* eſt tres *epaiſſe* & *forte* pour ſa *grandeur*, qui generalement eſt d'un quart à un demi pouce en *longeur*, d'une *forme* tout à fait effilée, mais pas fort deliée, point *luſtrée*, d'une *couleur* brune claire, & à *ſillons longitudinales* & *ſpirales* ou à *travers*, & ſerrés, qui entrecoupent l'une l'autre; formant un *ouvrage à treillis* ou *reticulé* conſiderable.

La *bouche* eſt à peu pres ronde. Elle a neuf *revolutions*, toutes appetiſſantes dans une pointe, de niveau ou ne debordant l'une l'autre, & ſeparées par une ſtrie forte & profonde.

Cette *eſpece* ſe trouve en *abondance* ſur les cotes de *Cornwall*, en la *Baie de Whitſand*, le *Land's End*, &c. et je l'ai trouvé, en quantité immenſe, ſur des bancs de ſable, formés par la mer, environ cincq miles de *Falmouth*. Elle ſe trouve auſſi en quelques *lieux* de la cote du comté de *Devon*.

LXXI.

COSTATUS. RIBBED.

Tab. VIII. fig. 14.

LXXI.

Pl. VIII. fig. 14.

Strombiformis parvus fuſcus, anfractibus coſtis elatis longitudinalibus inſignitis. Coſtatus. Tab. 8. fig. 14.

The *ſhell* is *thinniſh* and *ſemitranſparent*, totally of a cheſnut brown *colour*, hardly *gloſſy*, of a taper *ſhape*, but *thick* or *ſwell'd*, and generally about half an inch in *length*.

The *mouth* is large, and round the *outer lip* ſinuous or waved, and extends a little

La *coquille* eſt *quaſi mince* & *demitranſparente*, totalement d'une *couleur* brune chataigne, à peine *luſtrée*, d'une *forme* effilée, mais *groſſe* ou *enflée*, & generalement environ un demi pouce en *longeur*.

La *bouche* eſt *grande*, & autour de la *levre exterieure* ſinueuſe ou ondée, & s'etend

little into a *flap* or *wing*, thick at the edges, and *outwardly* has a thick prominent broad border.

The *spires* are ten; they taper to a sharp point, are greatly *rounded* and *swell'd*, and separated by a deep depression and furrow; they are wrought with many thick prominent longitudinal *ribs*, set close, in a very pretty manner.

This *curious small species* is found on the coasts of *Cornwall*.

tend un peu dans une *oreille* ou *aile*, epaisse aux bords, & *exterieurement* à bordure saillante, epaisse & large.

Les *revolutions* sont dix; elles appetissent dans une pointe aigue, sont tres *arrondies* & *enflées*, & separées par un enfoncement profond & une strie; elles sont travaillée à plusieurs *cotes* saillantes, epaisses & longitudinales, tres serrées, d'une maniere fort agreable.

Cette *curieuse petite espece* se trouve sur les cotes de *Cornwall*.

FAMILY OF BUCCINA. WILKS OR WHELKS,

Turbinated univalves, whose *mouths* are an oblong-oval, or near to it, the *upper part* whereof is *produced* or *lengthened* into a *gutter* or *beak*. The animal a *Slug*.

This *family* subdivides into *six genera*, viz. 1. *Buccina Canaliculata, Gutter'd Whelks.* 2. *Buccina Recurvirostra, Wry-mouth'd Whelks.* 3. *Buccina Longirostra, Beaked Whelks.* 4. *Buccina Umbilicata, Umbilicated Whelks.* 5. *Buccina Columella Plicata, Plaited Pillar Whelks.* And 6. *Strombi, Needles.* — Of this *family*, we have *shells* of the 1st, 2d, and 3d *genera*, in our *British seas*.

LA FAMILLE DE BUCCINA, OU BUCCINS.

Univalves contournées en spirale, dont les *bouches* sont oblongue-ovales, ou à peu près, le *haut* ou la *partie superieure* de laquelle est *allongé* en une *goutiere* ou *queue*. L'animal est une *Limace*.

Cette *famille* se soudivise en *six genres*, qui sont, 1. *Buccins à bouche en goutiere.* 2. *Buccins à bouche echancrée ou de travers.* 3. *Buccins à bouche garnie d'une longue queue.* 4. *Buccins umbiliquès.* 5. *Buccins à columelle plissée.* Et 6. *Strombi*, ou *Vis.* —De cette *famille*, nous avons des *coquilles* du 1. 2. & 3e. *genres*, dans nos *mers Britanniques*.

GENUS XIII. BUCCINA CANALICULATA, GUTTER'D WHELKS.

Whelks whose upper part of the mouth ends in a strait and somewhat produced gutter or channel.

GENRE XIII. BUCCINA CANALICULATA, BUCCINS en GOUTIERE.

Buccins dont le haut de la bouche est terminé dans un canal ou goutiere, droit & quelque peu allongé.

* MARINÆ. SEA.

LXXII.

MAGNUM. LARGE.

Tab. VI. fig. 4.

* MARINÆ. DE MER.

LXXII.

Pl. VI. fig. 4.

BUCCINUM canaliculatum magnum, crassum, striatum, album. Magnum. Tab. 6. fig. 4. *Buccinum album læve, maximum, septem minimum spirarum: A whelke.* Lift. H. An. Angl. p. 155. tit. 1. tab. 3. fig. 1.—And *Buccinum rostratum majus crassum, orbibus paulum pulvinatis.* Lift. H. Conch. tab. 913. fig. 4.

Wallace Orkn. p. 39.—*The greater smooth whelke.* Dale Harw. p. 381. 1.—Smith Cork, p. 318.—*The largest whelk, called Barnagh.* Rutty Dublin, p. 381.—*Murex despectus. Despised.* Penn. Brit. Zool. No. 98. tab. 78. fig. 98.

Buon.

Buon. Ricr. p. 214. fig. 190.—*Sipho lævis crassus.* Klein Ostrac. p. 53. § 154. No. 1. tab. 3. fig. 68.

Buccinum majus, canaliculatum, rostratum, ore simplici, læve, crassum, ponderosum, albidum. Gualt. tab. 46. fig. D.

Despectus. Murex testa patulo-subcaudata oblonga anfractibus octo: lineis duabus elevatis. Lin. S. N. p. 1222. No. 559.—F. Suec. II. No. 1266.

This *shell* is *large*, and found to above five inches in *length* and two one-half inches in *breadth* across the middle of the mouth, of a produced *shape*, *heavy*, *strong* and *thick*, yet *semitransparent*, totally *white*, not *glossy*, with some strong longitudinal *wrinkles*, and thickly set with transverse *striæ*, which are crossed by other slighter longitudinal ones, but do not form a very distinct *lattice work*.

The *mouth* is oval, and towards the top stretches out into a short and somewhat twirl'd *gutter* or *beak*; at the bottom it ends in an *angle*. The *outer lip* even; the *inner* or *pillar lip* twirl'd at the top, and spread down the side. *Within* it is perfectly smooth and glossy, in the *small shells* always *yellowish*, but in the *large* ones is of a beautiful deep *golden yellow colour*.

The *spires* are eight, all greatly *swell'd* and *rounded*, *prominent* from *each other*, and separated by a very broad tho' gentle slope. In the *turban*, or *lower spires*, this *declivity* verges on a *flat*, and in the *large shells* the *edge* of it runs in rude but large tumid *bumps*.

The *top* of the *body spire*, aside of the *gutter* and *mouth*, is a strong, thick, *prominent wreathed edge*, *curved* towards the *mouth*, between *which* and the *mouth* is a large gutter-like *depression*, which leads at

R the

La *coquille* est *grande*, & se trouve au delà de cincq pouces en *longeur* & deux & demi pouces en *largeur* à travers le milieu de la bouche, d'une *forme* allongée, *pesante*, *forte* & *epaisse*, cependant *demitransparente*, totalement *blanche*, point *lustrée*, à quelques *rides* longitudinales fortes, & à *stries* transversales tres serrées, qui sont croisées par d'autres longitudinales legeres, mais ne forment point un *ouvrage à treillis* fort distinct.

La *bouche* est ovale, & vers le haut se prolonge en une *goutiere* ou *queue*, quelque peu tortillée; & au bas elle finit dans un *angle*. La *levre exterieure* unie; l'*interieure* ou celle de la *columelle* tortillée au haut, & etendue le long du coté. En *dedans* elle est parfaitement lisse & lustrée, dans les plus *petites coquilles* toujours *jaunatre*, mais dans les *grandes* d'une tres belle *couleur jaune d'or* foncée.

Les *revolutions* sont huit, toutes *enflées*, *arrondies*, & *debordant l'une l'autre*, & separées par un penchant large mais leger. Dans la *clavicule*, ou les *revolutions inferieures*, cet *penchant* approche à etre *applati*, & dans les *grandes coquilles* le *bord* est à grandes *bosses* & fort enflées.

Le *haut* de la *revolution* du *corps*, à coté de la *goutiere* & de la *bouche*, est un *bord saillant*, epais, fort, *tortillé*, & *courbé* envers la *bouche*, entre lequel & la *bouche* se trouve un grand *enfoncement* comme une goutiere,

à

the end into a deep tho' narrow *hole*, *like to*, tho' not an *umbilicus*.

à la fin duquel est un *trou* profond quoique etroit, qui *ressemble à*, mais n'est point *un umbilic*.

This *large* and *fine shell*, to which the celebrated *Linné* and Mr. *Pennant* have *emphatically given* the *trivial name* of *Despectus*, *Despicable* or *Contemptible*, is the *largest* of the *turbinated univalves*, *natives* of the *British seas*.

Cette *grande* & *belle coquille*, à laquelle le celebre *Linné* & M. *Pennant* ont *avec emphase donné* le *nom de guerre* de *Despectus*, *Meprisable* ou *Vile*, est la plus *grande* des *univalves contournées en spirales*, *natives* des *mers Britanniques*.

It generally inhabits the deep sea, and is found at *Scarborough* and other places in *Yorkshire*, on the coasts of *Durham*, on the *Essex* shores, and is a *common shell* of that *sea*; on the *Sussex* shore, and *many other English shores*; in *Scotland*, in the *Orkneys*, in *Kincardinshire*, and *many other coasts*, and on *many* of the *Irish coasts*.

Elle habite generalement la haute mer, & se trouve a *Scarborough*, & autres lieux au comte de *York*, sur les cotes de *Durham*, sur les rivages de *Essex*, & est une *espece commune* dans cette *mer*; sur les cotes de *Sussex*, & *plusieurs autres de nos cotes Angloises*; en *Ecosse*, aux *Orcades*, au comté de *Kincardin*, & *plusieurs autres cotes*; & sur *plusieurs* des *cotes Irlandoises*.

It is eaten sometimes; however *Rutty notes* that the tail is said to be more fat and tender than a lobster. By our English fishermen it is chiefly used for baits to catch the fish.

Quelquefois elle sert de nourriture; cependant *Rutty remarque* que la queue est estimée d'etre plus grasse & tendre que une ecreviste de mer. Mais nos pecheurs Anglois l'usent principalement pour des amorçes à prendre les poissons.

LXXIII.
VULGARE. COMMON.
Tab. VI. fig. 6. 6.

LXXIII.
Pl. VI. fig. 6. 6.

Buccinum canaliculatum medium vulgare rufescens striatum, pluribus costis undatis distinctum. Vulgare. Tab. 6. fig. 6. 6.

Buccinum crassum rufescens, striatum & undatum. List. H. An. Angl. p. 156. tit. 2. tab. 3. fig. 2.—Et *Bucc. tenue, læve, striatum & undatum.* Id. p. 157. tit. 3. tab. 3. fig. 3.

Bucc. brevi rostrum tenuiter striatum, pluribus undatis sinubus distinctum. List. H. Conch. tab. 962. fig. 14.—Et *Bucc. brevi rostrum magnum, tenue, leviter striatum.* Id. tab. 962. fig. 15. 15. a.—Id. Exerc. Anat. Alt. p. 68.

Wallace Orkn. p. 39.—Martin W. Isles, p. 6. & alibi.—*Bucc. marinum nostras costis fasciatis & striatis.* Mus. Petiv. p. 83. No. 809.—*Rough*, and *our most common whelke.* Dale Harw. p. 382. No. 3. 4.—Wallis Northumb. p. 401. No. 35. 36.

Buccinum

Buccinum undatum, Waved; & *Bucc. striatum, Striated.* Penn. Brit. Zool. No. 90. & 91. tab. 73. & 74. fig. 90. & 91. et An *Murex antiquus, Antique*, No. 97?

Buon. Ricr. p. 213. fig. 189.—*Buccinum lacerum.* Klein Ostrac. p. 45. § 132. spec. 1. —An Gualt. I. Conch. tab. 46. fig. E?

Undatum. Buccinum testa oblonga rudi transversim striata: anfractibus curvato-multangulis. Lin. S. N. p. 1204. No. 475.—F. Suec. II. p. 2263.—et An *Murex antiquus.* Id. S. N. p. 1222. No. 558?

This *species* is generally about half the *size* of the last, rather *thin* and *semitransparent*, yet *strong* and *heavy*, not *glossy*, of a light yellowish brown *colour*, on the turban generally deeper *brown*, and some shells are almost *brown*. It is spirally *striated*, has many longitudinal *wrinkles*, and towards the bottom of the body spire is thickly and longitudinally *ribb'd*; but all the *turban* or *lower spires* are greatly ribb'd with very prominent, rounded, oblique waved *ribs*, also longitudinal.

The *mouth* is oval; at the *top* it runs into a broad *scoop* or *gutter*, which inclines downwards; and at the *bottom* it ends in an angle. The *outer lip* thin and even, but towards the bottom thickens very much, is spread, or *wing-like*, and somewhat *waved*. The *inner* or *pillar lip* is very thick, twirl'd at top, and spread. The *inside* of the *mouth* is smooth and glossy; and, in live fresh shells, white, or of a yellowish livid hue.

The *spires* are seven, all swell'd and rounded, and separated by a strong depression and furrow.

Some of these *shells* being only *striated*, or with *few* or *no ribs*, has led Dr. Lister to form *two species* of them, viz. *Tit.* 2 & 3. Other authors, as *Dale, Pennant, &c.* have followed

Cette *espece* est generalement environ la mi *grandeur* de la derniere, plutot *mince* & *demitransparente*, cependant *forte* & *pesante*, point *lustrée*, d'une *couleur* foible jaunatre brune, sur la *clavicule* generalement d'un *brun* plus foncé, & quelques coquilles sont presque *brunes*. Elle est à *stries* spirales, & a plusieurs *rides* longitudinales, & vers le bas de la revolution du corps est à *cotes* longitudinales fort serrées; mais toute la *clavicule* ou les *revolutions inferieures* sont fortement travaillées à *cotes* tres bombées, arrondies, obliques & ondées, comme aussi longitudinales.

La *bouche* est ovale; au *haut* elle s'etend en une *goutiere* ou *ecope*, qui penche en bas; & *au bas* elle finit dans un angle. La *levre exterieure* est mince & unie, mais vers le bas s'epaissit beaucoup, est etendue, ou *ailée*, & quelque chose *ondée*. L'*interieure* ou celle de la *columelle* est fort epaisse, tortillée au haut, & etendue. Le *dedans* de la *bouche* est lisse & lustrée; en les coquilles vivantes, blanc, ou d'un teint jaunatre livide.

Les *revolutions* sont sept, toutes bombées & arrondies, & separées par un enfoncement fort & une strie.

Quelques unes de ces *coquilles* etant seulement *striées*, ou avec *peu* ou *point* des *cotes*, à porté le Dr. Lister de les proposer comme *deux especes*, à *sçavoir*, *Tit.* 2 & 3. D'autres auteurs

followed him; but they are only *different growths* of the *same species*.

auteurs, comme *Dale* & *Pennant*, &c. l'ont suivi; mais elles sont seulement *differents ages* de la *même espece*.

This is the *common edible whelke* sold in the markets *throughout these kingdoms*, and is found in *plenty* on all our *shores*.

C'est le *buccin commun mangeable* qui se vend aux marchés *partout ces royaumes*, & se trouve en *abondance* sur toutes nos *cotes*.

LXXIV.

GRACILE. SLENDER.

Tab. VI. fig. 5.

LXXIV.

Pl. VI. fig. 5.

Buccinum canaliculatum medium, angustius, album, striatum, octo spirarum. Gracile. Tab. 6. fig. 5.

Buccinum angustius, tenuiter admodum striatum, octo minimum spirarum. List. H. An. Angl. p. 157. tit. 4. tab. 3. fig. 4.—App. H. An. Angl. p. 15. 16.—App. H. An. Angl. in Goedart. p. 24.—Et *Buccinum rostratum gracilius.* Id. H. Conch. tab. 913. fig. 5.

Wallace Orkn. p. 40.—*Lesser long and smooth whelke.* Dale Harw. p. 381. No. 2. and Smith Cork, p. 318.—*Narrow-mouth'd whelke with eight wreaths.* Wallis Northumb. p. 401. No. 7.—*Murex corneus. Horny.* Penn. Brit. Zool. No. 99. tab. 76. fig. 99.

Buon. Ricr. p. 186. fig. 53.—Gualt. I. Conch. tab. 46. fig. F. *quoad fig.*

Corneus. Murex testa oblonga rudi, anfractuum marginibus complanatis, apice tuberculoso, apertura edentula, cauda adscendente.—Lin. S. N. p. 1224. No. 565.

The *shell* is of a narrow slender *shape*, from two and a half to three inches *long*, and little better than about one inch *broad* across the middle of the mouth, rather *thin* and *semitransparent*, tho' *strong* and *heavy*. It is of a pure white *colour*, somewhat *glossy*, and, when *alive*, cover'd with a fine thin brown *film* or *epidermis*, like glue; it is *spirally striated*, or according to the run of the wreaths, with very fine slightly prominent *striae*, pretty thickly set.

La *coquille* est etroite & d'une *forme* deliée, de deux & demi à trois pouces en *longeur*, & un peu plus de environ un pouce en *largeur* ou au travers du milieu de la bouche, plutot *mince* & *demitransparente*, quoique *forte* & *pesante*. Elle est d'une *couleur* blanche pure, quelque peu *lustrée*, &, quand *vivante*, couverte d'une *peau deliée* ou *epiderme*, mince brune, comme de la colle; elle est *striée en spirale*, ou selon le tour des revolutions, par de *stries* fines, peu saillantes, & asses serrées.

The *mouth* is narrow and oblong-oval; it ends at *top* in a pretty produced *beak* or *gutter*, twirl'd or waved, and deep; at *bottom*

La *bouche* est etroite & oblongue-ovale; elle finit *au haut* dans une *queue* ou *goutiere*, asses allongée, tortillée & profonde; au *bas*

bottom in an angle. The *outer lip* thin and even; the *inner* or *pillar lip* thick, twirl'd and fpread. The *infide* of the *mouth* is fmooth, finely glazed, and pure white.

The *fpires* are eight, and taper to a point; they are all fwell'd and rounded, and feparated by a deep depreffion and furrow.

This *pretty fpecies* is found on feveral of our coafts, as *Yorkfhire*, *Northumberland*, *Effex*, &c. in the *Orkneys* and other fhores of *Scotland*, and on the *Irifh* coafts.

bas dans un angle. La *levre exterieure* eft mince & unie; l'*interieure*, ou celle de la *columelle*, epaiffe, tortillée & etendue. Le *dedans* de la *bouche* eft liffe, parfaitement luftré, & blanc pur.

Les *revolutions* font huit, & appetiffent en une pointe; elles font toutes bombées & arrondies, & feparées par un enfoncement profond & une ftrie.

Cette *jolie efpece* fe trouve fur plufieurs de nos cotes, comme aux comtés de *York*, *Northumberland*, *Effex*, &c. aux *Orcades* & autres cotes d'*Ecoffe*, & fur les cotes *Irlandoifes*.

LXXV.

Purpuro-Buccinum. Purple Whelke.

Tab. VII. fig. 1. 2. 3. 4. 9. 12.

LXXV.

Pl. VII. fig. 1. 2. 3. 4. 9. 12.

Buccinum canaliculatum minus, craffum, varicolor, ftriatum, feu Purpura Anglicana. Purpuro-buccinum. Tab. 7. fig. 1. 2. 3. 4. 9. 12.

Buccinum minus, albidum, afperum, intra quinas fpiras finitum. Lift. H. An. Angl. p. 158. tit. 5. tab. 3. fig. 5.—*White couvins.* App. H. An. Angl. p. 16. et App. ad Goedart. p. 25.—Et *Buccinum minus, ex albo fubviride, ore dentato, ecq; ex flavo leviter rufefcente.* Id. H. An. Angl. p. 159. tit. 6. tab. 3. fig. 6.—Exercit. Anat. Alt. p. 85.

Buccinum brevi roftrum fupra modum craffum, ventricofius, labro denticulato: Purpura Anglicana. Lift. H. Conch. tab. 965. fig. 18.—Et *B. brevi roftrum, album denticulo unico ad imam columellam. Purpura Anglicana.* Fig. 19.

Phil. Tranf. No. 178. tab. 3. fig. 3 a 7. and Lowthorp Abridg. vol. 2. p. 823. tab. 12. fig. 224 a 228.—Wallace Orkn. p. 40.—*Purpura allicans rugofa Anglicana.* Petiv. Gaz. tab. 93. fig. 16. Et *Buccinum cornubienfe, é caftaneo & albo fafciatum.* Id. tab. 18. fig. 5. —*Purpura Anglicana major & minor.* Dale Harw. p. 382. No. 5 & 6.—*The purple-marking whelke.* Borlafe Cornw. p. 277. tab. 28. fig. 11.—*Englifh purple.* Smith Cork, p. 318.—*Horfe wrinkles.* Smith Waterford, p. 272. Rutty Dublin, p. 382.—*Small purple whelk.* Wallis Northumb. p. 401.—*Buccinum lapillus. Mafiy.* Penn. Brit. Zool. No. 89. tab. 72. fig. 89.

Mem. Acad. Roy. de Sciences, 1711, tab. 6. fig. 5. 7.

Cochlea tefta craffa ovata utrinque producta; fpiris quinque fpiraliter fulcatis; aperturæ labro undulato.—Lin. F. Suec. I. p. 378. No. 1321. II. No. 2167.

Lapillus.

Lapillus. Buccinum testa ovata acuta striata levi, columella planiusćula.—Lin. S. N. p. 1202. No. 467.

This *shell* is extremely *thick* and *strong* for its *size*, which is *small*, generally from one to one and a half inch in *length*, and about three quarters of an inch in *breadth* across the middle of the mouth, of a full-bodied pyramidal *shape*, and terminates in a very sharp point.

La *coquille* est extremement *epaisse* & *forte* pour sa *grandeur*, qui est *petite*, generalement de un à un & demi pouce en *longeur*, & environ trois quarts d'un pouce en *largeur* au travers du milieu de la bouche, d'une *forme* grosse & pyramidale, & finit dans une pointe fort aigue.

The *mouth* is narrow and oblong-oval; it ends at the *top* in a wide, deep, and bent *gutter*; at *bottom* it runs into an angle. The *outer lip* is sharp edged, somewhat fluted or serrated, but from the *edge* it *slopes* down some way *within* the *mouth* to a *border*, where it becomes extremely thick in common with the rest of the shell; and on this *interior border* those *shells* that are *toothed* have the *teeth*, which are *five*, long, narrow, prominent and parallel.

La *bouche* est etroite & oblongue-ovale; elle finit au *haut* dans une *goutiere*, large, profonde, & tortillée; au *bas* elle s'etende dans un angle. La *levre exterieure* est à bord tranchant, & un peu canelé ou dentelé, mais dès le *borde* elle *penche* quelque chose en *dedans* la *bouche* à une *bordure*, ou elle devient extremement epaisse comme le reste de la coquille; & sur cette *bordure interieure* les *coquilles* qui sont à *dents* ont *cincq* dents, longues, etroites, saillantes & paralleles.

The *inner* or *pillar lip* is thick and very spread, and the whole *mouth within* is generally very smooth, glossy, white or livid.

La *levre interieure*, ou de la *columelle*, est epaisse et fort etendue, & toute la *bouche en dedans* est generalement fort lisse, lustrée, blanche ou livide.

The *spires* are five; the *body one* large and swell'd; the *others* diminish, but not quite gradually, to a sharp point; they are swell'd and rounded, and separated by a depression and furrow, spirally ridged, and pretty thickly set; the *ridges* are prominent, rounded, and like threads. The *turban* or *lower wreaths* have the *ridges notched* or *scaled*, and feel very rough.

Les *revolutions* sont cincq; *celle* du *corps* grande & bombée; les *autres* diminuent, mais non pas tout à fait, à une pointe aigue; elles sont bombées & arrondies, & separées par un enfoncement & une strie, à sillons spirales & asses serrés, saillants, arrondis, & comme des fils. La *clavicule*, ou les *revolutions inferieures*, ont les *sillons crenelés*, ou à *ecailles*, & sont fort rudes au toucher.

Aside of the *top* and *pillar lip* is a large prominent twirl'd *edge*, which in *many shells (not all)* has a *shallow* between *it* and the *mouth*, with a narrow *cavity*, *umbilicus like*, in the same manner as the *Buccinum Magnum*, No. 72. supra.

A coté du *haut* & de la *levre columelle* se trouve un grand *bord* saillant & tortillé, qui dans *plusieurs coquilles (non en toutes)* à un *enfoncement* entre *elle* & *la bouche*, avec une *cavité*, comme un *umbilic*, de la même maniere que le *Buccinum Magnum*, No. 72. sus dit.

The

Les

The *colours* of these shells are very various, *often* of a *single* or *uniform colour*, as pure *white*, pale *violet*, *brown* and *brownish*, *yellow*, *sandy*, *&c.* and as often finely *girdled* with broad *girdles* of fine *red*, *brown*, *&c.* on all the different *grounds*, sometimes with single, but generally with two or three girdles.

Les *couleurs* de ces coquilles varient beaucoup; *souvent* elle sont d'une *seule couleur uniforme*, comme *blanche* pure, *violet* pale, *brune* ou *brunatre*, *jaune*, *couleur de sable*, *&c.* & aussi souvent agreablement *fasciées* à *bandes* larges, d'un beau *rouge*, *brune*, *&c.* sur touts les differents *fonds*, quelquefois à une seule bande, mais generalement à deux ou trois bandes.

From its being *tooth'd*, or *not tooth'd*, (*only* the *effects* of *different growths*) Dr. *Lister* has *erroneously proposed it* as *two distinct species*.

Les coquilles etant *avec* des *dents*, ou *sans dents* (qui est *seulement* l'*effet* de leurs *differents ages*) à seduit le Dr. *Lister* de les *proposer*, *tres erronement*, comme *deux especes distinctes*.

The *shell*, when *fresh* from the *sea*, is generally cover'd with *filth*, but when clean'd, by its *fine colours*, makes a *beautiful appearance*.

It is found in *great plenty* on most of the coasts of *Great Britain* and *Ireland*.

La *coquille*, quand *recente* de la *mer*, est generalement couverte de *faletes*, mais quand nettoyée fait une *apparence elegante*, par ses *belles couleurs*.

Elle se trouve en *tres grande abondance* sur la plupart des cotes de la *Grande Bretagne* & l'*Irlande*.

This *species* yields the *purple juice*, *analagous* to the *Tyrian purple* of the *antients*. Mr. Cole has given the following account of it, in the Phil. Transf. 1684.—" It is a *white vein*, lying *transversely* in a *little furrow next* to the *head* of the *fish*, which must be digged out with the stiff point of a horse-hair pencil. The *letters*, *spots*, *&c.* to be *dyed*, will presently appear of a *pleasant light green colour*, and, *if placed in the sun*, will change into the following colours: first, a *deep green*, then a full *sea green*, after into a *watchet* or *pale blue*, then into a *purplish red*, and lastly, into a very deep *purple red*, beyond which the *sun* does no more. But the last and most beautiful *colour*, after washing in *scalding water* and *soap*, and dried in the *sun*, will be of a fair *bright crimson*, which will *continue*, though there is no use of any *styptic* to bind it.

Cette *espece* fournit la *liqueur pourpre*, *analogue* à la *pourpre Tyrienne* des *anciens*. M. Cole nous à donnè le recit suivant dans les Transactions Philosophiques de 1684. —" C'est une *veine blanche*, posée *à travers* dans une *petite fente proche* de la *tete* du *poisson*, qu'on doit oter avec la pointe roide d'un pinceau de poil de cheval. Les *lettres*, *taches*, *&c.* à *etre teintes*, paroitront bientot apres d'une *couleur agreable verte clair*, &, si *mise au soleil*, changeront aux couleurs suivantes: *premierement*, une couleur *verte foncée*, alors une *verte de mer parfaite*, apres dans une *bleu pale*, & alors dans une *couleur rouge pourprée* tres chargée, au delà de laquelle le *soleil* ne fait plus. Mais la derniere & la plus belle *couleur*, apres l'avoir lavée dans *l'eau bouillante* avec du *savon*, & sechée au *soleil*, sera une parfaite & *elegante couleur de cramoisi*, qui *continuera*, quoique l'on ne fait usage de *stiptique* à la fixer.

The

La

The *cochineal* dye, ſince the diſcovery of *America*, has deſtroyed the uſe of theſe *valuable purples*, and now they are *objects* of mere *curioſity*, for many of the *inhabitants* of our *ſea coaſts uſe it only* to mark the linen.

But before that *æra*, our *antient Engliſh* ſeem to have *uſed it* as a *Tyrian or valuable purple*. The *venerable Bede*, who flouriſhed about the *latter end* of the *ſeventh century*, is my *authority*, for in his *Eccleſiaſtical Hiſtory he ſays (moſt probably of this very ſpecies)* "There are ſnails in very great abundance, from which a ſcarlet or crimſon dye is made, whoſe elegant redneſs never fades, either by the heat of the ſun or the injuries of rain, but the older it is, it is more elegant."

La teinture de la *cochenille*, depuis la decouverte de l'*Amerique*, à detruit l'uſage de ces *pourpres precieuſes*, & à preſent elles ſont ſimplement des *objets* de la *curioſité*, car pluſieurs de *habitants* de nos *cotes maritimes* l'*uſent ſeulement* à marquer de linge.

Mais avant cette *ere*, nos *anciens Anglois* paroiſſent l'avoir *uſée* comme une *pourpre precieuſe* ou *Tyrienne*. *Bede le venerable*, qui floriſſoit vers *la fin* du *ſeptieme ſiecle*, eſt mon *authorite*, car dans ſon *Hiſtoire Eccleſiaſtique, il dit (fort probablement de cette eſpece même)* "Il y a de limaçons en fort grande abondance, de leſquels une teinture ecarlate ou cramoiſi ſe fait, dont la rougeur elegante jamais ſe ternit, ni par l'ardeur du ſoleil, ou par les injures de la pluie, mais plus elle viellit, elle eſt plus elegante.

Sunt cochleæ, ſatis ſuperque abundantes, quibus tinctura coccinei coloris conficitur. Cujus rubor pulcherrimus nullo unquam ſolis ardore, nulla valet pluviarum injuria palleſcere; ſed quo vetuſtior, eo ſolet eſſe venuſtior.—Hiſt. Eccleſ. (edit. opt.) l. 1. c. 1. p. 277.

LXXVI.

COSTATUM. RIBBED.

Tab. VIII. fig. 4. 4.

LXXVI.

Pl. VIII. fig. 4. 4.

Buccinum canaliculatum parvum, anfractibus coſtis longitudinalibus diſtinctis. Coſtatum. Tab. 8. fig. 4.

Murex coſtatus. Ribbed.—Penn. Brit. Zool. No. 100. tab. 79. fig. 1 & 4. *of the inner quadrangle.*

This *ſhell* is narrow and long, or of an oblong *ſhape*, and tapers to a fine point, about half an inch in *length*, very *thick* and *ſtrong*, of a white and cheſnut *colour*, the latter the *ground*; *ſmooth*, and very *gloſſy*. It is neatly *ribb'd* with about eight equidiſtant white *ribs*, that run its *length*, broad, thick, prominent, and not cloſe ſet.

Cette *coquille* eſt etroite & longue, ou d'une *forme* oblongue, & appetiſſante en pointe fine, environ un demi pouce en *longeur*, tres *epaiſſe* & *forte*, de *couleur* blanche & chataigne, la derniere etant le *fond*; *liſſe*, & fort *luſtrée*. Elle eſt nettement travaillée à *cotes*, environ huit, à diſtances egales, blanches, & qui courent ſa *longeur*, larges, groſſes, ſaillantes, & point ſerrées.

The

La

The *mouth* is narrow and oblong; it *ends* at *top* in a narrow ſtrait *gutter*, at the *bottom* in an acute angle. The *outer lip* very thick, and border'd outwardly with a thick, ſmooth, prominent ledge, which is only one of the ribs. The *interior* or *pillar lip* alſo very thick, and a little ſpread or faced.

The *ſpires* are ſix, all gradually tapering to a ſharp point, not jutting or prominent, and ſeparated by a ſtrong furrow.

I received this *ſpecies* from the coaſts of *Cornwall* and *Devonſhire*. Mr. Pennant notes it from *Angleſea*, and ſays it is alſo a *Norwegian* ſhell.

La *bouche* eſt etroite & oblongue; elle *finit* au *haut* dans une *goutiere* etroite & droite, au *bas* dans un angle aigu. La *levre exterieure* fort epaiſſe, & bordée en dehors par une bordure epaiſſe, liſſe & ſaillante, qui eſt ſeulement une des cotes. L'*interieure* ou celle de la *columelle* auſſi fort epaiſſe, & un peu etendue ou en face.

Les *revolutions* ſont ſix, toutes appetiſſantes à une pointe aigue, ne debordent l'une l'autre, & ſont ſeparées par une ſtrie forte.

J'ai receu cette *eſpece* des cotes de *Cornwall*, & du comté de *Devon*. M. Pennant la notifie de *Angleſea*, & dit qu'elle eſt auſſi une coquille *Norwegienne*.

GENUS GENRE

GENUS XIV.

Buccina Recurvirostra, Wry-mouth'd Whelkes.

Whelkes whose mouths are cut short at top, for the gutter or beak does not ascend, but bends and falls on the back, oblique or awry, exactly like the mouth of a soal or other flat fish.

GENRE XIV.

Buccina Recurvirostra, Buccins à Bouche echancrée.

Buccins dont les bouches sont comme coupées au haut, car la goutiere ou queue n'est pas allongée, mais se courbe & tombe sur le dos, obliquement ou de travers, exactement comme la bouche d'une sole ou autre poisson plat.

LXXVII.

Lineatum. Lineated.
Tab. VIII. fig. 5.

LXXVII.

Pl. VIII. fig. 5.

BUCCINUM recurvirostrum minimum pullum, lineis albidis spiraliter distinctum. Lineatum. Tab. 8. fig. 5.

The *shell* is *small, thick* and *strong*, about the *size* of a grain of wheat, but more swell'd at top, and sharp pointed at bottom, or of a pyramidal *shape*; *smooth, glossy*, of a dark brown near black *ground, thickly* set, in a *spiral* manner, with very fine *whitish lines* or *streaks*.

The *mouth* is oval, and ends at top in a slant *gutter* or cut, tending *outwards* on the *back*. The *outer lip* is thin and sharp; the *inner* or *pillar lip* thick and spread.

The *spires* are five, near on a level, or not prominent from each other, and are separated by a furrow. The *upper ones* are much swelled, or bulky; the *others* taper to a fine sharp point.

This *species* is found in great *abundance* on the coast of *Cornwall*, and also in great *plenty* in the *West Indies*.

La *coquille* est *petite, epaisse* & *forte*, environ la *grandeur* d'un grain de bled, mais plus bombée au haut, & en pointe aigue au bas, ou d'une *forme* pyramidale; *lisse, lustrée*, d'un *fond* brun chargé presque noir, à *stries* ou *lignes* fort fines, *blanchatres*, & *spirales*, tres *serrées*.

La *bouche* est ovale, & finit au haut dans une *goutiere* oblique ou coupée, aboutissant *exterieurement* sur le *dos*. La *levre exterieure* est mince & tranchante; l'*interieure* ou celle de la *columelle* epaisse & etendue.

Les *revolutions* sont cincq, presque de niveau, ou ne debordant l'une l'autre, & sont separées par une strie. Les *superieures* tres bombées ou enflées; les *autres* appetissent en une pointe aigue & fine.

Cette *espece* se trouve en grande *abondance* sur les cotes de *Cornwall*, & aussi en tres grande *quantité* aux *Indes Occidentales*.

LXXVIII.

Reticulatum. Reticulated.

Tab. VII. fig. 10.

LXXVIII.

Pl. VII. fig. 10.

Buccinum recurviroſtrum cancellatum, columella ſinuoſa, labro dentato. Reticulatum. Tab. 7. fig. 10.

Buccinum brevi roſtrum cancellatum, denſe ſinnoſum, labro dentato.—Liſt. H. Conch. tab. 966. fig. 21.

Buccinum marinum cancellatum. Small latticed whelke. Petiv. Gaz. tab. 75. fig. 4.—Dale Harw. p. 283. No. 7. & p. 285. No. 3.—*Smooth chequer'd whelke.* Smith Cork, p. 318.

B. reticulatum. Reticulated.—Penn. Brit. Zool. No. 92. tab. 72. fig. 92.

Buon. Ricr. fig. 62.—*Reticulatum. Buccinum teſta ovato-oblonga tranſverſim ſtriata, longitudinaliter rugoſa, apertura dentata.* Lin. S. N. p. 1204. No. 476.

This *ſhell* is very *thick* and *ſtrong*, yet *ſemitranſparent*, of an oblong-oval *ſhape*, about the *ſize* of a filberd, and often of double that bigneſs. The *ground colour* is whitiſh, greatly tinted with pale brown, and a very pale *bluiſh* or *livid girdle*, pretty broad and regular, runs *ſpiral* near to the *bottom* of each *ſpire*. It is thickly wrought with broad, rounded, prominent *longitudinal ribs* and *intermediate furrows*, croſſed by *tranſverſe* or *circular ſtriæ*. Theſe *ſtriæ*, on croſſing the *ribs*, cut and notch them in ſuch a manner, that they appear as made up of little beads or bumps.

La *coquille* eſt tres *epaiſſe* & *forte*, cependant *demitranſparente*, d'une *forme* oblongue-ovale, environ la *grandeur* d'une noiſette, & ſouvent de double cet volume. Le *fond* eſt blanchatre, beaucoup teint de brun pale, & une *bande bleuatre* fort pale, ou *livide*, aſſes large, reguliere & *ſpirale*, ſe trouve près du *bas* de chaque *revolution*. Elle eſt travaillée à *cotes* larges, arrondies, faillantes & *longitudinales*, tres ſerrées, & des *canelures intermediaires*, traverſées par des *ſtries circulaires* ou *tranſverſales*. Ces *ſtries*, en traverſant les *cotes*, les coupent & les entaillent en telle maniere qu'elles paroiſſent comme compoſées de petits grains ou boſſes.

The *mouth* is narrow and oval; at the top it is cut into a large ſemicular *notch*, *oblique* or *awry*, turns *downwards* on the *back*, and exactly reſembles the mouth of a *ſoal*, or other flat fiſh. The *outer lip* is thick, crenated on the edges, and ſlopes a little ſome way within the mouth, where it forms a *ledge*, which is ſet with ſmall prominent *teeth*, leſs than a pin's head, and *generally ſix* in number. The *inner* or *pillar lip*

La *bouche* eſt etroite & ovale; au haut elle eſt coupée dans une grande *entailleure* ſemicirculaire, *oblique* ou *de travers*, tournant *en bas* ſur le *dos*, & exactement reſſemble la bouche d'une *ſole*, ou autre poiſſon plat. La *levre exterieure* eſt epaiſſe, crenelée ſur les bords, & penche un peu pour quelque eſpace au dedans de la bouche, ou elle forme une *bordure*, qui eſt garnie de quelques petites *dents* faillantes, moins

lip is greatly twirl'd, and, by the *notch* at top, is form'd into a broad rounded *neck*, quite diſtinct from the body ſpire, which is ſet on the *outſide* with ſtrong *circular furrows*; thence it curves and ſlopes down to *within* the *mouth*, where are *two ſmall* knobs or *teeth*; and then ſpreads extremely much all *down* and on the *body ſpire*. The *whole inſide* of the *mouth* is *pure white*, and finely *glazed*.

The *ſpires* are ſeven, ſcarcely raiſed or prominent from each other, and are ſeparated by a *folding furrow*. The body one is pretty much ſwell'd; the others taper to a very ſharp point.

This *pretty ſpecies* is common in *Suſſex*, and on the *Kentiſh* ſhores, as at *Shellneſs* in *Sheerneſs*, *Margate*, and *Deal*; the *Eſſex* ſhores; the coaſts of *Cornwall*, *Devonſhire*, and *Dorſet*; but *it is not* an *inhabitant* of the *northern* coaſts of *England*. It is alſo found on ſome of the coaſts of *Wales* and *Ireland*.

This *kind* likewiſe inhabits the *Mediterranean*, and the coaſt of *Senegal* in *Guinea*.

moindre que la tete d'une epingle, & *generalement* au nombre de *ſix*. La *levre interieure*, ou celle de la *columelle*, eſt beaucoup tortillée, & par l'*entailleure* au haut ſe forme dans un *col*, arrondi & large, tout à fait diſtinct de la revolution du corps, & *exterieurement* à *ſtries circulaires* fortes; de là il courbe & penche en *bas* à la *bouche*, ou il eſt garni avec *deux petites* boſſes ou *dents*; alors il s'etend extremement *en bas* & ſur la *revolution* du *corps*. *Tout le dedans* de la *bouche* eſt *blanc pur*, & parfaitement bien *verni*.

Les *revolutions* ſont ſept, à peine faillantes ou debordant l'une l'autre, & ſont ſeparées par une *ſtrie pliſſée*. Celle du corps eſt aſſes bombée; les autres appetiſſent à une pointe fort aigue.

Cette *jolie eſpece* eſt commune au comté de *Suſſex*, & ſur les cotes du comté de *Kent*, comme à *Shellneſs* en *Sheerneſs*, à *Margate*, & *Deal*; les cotes de *Eſſex*; celles de *Cornwall*, & des comtés de *Devon* & *Dorſet*; mais *elle n'eſt pas habitante* des cotes du *nord* d'*Angleterre*. Elle ſe trouve pareillement ſur quelques unes des cotes de *Galles* & d'*Irlande*.

Cette *ſorte* eſt auſſi habitante de la *Mediterranée*, & de la cote de *Senegal* en *Guinée*.

GENUS XV.
BUCCINA LONGIROSTRA, LONG-BEAKED WHELKES.

Whelkes that have a produced or ascending beak.

GENRE XV.
BUCCINA LONGIROSTRA, BUCCINS à BOUCHE GARNIE D'UNE LONGUE QUEUE.

Buccins qui ont une queue allongée ou elevée.

LXXIX.

PORCATUM. RIDGED.
Tab. VIII. fig. 7. 7.

LXXIX.

Pl. VIII. fig. 7. 7.

BUCCINUM *longiroſtrum medium ſubangulatum, porcis ſpiralibus diſtinctum. Porcatum.* Tab. 8. fig. 7. 7.

An *Buccinum roſtratum labro duplicato, ſtriis latis & eminentibus diſtinctum, ſinuoſum.*—Liſt. H. Conch. tab. 942. fig. 38?—Leigh Lancaſh. tab. 3. fig. 9 & 11.—*Murex erinaceus. Urchin.* Penn. Brit. Zool. No. 95. tab. 76. fig. 95.

Buccinum majus canaliculatum, roſtratum, ore labioſo, fimbriatum, umbilicatum, ore anguſto, oblongo, rugoſum, coſtulatum, ſtriis eminentibus reticulatim exaſperatum, albidum.—Gualt. I. Conch. tab. 49. fig. H.

Erinaceus. Murex teſta, multifariam ſubfrondoſo-ſpinoſa, ſpiræ anfractibus retuſo-coronatis, cauda abbreviata.—Lin. S. N. p. 1216. No. 526.

The *ſhell* is *thick*, very *rugged*, *ſtrong* and *opake*, *except* the *interſtices* between the *ridges*, which, when held to the light, are *thin* and *ſemitranſparent*. It is, on the fore or mouth ſide, of an *oval ſhape*, pointed at both ends; on the *back* part more *irregular*; but the *whole ſhell*, though of an *uncouth appearance*, is *ſubangular*, or *divided* into *ſix longitudinal angles*. It is entirely of a yellowiſh brown or ſandy *colour*, and the moſt general *ſize* is that of a walnut, or about two inches *long*, and about an inch *broad* juſt below the mouth; but *larger* ones are found, and *frequently ſmaller*.

La *coquille* eſt *epaiſſe*, *raboteuſe*, *forte* & *opaque*, *excepté* les *intervalles* entre les *ſillons*, qui, quand tenus au jour, ſont *minces* & *demitranſparentes*. Au devant ou du coté de la bouche, elle eſt d'une *forme ovale*, pointue aux deux bouts; ſur le *dos* plus *irreguliere*; mais la *coquille entiere*, quoique d'une *apparence rude*, eſt quaſi *angulaire*, ou *diviſée* en *ſix angles longitudinales*. Elle eſt totalement jaunatre brune, ou *couleur* de ſable, & ſa *grandeur* generalement eſt celle d'une noix, ou environ deux pouces en *longeur*, & environ un pouce en *largeur* juſtement au deſſous de la bouche; mais des plus *grandes* ſe trouvent, & *frequemment* des plus *petites*.

The

Toute

The *whole shell* is wrought with *circular ridges*, not set *very close*, but *equidistant*. These *ridges* rise at the border of the mouth, where they are thick and knobbed, and also on the turn of the back they are broad, but less swell'd. They run thence quite *across* the *shell*, *pretty broad*, *rounded*, very *prominent*, and in many places are a little rough, being set with minute points or scales. *Between each* is a *smaller* thread-like *ridge*, *bounded* by two *striæ* or *furrows*, which regularly accompany the larger. These *smaller ridges* are very rough, being all over cover'd with points or raised scales. The *shell*, as *before observed*, is *subangular*, or *divided* into *six* very high and *longitudinal angles* or *ridges*, like a chain of hills, three large and three less, with intermediate depressions. These *six longitudinal angles* are formed by the *circular* or *transverse ridges*, which at crossing them are excessively swell'd and bumped, as they are also on the edges of the spires. All these *circular ridges* are *hollow* or *tubular*, as *appears* by their *ends* on the *border* of the *mouth*.

Toute la *coquille* est à *sillons circulaires*, point *fort serrés*, mais à *distances egales*. Ces *sillons* s'elevent de la bordure de la bouche, ou ils sont gros & en bosse, & aussi sur le tour au dos ils sont larges, mais moins enflés. De là ils courent tout *au travers* de la *coquille*, *asses larges*, *arrondis*, tres *saillants*, & dans plusieurs places un peu aspre, etant chargès de petites pointes ou ecailles. *Entre chacun* se trouve un autre *sillon*, *plus petit*, & gros comme un fil, *borné* par deux *stries* ou *canelures*, qui accompagnent regulierement les grands. Ces *petits sillons* sont tres aspre, etant tout à fait couverts des pointes ou ecailles saillantes. La *coquille*, comme *il a été dit*, est *quasi angulaire*, ou *divisée* en *six angles* ou *sillons*, tres elevés & *longitudinales*, comme une lisiere de montagnes, trois grands & trois moindres, avec des enfoncements intermediats. Ces *six angles longitudinales* sont formés par les *sillons circulaires* ou *transversales*, qui en les traversant sont excessivement enflès & bossus, comme ils sont aussi aux bords des revolutions. Touts ces *sillons circulaires* sont *creux* ou *tubulaires*, comme il *paroit* par leurs *bouts* sur la *bordure* de la *bouche*.

The *mouth* is perfectly oval, and *ends* at *top* in a *long* or *produced* and somewhat *twirl'd beak*, quite *tubular*, for it is cover'd from the mouth upwards with a rugged scale-work, but is *open* at the very *top*. The *outer lip* is notched, and bounded by a large, thick, and swell'd *border*, formed by the ends of the circular ridges; and here they shew their tubular structure, and are likewise finely wrought in a *curious* cancellated and plated *work*, like some *Madrepore corals*. The *inner* or *pillar lip* is

La *bouche* est tout à fait ovale, & *finit* au *haut* dans une *queue allongée*, & un peu *tortillée*, absolument *tubulaire*, car elle est couverte dès la bouche par haut d'un ouvrage ecailleux & raboteux, mais elle est *ouverte* au *haut* même. La *levre exterieure* est crenelée, & bordée par une *bordure* grande, epaisse & fort enflée, formée par les bouts des sillons circulaires; & ici ils demontrent leur structure tubulaire, & sont aussi tres agreablement travaillés en un *ouvrage curieux*, celluleux & à lames comme

is somewhat spread, and the *inside* of the *mouth* is *glossy* and *pure white*.

comme quelques *coraux Madrepores*. L'*interieure* ou *celle* de la *columelle* est un peu etendue, & tout le *dedans* de la *bouche* est *lustré* & *blanc pur*.

The *spires* are eight, all greatly prominent from each other, by a very broad depression, and by the large swell'd bumps of the circular ridges at their edges, and they gradually taper to a very fine point.

Les *revolutions* sont huit, toutes fortement debordant l'une l'autre, par un enfoncement large, & par les grandes bosses tres enflées des sillons circulaires de leurs bords, & elles appetissent par degrès à une pointe tres fine.

This *species* is found on the coasts of *Cornwall* and *Dorsetshire*; and Leigh notes that it is got on all the coast near and on *Hilbree island*, in *Cheshire*.

It is also an inhabitant of the Mediterranean sea.

Cette *espece* se trouve sur les cotes des comtés de *Cornwall* & de *Dorset*; & Leigh notifie qu'elle se trouve sur toute la cote & sur l'*isle de Hilbrée*, au comté de *Chester*.

Elle est pareillement une habitante de la mer Mediterranée.

FAMILY OF MURICES, ROCKS.

Turbinated univalves, whose *mouths* are oblong, narrow, and *end* in a *gutter* or *beak*. The animal a *Slug*.

This *family* subdivides into *four genera*, viz. 1. *Murices*, or *Rocks*. 2. *Rhombi*; of a rhombic shape, or contour. 3. *Alatæ*, *winged*; whose outer lip expands into a large entire flap or wing. And 4. *Aporrhaidæ*, *digitated*; whose flap or wing is cut or flash'd into spikes or fingers.—Of this *family*, one *species only*, of the *4th genus*, has as yet been discovered in our *British seas*.

LA FAMILLE DE MURICES, OU ROCHERS.

Univalves contournées en spirale, dont les *bouches* sont oblongues, etroites, & *finissent* dans une *goutiere* ou *queue*. L'animal est une *Limace*.

Cette *famille* se soudivise en *quatre genres*, qui sont, 1. *Murices*, ou *Rochers*. 2. *Rhombi*, ou *Rochers*, d'une forme ou contour de rhombe. 3. *Alatæ*, *Rochers ailés*, dont la levre exterieure se deploye dans une aile grande & entiere. Et 4. *Aporrhaidæ*, *Rochers ailés & a pattes*.—De cette *famille*, *seulement une espece*, du 4e *genre*, jusques à present a été decouverte dans nos *mers Britanniques*.

GENUS XVI. APORRHAIS.

Murices or Rocks, whose wing is cut into spikes or fingers.

GENRE XVI. APORRHAIS.

Murices ou Rochers, dont l'aile est decoupée en pattes ou doigts.

* MARINÆ. SEA.

LXXX.

QUADRIFIDUS. FOUR-FINGER'D.

Tab. VII. fig. 7.

* MARINÆ. DE MER.

LXXX.

Pl. VII. fig. 7.

APORRHAIS subfuscus, anfractibus nodosis, labro palmato quadrifido. Quadrifidus. Tab. 7. fig. 7.

Buccinum bilingue striatum labro propatulo digitato.—List. H. Conch. tab. 865. fig. 20.

Aporrhais Edinburgicus minor nodosa. Petiv. Gaz. tab. 79. fig. 6. & tab. 127. fig. 11. —*Buccinum dactyloides labrosum.* Memoirs for the Curious, October 1708, p. 386. No. 12. Wallace Orkn. p. 39.—*Strombus Pes Pelecani. Corvorant's foot.* Penn. Brit. Zool. No. 94. tab. 75. fig. 94.

Turbo

Turbo pentadactylus. Buon. Ricr. p. 192. fig. 87.—*Pes Anserinus.* Klein Ostrac. p. 32. § 84. spec. 1. tab. 2. fig. 41.—*Rocher.* Argenv. Conch. I. p. 293. tab. 17. fig. M. II. p. 251. tab. 14. fig. M.—*Aile de Chauve Souris femelle, Patte d'Oye,* ou *Hallebarde.* D'Avila Cab. p. 191. No. 344.

Strombus canaliculatus, rostratus, ore labioso, striatus, papillosus, auritus aure admodum crassa, & in quatuor appendices breviores expansa, ex candido cinereus.—Gualt. I. Conch. tab. 53. fig. A.

Cochlea testa longa acuminata, aperturæ labro dilatato, duplici stria antice sinuato. Lin. F. Suec. I. p. 378. No. 1323. II. No. 2164.—Id. Mus. Reg. p. 615. No. 273.—*Pes Pelecani. Strombus testæ labro tetradactylo palmato digitis angulatis, fauce lævi.* Id. S. N. p. 1207. No. 490.

This *shell* is extremely *thick* and *strong.* The *body* is *narrow, slender,* and *tapers* to a very *fine point*; from top to bottom it generally measures from one inch and a half to two inches in *length,* and from the body to the middle of the wing, between the fingers or prongs, one inch and a quarter *broad*; and the *colour* is very pale yellowish brown.

The *mouth* is narrow and oblong. The *outer lip expands* into a very thick *wing,* swelling to a greater thickness at the *edge,* which *slopes* slightly *downwards,* and is *sinuous* or *waved.* It is *cut* or *divided* into *four fingers* or *prongs,* whereof *two* are in the *middle:* the *upper* of these is nearly *horizontal,* or slightly *ascending,* the *other* very *obliquely descending.* The remaining *two,* which are on the *body,* are *both* very *perpendicular*; the *upper* one, directly on the top of the mouth, *forms* a pretty long waved pyramidal sharp *beak*; the *other,* that runs *downwards along* the *body* to almost the end of the turban, is very valid, spread, and sharp pointed. These *two* latter *prongs* should more properly be said to belong to the *pillar* than the *outer lip.* The *interstices* between the prongs are cut into *deep semilunar cuts* or *scollopings,* and each

Cette *coquille* est extremement *epaisse* & *forte.* Le *corps* est *etroit, delié,* & *appetisse* à une *pointe* tres *fine*; du haut au bas sa mesure est généralement d'un pouce & demi à deux pouces en *longeur,* & du corps jusques au milieu de l'aile, entre les doigts ou pattes, un pouce & un quart en *largeur*; & sa *couleur* est pale jaunatre brune.

La *bouche* est etroite & oblongue. La *levre exterieure se deploye* dans une *aile* fort epaisse, grossissant à une grande epaisseur au *bord,* qui *penche* legerement *en bas,* & est *sinueuse* ou *ondée.* Elle est *decoupée* ou *partagée* en *quatre doigts* ou *pattes,* dont *deux* sont au *milieu:* le *superieur* de ceux ci est à peu pres *horizontal,* ou peu *elevé,* l'*autre descend* fort *obliquement.* Les *deux* autres, qui se tiennent au *corps,* sont *tous deux* fort *perpendiculaires*; le *superieur,* directement au haut de la bouche, *forme* une asses longue *queue,* pyramidale, pointue & ondée: l'*autre,* qui *descend le long* du *corps* à presque la fin de la clavicule, est extremement fort, etendu, & en pointe aigue. Ces *deux* derniers *doigts* doivent etre dit, à proprement parler, de appartenir a la *columelle* plutot que à la *levre exterieure.* Les *intervalles* entre les doigts sont decoupés à *decoupures profondes* semi-

each *prong* on this side has a *channel* along the middle.

The *pillar lip* is very thick, strong, and spread; and all the *mouth*, *wing*, and *pillar*, *within*, are white, smooth, and finely glazed, except towards the ends of the prongs, which are tinctured with shades of a chesnut colour.

The *spires* are about eight, *tapering* to a very fine point, and *separated* by a deep slope and furrow. Each *spire* has a prominent *ridge*, *studded* or bumped with round knobs, size of a pin's head, that runs *circular* along the *middle* of it. From this ridge it slopes upwards and downwards. But the *body spire* has *two ridges*, the *lowermost* whereof is greatly *knobb'd*, and extends (but then plain, very sharp, prominent and oblique) to the border of the wing, exactly to the *descending prong*, and *answers* to its *gutter* on the inside; the *upper ridge* is not so bumped, and continues also sharp, *ascending* along the wing, and answers to the inside *gutter* of the *ascending prong*. Besides, there is a third slightly *prominent plain ridge* just above the *latter*, which also continues on the wing, but slightly, and does not reach to the edge of it. The *spires*, and also the *wing*, are besides *finely* and *circularly striated*.

This *species* is found in *Cornwall*, *Devonshire*, *Durham*, *Sussex*, and some other *shores*; in *Carnarvonshire* and *Merionethshire*, *Wales*; and is common in the *Orkneys*, at *Leith*, and several other coasts of *Scotland*.

It inhabits also the Norwegian, Mediterranean, and American seas.

semilunaires ou à *languettes*, & chaque *doigt* de cet coté est creusé en *canal* le long du milieu.

La *levre* de la *columelle* est tres epaisse, forte, & etendue; & toute la *bouche*, l'*aile*, & la *columelle*, *en dedans*, sont blanches, lisses, & bien vernissées, excepte vers les bouts des doigts, qui sont teints de nuances de couleur chataigne.

Les *revolutions* sont environ huit, *appetissantes* à une pointe tres fine, & *separées* par un penchant profond & une strie. Chaque *revolution* a un *sillon* saillant, *tuberculeux* ou à petits tubercules arrondis, de la grandeur d'une tete d'epingle, qui courent *circulairement* le long de son *milieu*. De cet sillon il penche par haut & par bas. Mais la *revolution* du *corps* est à *deux sillons*, l'*inferieur* desquels est fort *tuberculeux*, & s'etend (mais alors uni, fort saillant, tranchant, & oblique) jusques au bord de l'aile, exactement au *doigt* qui *descend*, & *repond* a son *canal* interieur; le *sillon superieur* n'est pas si tuberculeux, & continue aussi tranchant, *monte* le long de l'aile, & repond au *canal* interieur du *doigt elevé*. Outre ceux ci, il se trouve un troisieme *sillon*, peu *saillant* & *uni*, justement au dessus du *dernier*, qui continue aussi sur l'aile, mais legerement, & n'atteint pas à son bord. Les *revolutions*, comme aussi l'*aile*, de plus, sont à *stries fines* & *circulaires*.

Cette *espece* se trouve aux comtés de *Cornwall*, *Devon*, *Durham*, *Sussex*, & quelques autres *cotes*; aux comtés de *Carnarvon* & *Merioneth*, en *Galles*; & elle est commune aux *Orcades*, à *Leith*, & plusieurs autres cotes de l'*Ecosse*.

Elle est pareillement habitante des mers de la Norwege, la Mediterranée, & de l'Amerique.

END OF THE UNIVALVES.

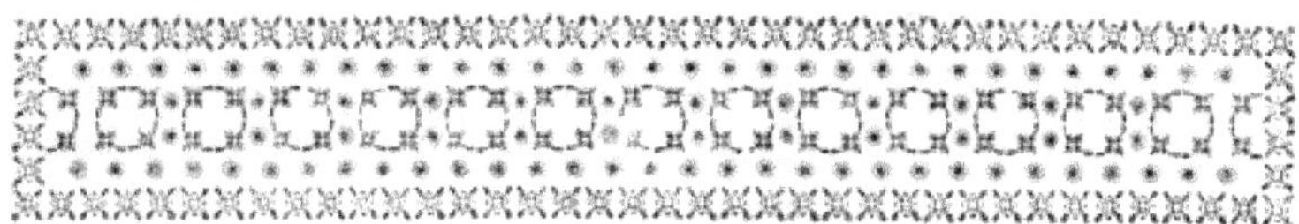

A NATURAL HISTORY OF BRITISH SHELLS.

ORDER II.

BIVALVES.

Double shells, or composed of two parts or pieces, simple, or without any sort of convolution called the *valves*, which are connected and joined together by means of a hinge or articulation.

The *Bivalves* may be divided into *Part I.* most generally with unequal valves, and shut close; *Part II.* with equal valves, and shut close; and *Part III.* with valves that never shut close, or gape.

HISTOIRE NATURELLE DES COQUILLES BRITANNIQUES.

ORDRE II.

BIVALVES.

Coquilles doubles, ou composées de deux pieces ou parties, simples, & en aucune maniere contournées, ou qu'on puisse regarder comme des spires, elles sont nommées les *battans* ou *valves*; & se joignent ensemble par le moyen d'une charniere ou articulation.

Les *Bivalves* peuvent etre divisées en *Premiere Partie*, Celles qui generalement ont les battans inegaux, quoiqu'elles se ferment exactement; *Seconde Partie*, Celles avec les battans egaux, & se ferment exactement; & *Troisieme Partie*, Celles dont les battans ne se ferment jamais exactement, ou sont beantes.

PART I.

Bivalves, most generally with unequal valves, and shut close.

PREMIERE PARTIE.

Bivalves, qui ont generalement les battans inegaux, quoiqu'elles se ferment exactement.

GENUS I.

PECTEN. The ESCALLOP.

Bivalves, most generally with *unequal valves*, that shut close, and are *eared*. The *hinge* is toothless, being only a trigonal cavity in the very center of the commissure or summit of the shell, which runs on a strait horizontal line. —The animal, a *Tethys*.

GENRE I.

LE PEIGNE.

Bivalves, qui ont generalement les *battans inegaux*, qui se ferment exactement, & sont *à oreilles*. La *charniere* n'a point de dents, mais a seulement une cavité triangulaire, dans le centre même de la commissure ou sommet de la coquille, qui court sur une ligne droite horizontale.—L'animal est un *Tethys*.

* MARINÆ. SEA.

I.

P. VULGARIS. COMMON.

Tab. IX. fig. 3. 3.

* DE MER.

I.

LE PEIGNE COMMUN.

Pl. IX. fig. 3. 3.

P. *Maximus inæquivalvis, costis circiter quatuordecim rotundatis, & admodum crassis. Vulgaris, the common escallop.* Tab. 9. fig. 3. 3.

P. maximus, circiter quatuordecim striis, admodum crassis & eminentibus, et iisdem ipsis striatis insignitus. A scallop. List. H. An. Angl. p. 184. tit. 29, tab. 5. fig. 29.—App. H. An. Angl. 4to. p. 18.—App. H. An. Angl. in Goedart. p. 32.—Id. Exercit. Anat. 3. p. 41. tab. --- & Edit. 2. H. Conch. App. tab. 17.

P. magnus, albidus, circiter duodecim striis, multis minutisque incisuris exasperatis donatus. List. H. Conch. tab. 163. fig. 1.

Escallop, or Scollop. Merret Pin. 193. Grew Muf. 142. Sibbald Muf. 162.—*Scollop or Clam-shell.* Wallace Orkn. p. 43. &c.—*Pecten major nostras vulgaris.* Muf. Petiv. p. 86. No. 829. Smith Cork, p. 318. Smith Waterford, p. 272. Nat. Hist. Down, p. 239. Rutty Dublin, p. 378.—*Frill or Scallop.* Hutchins Dorset, p. 77.

P. maximus, Great. Penn. Brit. Zool. No. 61. tab. 59. fig. 61.

Rondel de Testac. l. 1. c. 14. p. 15.—Buon. Ricr. p. 152. fig. 8.—Argenv. Conch. II. Zoom. p. 55. pl. 6. fig. A. B.

Concha

Concha pectinata major, in parte plana striis eminentibus asperis, in parte vero convexa striis striatis insigniter notata, nonnullisque lineis fasciata, ex albido rufescens. Gualt. I. Conch. tab. 98. fig. A. B. & tab. 99. fig. A.

Concha testa aurita, striis quatuordecim. Lin. F. Suec. I. p. 383. No. 1343. II. No. 2148.

Maxima. Ostrea testa inæquivalvi radiis rotundatis longitudinaliter striatis. Lin. S. N. p. 1144. No. 185.—Muf. Reg. p. 522. No. 98.

The *shell* is pretty thick and strong; the *shape* narrow, and eared at top, and thence extends into a round circumference at bottom: the *valves* very unequal, the *upper* flat, the *under* concave.

The most general *size* is from three to four inches in length, or from the hinge to the bottom; and somewhat more in *breadth*, or across the shell. Dr. Lister says he got some, at *Scarborough*, five inches long and six inches wide; an uncommon size.

The *commissure* or summit of the hinge lies on a strait line, and the edges of it are thick and reflected inwards. The *ears* are large, equal, rectangular, and strongly striated in a transverse manner. The *beaks* of the valves are pointed, but do not over-top the commissure; that of the upper is always depressed or hollow'd.

The *under* or concave *valve*, on the *outside*, is whitish, generally, with a large circular space round the top, of a rosy or reddish hue, and a few scatter'd brownish streaks or lines. The *ribs* or ridges are about fourteen, longitudinal or the length of the shell, broadish, rounded, and strongly striated lengthways. The interstices between them are striated in the same manner, but they are also wrought with fine transverse striæ; and the whole shell is likewise crossed by transverse concentric wrinkles,

La *coquille* est asses epaisse & forte; la *forme* etroite, & à oreilles au haut, & de là s'elargit pour prendre une figure arrondie au bas; les *valves* sont fort inegales, la *superieure* platte, l'*inferieure* concave.

La *grandeur* la plus generale est de trois a quatre pouces en longeur, ou dès la charniere au bas, & quelque chose plus en *largeur*, ou à travers la coquille. Le Dr. Lister dit qu'il a obtenu quelques uns, a *Scarborough*, qui portoient cincq pouces de long, sur six pouces de large; une grandeur qui n'est pas commune.

La *commissure* ou sommet de la charniere est coupée en ligne droite, & ses bords sont epais & recourbés en dedans. Les *oreilles* sont grandes, egales, rectangulaires, & fortement striées à travers. Les *becs* des valves sont pointus, mais ils ne surmontent la commissure; celui de la valve superieure est toujours enfoncé ou comprimée.

La *valve* inferieure ou concave, à l'*exterieur*, est communement blanchatre, avec un grand espace circulaire autour de la tete, de couleur rougeatre ou de rose, avec quelques rayes ou lignes brunatres dispersées. Les *cotes* sont environ quatorze, longitudinales, larges, arrondies, & fortement striées selon leur longeur. Les intervalles ou cannelures entre les cotes sont striés de la même maniere, mais ils sont aussi finement striés à travers; & toute la coquille est pareillement traversée par

wrinkles, which become ſtronger at the bottom.

The *upper* or *flat ſhell* is of a reddiſh hue, with a circular ſpace of white from the beak to about three quarters of an inch down; and below it very often are ſeveral pretty ſcatter'd ſtreaks of white and brown, like arrow-heads. The *ribs*, alike in number with the under ſhell, are narrower or more contracted, very ſtrongly ſtriated or rather fluted, and the interſtices between them are very finely and tranſverſely ſtriated.

The *inſide* of both valves, in the *young ſhells*, are entirely white, ſmooth, and gloſſy, but in the *full grown* ones the center is only *white*, for the *top*, the *ears*, and a very broad regular liſt or *border* all round the margins, are of a fine bright *bay*, or deep *cheſnut colour*. The outer ribs form broad *channels withinſide* the ſhells. On each ſide the *hinge* are ſome very oblique, parallel, ſtrong furrows and ridges. The *ears* are ſomewhat depreſſed or hollow'd; and at the external bottom or root of them, where they join the body of the ſhell, they riſe into a remarkable thick prominency or knob.

This *ſpecies* is found on moſt of the coaſts of *Great Britain* and *Ireland*, but one of the principal places is at the iſlands of *Portland* and *Purbeck*, in *Dorſetſhire*.

The eſcallop has always been eſteemed a principal food among the teſtaceous animals.

It is affirmed by both antient and modern writers, that eſcallops will move ſo ſtrongly as even to leap out of the catcher wherein

par des rides concentriques tranſverſales, qui deviennent plus fortes au bas.

La *valve ſuperieure* ou *platte* eſt d'une couleur rougeatre, avec une eſpace circulaire blanche dès le bec à environ trois quarts d'un pouce plus bas, & au deſſous communement pluſieurs jolies rayes diſperſées blanches & brunes, comme les fers d'une fleche, ſe trouvent. Les *cotes* ſont de même nombre que dans l'inferieure, mais plus etroites ou retrecies, fortement ſtriées ou plutot cannelées, & les intervalles entre elles ſont tres finement & tranſverſalement ſtriés.

Le *dedans* ou l'*interieur* des deux valves, dans les *coquilles jeunes*, eſt entierement blanc, liſſe & luſtré; mais dans *celles* qui ſont *adultes* le centre ſeulement eſt *blanc*, car le *ſommet*, les *oreilles*, & une *liſiere* ou *bordure* tres large & reguliere, qui environne les bords, ſont d'une belle *couleur rouge-brun* ou de *chataigne* foncée. Les cotes exterieures forment des *cannelures* larges *en dedans* les coquilles. De chaque coté de la *charniere* ſe trouvent quelques ſillons & ſtries fortes, paralleles, & tres obliques. Les *oreilles* ſont quelque peu enfoncées ou comprimées, & à leur bas exterieur, ou élles joignent le corps de la coquille, elles s'elevent en une boſſe epaiſſe & remarquable.

Cette *eſpece* ſe trouve ſur la plupart des cotes de la *Grande Bretagne* & l'*Irlande*, mais une des places principales eſt aux iſles de *Purbeck* & *Portland*, au comte de *Dorſet*.

Le peigne a toujours eté eſtimé comme un aliment principal entre les animaux teſtacés.

Les auteurs tant anciens que moderns nous aſſurent, que les peignes ſe remueront ſi fortement que même de ſauter hors des paniers

wherein they are taken: their way of leaping, or raising themselves up, is by forcing their under valve against the body whereon they lie.

Escallop shells have always been a mark of *Christian devotion*, pilgrims wearing them on their hats and cloaks, as badges of their having been in crusades, visited the Holy Land, or performed some other pious acts. The name it bears, in several modern languages, of *St. James's shell*, is derived thence.

paniers ou autres choses dans lesquels il sont pris: leur maniere de sauter ou s'elever ainsi, est en forcant leur valve inferieure contre le corps ou ils se trouvent.

Les *coquilles peignes* ont toujours eté une marque de *devotion Chretienne*, les pelerins les portant sur leurs chapeaux ou manteaux, comme des symboles d'avoir entrepris des crusades, visité la Terre Sainte, ou accompli quelques actes religieux. Le nom qu'elle porte en plusieurs langues modernes, de *pelerines*, ou *coquilles de Saint Jaques*, y est derivè.

II. P. JACOBÆUS. THE PILGRIM.

II. PELERIN.

P. magnus inæquivalvis operculo subrufo, fornix vero albus costis angulatis & canaliculatis. Jacobæus.

P. magnus, subrufus, variegatus, sexdecem striis, ipsisq; striatis distinctus; at canaliculi leviores sunt. List. H. Conch. tab. 165. fig. 2. 3.

P. Jacobæus. Lesser. Penn. Brit. Zool. No. 62. tab. 60. fig. 62.

Pettine. Buon. Ricr. p. 152. fig. 4.

Concha pectinata, sed striis magis profunde striatis, seu magis canaliculatis distincta, candidissima. Gualt. I. Conch. tab. 99. fig. B.

Jacobæa. Ostrea testa inæquivalvi radiis quatuordecim angulatis longitudinaliter striatis. Lin. S. N. p. 1144. No. 186.—Mus. Reg. p. 522. No. 99.

This *species* is rather *smaller*, but in *shape* and *other particulars* like the foregoing.

The *upper valve* is flat; the *outside* red; but the *beak*, which is greatly depressed or hollow'd, is surrounded by a small circular space of white, mottled and streaked with red, and below it the ground is often prettily set with transverse streaks of dark purple red.

The *ears* are equal, and on this side are much depress'd or hollow'd, and finely striated

Cette *espece* est quelque peu plus *petite*, mais en *forme* & *autres particuliers* ressemble à la precedente.

La *valve superieure* est platte; le *dehors* rouge; mais le *bec*, qui est fort enfoncé ou concave, est environné par un petit espace circulaire blanc, tacheté & rayé de rouge, & en dessous, la robe est souvent tres joliment parsemée par des rayes transversales de pourpre foncé & rouge.

Les *oreilles* sont egales, & de cet cote beaucoup enfoncées ou concaves, & fine-

ſtriated lengthways. The *ribs* are about fifteen, obtuſe or rounded, tho' narrow and prominent; they are longitudinally and alſo very finely ſtriated in a tranſverſe manner, or acroſs, as are the interſtices between them. The *inſide* is quite white, gloſſy and ſmooth.

finement ſtriées leur longeur. Les *cotes* ſont environ quinze, obtuſes ou arrondies, quoique etroites & ſaillantes; elles ſont chargées de ſtries longitudinales, & auſſi fort finement ſtriées à travers, comme ſont pareillement les intervalles entre elles. Le *dedans* eſt entierement blanc, liſſe & luſtré.

The *under valve* is quite *white, within* and *without*; has the ſame number of *ribs* as the upper, but they are very ſtrong, prominent, narrow and angular, for their ſides are perpendicular; they are longitudinally fluted, the *flutings* deep, and about *five* on each *rib*. The *interſtices* between the ribs are tranſverſely ſet with very fine ſtriæ.

La *valve inferieure* ou concave eſt entierement *blanche* tant en *dedans* que en *dehors*, elle a le même nombre des *cotes* que la valve ſuperieure, mais elles ſont tres fortes, ſaillantes, etroites & angulaires, car leurs cotés ſont perpendiculaires; elles ſont cannelées longitudinalement par des *cannelures* profondes, & environ *cincq* ſur chaque *cote*. Les *intervalles* entre les cotes ſont finement ſtriés à travers.

This *ſpecies* is *ſometimes*, not *commonly*, found on *our coaſts*. I have received it from *Scarborough* in *Yorkſhire*, and from the coaſts of *Cornwall* and *Dorſetſhire*.

It is alſo a native of the Mediterranean ſea, and is probably the very ſpecies worn by pilgrims as a mark of devotion; I have therefore given it the *trivial name* of the *Pilgrim*.

Cette *eſpece* eſt *quelquefois*, pas *communement*, trouvée ſur *nos cotes*. Je l'ai recu de *Scarborough*, au comté de *York*, & des cotes des comtés de *Cornwall* & *Dorſet*.

Elle eſt auſſi native de la Mediterranée, & probablement l'eſpece même portée par les pelerins, comme une marque de devotion; pour cette raiſon je l'ai donné le *nom de guerre* de *Pelerin*.

III.

P. Pictus. Painted.

Tab. IX. fig. 1. 2. 4. 5.

III.

Pl. IX. fig. 1. 2. 4. 5.

P. mediocris, fere æquivalvis, tenuis, variis coloribus perbelle variegatus. Pictus. Tab. 9. fig. 1. 2. 4. 5.

P. tenuis, ſubrufus, maculoſus, circiter viginti ſtriis majoribus, at lævibus, donatus. Liſt. H. An. Angl. p. 85. tit. 30. tab. 5. fig. 30.

P. mediocris, latus, ex rufo variegatus, circiter viginti ſtriis, tenuiter admodum ſtriatis diſtinctus. Liſt. H. Conch. tab. 190. fig. 27. & tab. 191. fig. 28.—Wallace Orkn. p. 43.

p. 43. Martin W. Isles, p. 38.—*Pecten media lata compressa.* Petiv. Gaz. tab. 94. fig. 21. Wallis Northumb. p. 398. 399. No. 22. 23. 25. a 29.—*P. subrufus. Red.* Penn. Brit. Zool. No. 63. tab. 60. fig. 63.

Pectunculus pennatus striis dense notatus, luteo purpurascens. Pecten altis striis albo purpureis transverse variegatis insignis; & Pectunculus purpurascens vittis albis circularibus variegatus. Borlase Cornw. p. 277 & 278. tab. 28. fig. 18. 21 & 22.

Buon. Ricr. p. 152. fig. 6.—Gualt. I. Conch. tab. 73. fig. F. R. tab. 74. fig. A. B. C. D. E.

Concha testa aurita rufescens; striis viginti. Lin. F. Suec. I. p. 384. No. 1344.—An *Opercularis. Ostrea testa inequivalvi radiis viginti, subrotunda decussatim striato-scabra, operculo convexiore.* Id. S. N. p. 1147. No. 202? Mus. Reg. p. 530. No. 115?

The *shell* is thinnish and fine; both *valves* are somewhat convex, but not equally so. The general *size* is about two inches long, and two inches one quarter broad or across; and the *shape* as usual of this genus.

La *coquille* est mince & fine; les deux *valves* sont un peu convexes, mais point egalement. La *grandeur* generale est environ deux pouces en longeur, & deux pouces un quart en largeur, ou à travers; & la *forme* comme usuelle à cet genre.

The *ears* are unequal, one being as it were cut or separated from the body, and forming a slip, the outer margin whereof is waved or scollop'd; the other is rectangular, and joins the body as usual. They are finely striated in a transverse manner.

Les *oreilles* sont inegales, une etant comme coupée ou separée du corps, & forme une languette, dont les bords exterieurs sont ondés ou decoupés; l'autre est rectangulaire, & joint au corps comme à l'ordinaire. Elles sont finement & transversalement striées.

The *insides* of the shells are most generally white, smooth and glossy, but sometimes brownish, fainter in the middle, and deeper colour'd at the margins; at other times a deep bay colour appears at the ears, and under them. The *ears* are slightly concave, and smooth. The *commissure* of the hinge is very thick, inflected, and the outer ribs form very shallow *furrows*.

L'*interieur* des coquilles est plus communement blanc, lisse & lustré, mais quelquefois brun, plus pale au centré, & plus foncé vers les bords; d'autres paroissent d'une couleur rouge-brune foncée aux oreilles, & en dessous. Les *oreilles* sont un peu concaves, & lisses. La *commissure* de la charniere est fort epaisse, & tournée en dedans, & les cotes exterieures forment des *sillons* peu profonds.

Both *valves* are convex, but not much so, and the convexity of one exceeds by near as much again the convexity of the other. The *beaks* of both valves also are convex,

Les deux *valves* sont convexes, mais pas beaucoup; la convexité de l'une surpasse presque d'autant plus la convexité de l'autre. Les *becs* des deux valves sont aussi convexes,

convex. The *ribs* are about twenty, longitudinal, and finely ſtriated their length: on the moſt convex ſhell they are obtuſe or rounded, and broad; on the other ſhell, more narrow, prominent, and riſe to a ſharp edge, which is very perceivable from the upper part to the middle of the ſhell, but from thence to the bottom is leſs obſervable. The *interſtices* between the ribs are very finely ſtriated net-ways, or both longitudinally and tranſverſely.

The *colours* of this ſpecies are very beautiful and various, and the two *valves* are moſt generally of different colours and variegations. The chief are, 1. *almoſt white*, and *white ground*, charged with *brown*, *red*, or *purple*; 2. *uniform* bright *yellow*, and pale *yellow*, with *white*; 3. *uniform brown*, and *brown*, *red*, or *purpliſh* grounds, with *white*, &c. All theſe *colours* are elegantly *blended* and *variegated*, ſometimes *marbled* or *mottled*, ſometimes in tranſverſe *zones* or *girdles*, and ſometimes in longitudinal and broad *rays*.

This *ſpecies* is found on ſeveral of the *coaſts* of *Great Britain* and *Ireland*; but thoſe of the ſhores of *Cornwall*, *Dorſet*, *Northumberland*, and *Ireland*, are eſpecially *very beautiful*, by the richneſs of their colours and variegations.

vexes. Les *cotes* ſont environ vingt, longitudinales, & finement ſtriées ſelon leur longueur. Sur la valve la plus convexe elles ſont obtuſes ou arrondies, & larges; ſur l'autre valve plus etroites, ſaillantes, & s'elevent en vive arrête, qui eſt tres viſible dès la tete juſques au milieu de la coquille, mais de là au bas eſt moins remarquable. Les *intervalles* entre les cotes ſont tres finement ſtriés en reſeau, ou longitudinalement & tranſverſalement.

Les *couleurs* de cette eſpece ſont fort belles & diverſes, & les deux *valves* ſont communement de differentes couleurs & bigarrures. Les chefs ſont, 1. *entierement blanche*, & à *robe blanche*, chargée de *brun*, *rouge*, ou *pourpre*; 2. de couleur *uniforme jaune* eclatante, & *jaune* pale, avec *blanche*; 3. *brune uniforme*, & à robe *brune*, *rouge*, ou *pourprée*, variée de *blanche*, &c. Toutes ces *couleurs* ſont elegamment *entremelées* & *bigarrées*, tantot *marbrées* ou *tachetées*, tantot en *bandes* ou *zones* tranſverſales, & tantot en des *rayes* larges & longitudinales.

Cette *eſpece* ſe trouve ſur pluſieurs des *cotes* de la *Grande Bretagne* & de l'*Irlande*; mais celles des cotes des comtés de *Cornwall*, *Dorſet*, *Northumberland*, & d'*Irlande*, ſont ſpecialement *tres belles*, par la richeſſe de leurs couleurs & bigarrures.

IV. IV.

IV.

P. Lineatus. Streaked.
Tab. X. fig. 8.

P. mediocris, fere æquivalvis, tenuis, valva una alba, altera vero cum linea purpurascente in summitate unaquæque costæ. Lineatus. Tab. 10. fig. 8.

Shell thinnish; general *size* about one and a half inch long, and near the same breadth. *Both valves* are slightly convex. The *ears* are large, and near equal; one rectangular, the other crumpled or sinuous down the perpendicular edge, but both transversely striated. The edge of the *commissure* on one valve is smooth, but on the other very thick, and finely serrated.

Outside. About twenty longitudinal *ribs*, transversely striated, and on one valve rough, or set with points, but on the other narrow, and rise into a ridge or sharpish edge. The *interstices* or furrows are broad, deep, and longitudinally striated.

Colours on the outside. *One valve* is entirely white; the *other* is also white, but the *ridge* or *summit* of each *rib* is adorn'd it's length with a fine reddish purple *line*.

Within it is milk white, and glossy.

This *pretty species* is very *rare*; I have only seen it from *Cornwall*, but am informed it has also been fish'd about *Weymouth*, in *Dorsetshire*.

IV.

Pl. X. fig. 8.

La *coquille* est mince; la *grandeur* generale environ un & demi pouce en longeur, & près de la même largeur. Les *deux valves* sont un peu convexes. Les *oreilles* grandes, & presque egales; une rectangulaire, l'autre chiffonnée ou sinueuse le long du bord perpendiculaire, mais toutes les deux sont striées transversalement. Le bord de la *commissure* dans une valve est uni, mais dans l'autre il est fort epais, & joliment dentelé.

L'exterieur. Environ vingt *cotes* longitudinales, striées à travers, & dans une valve apres ou chargées des pointes, dans l'autre etroites, & s'elevent en vive arrete. Les *intervalles* ou sillons sont larges, profonds, & striés selon leur longeur.

Couleurs à l'exterieur. *Une valve* est entierement blanche; l'*autre* aussi est blanche, mais l'*arrete* ou *sommet* de chaque *cote* est orné sa longeur d'une belle *ligne* rouge pourpre.

Le *dedans* parfaitement blanc, & lustré.

Cette *jolie espece* est fort *rare*; je l'ai vu seulement de *Cornwall*, mais je suis informé qu'on la aussi pesché autour de *Weymouth*, au comté de *Dorset*.

V.

P. Distortus. Distorted.
Tab. X. fig. 3. 6.

V.

Pl. X. fig. 3. 6.

P. parvus inæquivalvis, informis, ſtriatus. Diſtortus. Tab. 10. fig. 3. 6.

P. minimus anguſtior, inæqualis fere & aſper, ſinu ad cardinem cylindraceo, creberrimis minutiſſimiſque ſtriis donatus. Liſt. H. An. Angl. p. 186. tit. 31. tab. 5. fig. 31.

Pecten minor ex croceo variegatus, aſper & fere ſinuoſus & inæqualis, leviter & admodum crebro ſtriatus. Liſt. H. Conch. tab. 172. fig. 9.

Twiſted pectines of Stroma. Wallace Orkn. p. 43. 44.—*Pecten minor alba contuſa proteiformis.* Petiv. Gaz. tab. 94. fig. 2.—*Puſio. Writhen.* Penn. Brit. Zool. No. 65. tab. 61. fig. 65.

Concha teſta aurita, ſtriis circiter quadraginta. Lin. F. Suec. I. p. 384. No. 1345.—An *Puſio.* Lin. S. N. p. 1146. No. 200. Muſ. Reg. p. 529. No. 113?

The *ſhell* very thick and ſtrong. The moſt general *ſize* is that of a large walnut, or about one and a half inch in *length*, and ſomewhat leſs in *breadth*.

The *whole ſhell* is vaſtly irregular, but tends to a globoſe *ſhape*; however, it is always ſo *writhen*, *crumpled*, *diſtorted*, and ſeemingly *bruiſed*, that it appears always like a ſhell greatly hurt by *accident*, and hinder'd in its *natural growth*, for though conſtantly *diſtorted*, yet two ſhells hardly wear the ſame diſtortions.

Both valves are *convex*, ſome greatly ſo, others leſs, and others ſo very little as to be near flattiſh. The *outſide* of the *flattiſh* or *upper* ſhells is generally leſs uneven or warped than the other, for the bruiſings or ſinuoſities are but ſlight on their upper or head part; on the lower they are larger and ſtronger, ſo that the margins are much depreſſed, and thrown out of their ſeeming natural poſition. The *ribs* or *ſtriæ* on this *valve* are always very viſible,

La *coquille* eſt tres epaiſſe & forte. La *grandeur* la plus generale eſt celle d'une grande noix, ou environ un & demi pouce en *longueur*, & un peu moins en *largeur*.

Toute la coquille eſt extremement irreguliere, mais tend a une *forme* ronde; cependant, elle eſt toujours ſi *pliée*, *froiſſée*, *difforme*, & en apparence *concaſſée*, qu'elle paroit comme une coquille beaucoup dommagée par *accident*, & detournée de ſon *accroiſſement naturel*, car quoiqu'elle eſt conſtamment *difforme*, cependant deux coquilles ne portent jamais les mêmes contorſions.

Les *deux valves* ſont *convexes*, quelques unes beaucoup, d'autres moins, & d'autres ſi peu que de paroitre preſque plattes. L'*exterieur* de celles qui ſont *preſque plattes*, ou les valves *ſuperieures*, eſt generalement moins raboteux ou froiſſé que l'autre, car les contuſions & les ſinuoſités ne ſont que legeres ſur la partie ſuperieure ou de la tete; au bas elles deviennent plus grandes & plus fortes, de ſorte que les bords ſont beaucoup enfoncés & jettés hors de leur ſituation

ſible, and generally not interrupted or broken by the warpings. This *upper ſhell* ſeems, when leaſt diſtorted, to be near flattiſh, or at moſt very ſhallow.

The *under valve* is very deep or concave, and is the *valve* that ſuffers the greateſt diſtortions or irregularities. The *ribs* or *ſtriæ* are always interrupted and miſplaced. It is impoſſible to deſcribe the various miſhapings of this valve. The top or head part generally ſlopes very obliquely, and is prettily impreſt with a ſmall fine ſtriated eſcallop; thence it riſes into a very high irregular hump, and the impreſſion there is generally quite loſt or interrupted; and from thence it takes an almoſt perpendicular declivity, with many ſinuoſities or warpings, quite to the lower margins, and regains the ribs or ſtriæ, which are much thicker and ſtronger.

The *ribs* are numerous, cloſe ſet, longitudinal and prominent, but very fine, or like mere threads, ſo that the ſhell appears only as thickly ſtriated. The *colours* are moſtly ſullied white, pale dull violet, dull ruſſety or ſaffrony; ſometimes the upper or flattiſh ſhells are of a pale violet colour, or ruſſety, and ſet with browniſh tranſverſe ſtreaks, and the whitiſh ones are much mottled or variegated with red.

The *inſide* of the *concave lower valves* is white and ſmooth when *alive*, but ruſſety, livid and dull, when *dead* or *worn*. The *upper* or *flattiſh ſhells*, when *alive*, have a

 ſtrong

ſituation naturelle apparente. Les *cotes* ou *ſtries* ſur cette *valve* ſont toujours fort viſibles, & generalement point interrompues ou coupées par les froiſſures. Cette *valve ſuperieure* paroit etre, quand elle eſt que peu froiſſée, preſque applatie, ou tout au plus tres peu concave.

La *valve inferieure* ou *concave* eſt tres profonde, & c'eſt la *valve* qui ſouffre les plus grandes contorſions & irregolarités. Les *cotes* ou *ſtries* ſont toujours interrompues & deplacées. C'eſt même impoſſible de decrire les difformites differentes de cette valve. La partie du ſommet ou de la tete penche fort obliquement, & ſe trouve joliment empreinte d'un petit peigne elegamment ſtrié; de là elle s'eleve dans une boſſe fort haute & irreguliere, & l'empreinte là eſt generalement entierement effacée ou interrompue; & de là elle prend une pente preſque perpendiculaire, avec pluſieurs ſinuoſités ou froncis, tout à fait juſques aux bords inferieurs, & regagne les cotes ou ſtries qui ſont plus epaiſſes & fortes.

Les *cotes* ſont nombreuſes, ſerrées, longitudinales & ſaillantes, mais tres fines, ou comme de ſimples fils, ainſi que la coquille paroit ſeulement comme ſerrement ſtriée. Les *couleurs* ſont, pour la plupart, blanche terne, violette pale triſte, rougeatre ſale ou ſaffranée; quelquefois les valves plattes ou ſuperieures ſont de couleur pale violette, ou rougeatre, chargées de rayes brunes tranſverſales, & les blanches ſont tres tachetées & variées de rouge.

Le *dedans* des *valves concaves* ou *inferieures* eſt blanc & liſſe quand *vivantes*, mais rougeatre, terne & ſale, quand *mortes* ou *roulées*. Les *valves ſuperieures* ou *applaties*,

ſtrong pearly gloſs. The lower *margins*, within, are finely notch'd or crenated from the outſide ribs.

The *ears* are large for the ſize of the ſhell, and nearly equal, when not loſt in the irregularities or diſtortions; in that caſe they are greatly unequal, and one is ſometimes even not very apparent. They are very ſinuous, or bent to and fro, and are tranſverſely ſtriated.

This *odd ſpecies* is found on ſome of our Engliſh ſhores, as in *Yorkſhire*, at *Scarborough* and the *mouth of the river Tees*, the coaſts of *Dorſetſhire*, &c.

Dr. Wallace got it on the ſhores of the *Orkneys*. At *Stroma*, a little iſland that lies in *Pightland Firth*, and in ſome places in *Orkneys*, where there went extraordinary croſs and ſtrong tides, he obſerved the *pectens* ſo *twiſted*, and of ſuch an *irregular ſhape*, that he was ſurprized to ſee it. He adds "I cannot think the odd ſtrange tumbling the tides make there can contribute any thing to that frame; yet, after all, I never ſee them in any other place."

Obs.—This kind is pretty frequently found foſſil in the chalk-pits of *Kent* and *Surrey*.

platies, quand *vivantes*, ont un grand luſtre de nacre. Les *bords* inferieurs, du dedans, ſont finement crenelés par les cotes exterieures.

Les *oreilles* ſont grandes pour le volume de la coquille, & preſque egales quand elles ne ſont abymées dans les ſinuoſites ou contorſions; en cet cas elles ſont tres inegales, & une même quelquefois n'eſt pas tres viſible. Elles ſont fort ſinueuſes, ou courbées ça & la, & ſont ſtriées tranſverſalement.

Cette *eſpece bizarre* ſe trouve ſur quelques de nos cotes Angloiſes, comme au comté de *York*, a *Scarborough* & à l'*embouchure du Tees*, les cotes du comté de *Dorſet*, &c.

Le Dr. Wallace l'a acquis ſur les cotes des *Orcades*. A *Stroma*, une petite iſle en *Pightland Firth*, & en autres lieux des *Orcades*, ou il y avoit des marées extraordinaires fortes & de travers, il obſerva ces *peignes* ſi *pliés* & de *forme ſi irreguliere*, qu'il etoit fort ſurpris. Il ajoute, "Je ne peus croire que cet bizarre & etrange roulement des marées en cet lieu peut contribuer a cette formation; cependant, je n'ay les ai jamais vu en aucune autre place."

Obs.—Cette eſpece ſe trouve aſſes frequemment dans les puits à craie des comtés de *Kent* & *Surrey*.

VI.

P. Monotis. One-Eared.

Tab. X. fig. 1. 2. 4. 5. 7. 9.

VI.

Pl. X. fig. 1. 2. 4. 5. 7. 9.

P. parvus angustior, æquivalvis, inæqualiter auritus, strigis echinatis. Monotis. Tab. 10. fig. 1. 2. 4. 5. 7. 9.

P. subrufus, striis viginti quatuor, ad minimum donatus.—P. parvus, ex croceo variegatus, tenuiter admodum striatus, alternis fere striis paulo minoribus. List. H. Conch. tab. 180. fig. 17. tab. 181. fig. 18. & tab. 189. fig. 23.

Pecten minor nostras, striis plurimis minoribus. Muf. Petiv. p. 86. No. 830.—*Pectunculus echinatus fusco purpureus.* Borlafe Cornw. p. 277. tab. 28. fig. 19. Wallis Northumb. p. 399. No. 24.—*P. varius. Variegated.* Penn. Brit. Zool. No. 64. tab. 61. fig. 64.

Buon. Ricr. p. 152. 153. fig. 5. 15.—Argenv. Conch. I. p. 342. pl. 27. fig. H. II. p. 304. pl. 24. fig. H.—Gualt. I. Conch. tab. 73. fig. I. N. tab. 74. fig. G. H. M. N. O. P. Q. R. S. V. X.

Varia. Ostrea testa æquivalvi radiis triginta, scabris compressis echinatis, uniaurita. Lin. S. N. p. 2146. No. 199. Muf. Reg. p. 529. No. 112.

The *shell* is thinnish and femitranfparent, of an oval *shape*, or narrow and long, and its *size* is from one and a half to two inches in length.

The *valves* are almoft equally convex, but the convexity is flight or compress'd.

The *ears* are very unequal; *one* vaftly large and extended, often rectangular and broad, but at other times fcallop'd and rounded on the bottom part; the *other* is very fmall and rectangular, not one quarter part as big, and of fuch difparity as hardly to be call'd an ear. They are *both* tranfverfely ftriated with ftrong prickly ftriæ; and the *commissure* is alfo fet with a row of fmall but ftrong prickles.

The *ribs* are about thirty, longitudinal, rounded, and thick as coarfe threads, not ftriated, but imbricated or fet with a tiled-like work of tranfverfe fcales, that, toward

La *coquille* eft plutot mince & demi-tranfparente, d'une *forme* ovale, ou etroite & longue, & fa *grandeur* eft de un & demi à deux pouces en longueur.

Les *valves* font prefque egalement convexes, mais la convexite eft legere ou comprimée.

Les *oreilles* font fort inegales; *une* extremement grande & etendue, fouvent rectangulaire & large, mais tantot decoupée & arrondie au bas; l'*autre* eft fort petite & rectangulaire, pas la quatrieme partie fi grande, & de telle inegalité que à peine on la peut nommer une oreille. Elles font toutes *deux* ftriées tranfverfalement à ftries fortes & epineufes; & la *commissure* eft auffi chargée d'une ranges des epines petites, mais fortes.

Les *cotes* font environ trente, longitudinales, arrondies, & epaiffes comme un fil groffier, point ftriées, mais tuilées par des ecailles tranfverfales, qui, vers les cotés & au

toward the ſides and bottom, riſe high and ſharp, like plates, in ſuch manner as to render thoſe parts very ſcabrous and prickly, and ſometimes the ſhell is all over in like manner ſcabrous or echinated. This roughneſs or pricklineſs is more conſtant and apparent on the deeper or under valve. The *interſtices* between the ribs are finely and tranſverſely ſtriated.

Within the valves are ſmooth, gloſſy, and moſt generally white, which ſeems to be the true natural colour; but ſome are tinted ſordid red, yellowiſh, ſtrong purple or violet. All theſe latter ſeem to owe their tinge to the ſtrong colours of their outſides.

Outſide of various colours; frequently of a ſordid red, and uniform, being but little waved or mottled with white or purple, and then only on the upper part, or near the hinge; ſometimes of a coral or bright red colour, but white on the upper part; others of an uniform bright yellow; others whitiſh, with tranſverſe zones and tints of waſhy or pale red; others deep brown, and others of a fine purpliſh or violet colour; and great numbers are beautifully variegated with different colours, as white, red, purpliſh, deep brown, &c.

This *ſpecies* is frequent on moſt of the ſhores of *England*; as in *Wales*; at *Deal*, *Margate*, and *Shellneſs*, in *Kent*; in *Suſſex* and *Dorſetſhire*; in *Devonſhire*; at *Lelant* and *Whitſand Bay*, &c. in *Cornwall*; the *oſtium* of the river *Aln* in *Northumberland*, and many other places.

au bas ſont elevées & aigues, comme des lames, de ſorte que de rendre ces parties tres raboteuſes & epineuſes, & quelquefois la coquille eſt tout a fait raboteuſe ou epineuſe de la même maniere. Cette aſpreté & heriſſement eſt toujours plus conſtante & apparente ſur la valve plus concave ou inſerieure. Les *intervalles* entre les cotes ſont finement & tranſverſalement ſtriés.

Le *dedans* des valves ſont liſſes, luſtrées, & generalement blanches, qui ſemble etre la couleur naturelle; mais quelquefois elles ſont teintes de rouge triſte, jaunatre, pourpre ou violette. Toutes ces dernieres paroiſſent devoir ces teintes aux couleurs fortes de leur exterieures.

L'*exterieur* eſt de differentes couleurs; frequemment d'un rouge morne, & uniforme, etant que tres peu ondé ou bigarré de blanc ou pourpre, & alors ſeulement ſur la partie ſuperieure, ou près de la charniere; tantot d'une couleur de corail ou rouge vive, mais blanche ſur la partie ſuperieure; d'autres d'une couleur jaune eclatante uniforme; d'autres blanchatres, avec des zones tranſverſales & des teints rougeatres ou foibles; d'autres brun foncé, & d'autres d'une belle couleur pourpre ou violette; & un grand nombre ſont elegamment bigarrées de differentes couleurs, comme blanche, rouge, pourpre, brune foncée, &c.

Cette *eſpece* eſt frequent ſur la plupart des cotes d'*Angleterre*; comme dans la *Principauté* de *Galles*; à *Deal*, *Margate*, & *Shellneſs*, en *Kent*; aux comtés de *Suſſex*, *Dorſet*, *Devon*, &c. à *Lelant* & *Whitſand Bay*, &c. en *Cornwall*; l'*embouchure* de la riviere *Alne* en *Northumberland*, & pluſieurs autres lieux.

VII.

P. Parvus. Small.

P. parvus fuſcus longitudinaliter ſtriatus. Parvus.
P. obſoletus. Worn. Penn. Brit. Zool. No. 66. tab. 61. fig. 66.

A *ſmall ſpecies,* about three quarters of an inch *long*, and near the ſame *breadth*. The *valves* are equal and ſhallow, or near flat, and the *ſhell* is very thin and ſemi-tranſparent.

The *ears* are very unequal, *one* vaſtly large, the *other* ſo ſmall as to be ſcarcely apparent. The *large* one is white in my ſhells (query, if a *luſus*, or *conſtantly ſo?*) rectangular, ſpiked or jagged down the perpendicular ſide, and ſtrongly ſtriated in a tranſverſe manner. The *ſmall* ear ſeems alſo tranſverſely ſtriated, and is of the ſame colour as the ſhell.

Within ſmooth and browniſh, with a very pearly hue.

Outſide of a ſtrong uniform dull purpliſh brown colour, with numerous longitudinal fine *ſtriæ*, very cloſe ſet, and at irregular intervals they riſe much more prominent into ſome eight or ten rays, which are the *obſolete rays of Mr. Pennant.* The ſtriæ notch the margins very finely, and ſome few ſlight ſtriæ run acroſs the ſhell.

It is a *very rare ſpecies*; I received mine from the coaſt of *Cornwall*. Mr. *Pennant* is *ſilent* as to the *natal place* of his ſhell.

The *figure* ſaid author gives is good, except as to the outſide work of the ſhell.

GENUS

VII.

Une *petite eſpece,* environ trois quarts d'un pouce en *longeur*, & à peu près de la méme *largeur*. Les *valves* ſont egales, tres peu profondes ou preſque applaties, & la *coquille* eſt fort mince & demitranſparente.

Les *oreilles* ſont fort inegales, *une* extremement grande, l'*autre* ſi petite que à peine eſt elle viſible. La plus *grande* eſt blanche dans mes coquilles (demande, ſi c'eſt un *jeu*, ou *conſtamment ainſi?*) rectangulaire, decoupée ou crenelée le long de ſon coté perpendiculaire, & fortement ſtriée à travers. La *petite* oreille ſemble etre auſſi ſtriée à travers, & eſt de la même couleur que la coquille.

Le *dedans* liſſe & brunatre avec un luſtre fort nacré.

L'*exterieur* d'une couleur pourpre brune foncée, morne & uniforme, chargé de *ſtries* nombreuſes, fines & longitudinales, tres ſerrées, & à intervalles irreguliers, elles s'elevent plus hautes en quelques huit ou dix rayes, qui ſont les *rayes peu marquées de M. Pennant.* Les ſtries crenelent tres finement les bords, & quelques ſtries legeres courent à travers de la coquille.

C'eſt *une eſpece tres rare.* J'ay reçu mes echantillons de la cote de *Cornwall*. M. *Pennant ne dit mot* du *lieu natal* de ſa coquille.

La *figure* que le dit auteur donne eſt bonne, excepté a l'egard de l'ouvrage exterieur de la coquille.

GENRE

GENUS II.

OSTREUM. The OYSTER.

Bivalves, most generally with *unequal valves*, that shut close. The *hinge* is toothless, and consists of a large transversely striated gutter, running the length of the summit of the shells.—The animal, a *Tethys*.

GENRE II.

L'HUITRE.

Bivalves, qui ont generalement les *battans inegaux*, & se ferment exactement. La *charniere* n'a point des dents, mais consiste d'une grande goutiere, striée à travers, qui court le long du sommet des coquilles.—L'animal est un *Tethys*.

* MARINÆ. SEA.

VIII.

O. VULGARE. COMMON.

Tab. XI. fig. 6.

* DE MER.

VIII.

L'HUITRE COMMUN.

Pl. XI. fig. 6.

O. *Vulgare subrotundum rugosum & crassum, lamellis foliaceis imbricatum, intus perlaceo quodam splendore albescens. Vulgare, the common oyster*. Tab. 11. fig. 6.

Ostreum vulgare maximum, intus argenteo quodam splendore albescens. List. H. An. Angl. p. 176. tit. 26. tab. 4. fig. 26.

Ostrea major sulcata, inæqualiter utrinq; ad cardinem denticulata. List. H. Conch. tab. 193. 194. fig. 30. 31. & tab. Anat. 195. 196.—In 2da vero Edit. Huddesford, App. tab. 15 & 16.

Mus. Sibbald. p. 161.—Merret Pin. p. 193.—Wallace Orkn. p. 42.—Martin W. Isles, p. 55. 343. &c.—Mus. Petiv. p. 85. No. 822.—Leigh Lancash. p. 134—Dale Harw. p. 383.—Morant Colchester, p. 85.—Borlase Cornw. p. 274.—Hutchins Dorset. p. lxxvii. & p. 10.—Smith Cork, p. 318.—Id. Waterford, p. 272.—County Down, p. 22. & 239.—Rutty Dublin, p. 375.—Wallis Northumb. p. 394.—*O. edulis, Edible*. Penn. Brit. Zool. No. 69.—Jacob Hist. of Faversham, p. 75.—Hist. of Rochester, p. 84.

Gesner de Aquat. 2. p. 645.—Rondelet de Aquat. 2. p. 37.—Aldrov. Exsang. p. 482. —Buon. Ricr. p. 167. fig. 70.—Klein Ostrac. tab. 8. fig. 21.—Argenv. Conch. II. p. 48. Zoom. pl. 5. fig. A.—Gualt. I. Conch. tab. 102. fig. A.

Concha testa subrotunda rugosa substriata: Valvis inæqualibus, cardine obliterato. Lin. F. Suec. I. p. 382. No. 1338. II. No. 2149.—*Edulis. Ostrea testa inæquivalvi semiorbiculata membranis imbricatis undulatis, valva altera plana integerrima*. S. N. p. 1148. No. 211.—Mus. Reg. p. 534. No. 123.

This *shell* is thick, strong, heavy and rugged, commonly of a round or oval *shape*, but narrowing towards the hinge, and on that part is of dissimilar sides, one side forming an oblique line, the other is rounded. The *valves* are very unequal, the *upper* flat, the *other* very concave.

The *size*, in common, is about three inches *long*, and less in *breadth*; but the *sizes* as well as the *shapes*, from their adhesion to other bodies, vary in different places.

Within the *valves* are smooth, white, and generally of a pearly glare, which oftentimes is very radiant or orient. The summit or *commissure* of the hinge is slightly notched, and from the *hinge* down each *side* runs a *row* of many small *knobs* or *teeth*, the largest about the size of a small pin's head; they are generally more distinct on the *under* than the *upper* valve.

Outside of the *upper* or *flat valve* is of a dirty brownish hue, and roughly plated or made up of transverse flakes, exceeding thin, like paper or leaves, imbricating or partly lying over one another, similar to a tiled-work; these *flakes* or *leaves* lie close, compact, and strongly set together on the upper part, till towards the middle of the shell, and from thence to the bottom they are looser set, become finer, thinner, and more extended beyond one another, and as they approach to the bottom margin, are generally so loose, so separate, and so extended, as to *foliate* the shell very *finely*, and even much beyond the edges of it.

Cette *coquille* est epaisse, forte, pesante & raboteuse, communement d'une *forme* ronde ou ovale, mais etrecissant vers la charniere, & en cette partie les cotés sont inegaux, l'un tendant en ligne oblique, l'autre est arrondi. Les *valves* sont inegales, la *superieure* est applatie, l'*autre* tres concave.

La *grandeur* generale est environ trois pouces en *longeur*, & un peu moins en *largeur*; mais les *grandeurs* tant que les *formes*, de leur proprieté de s'attacher à d'autres corps, varient en differentes places.

Le *dedans* des *valves* sont lisses, blancs, & generalement d'un lustre nacré, qui souvent est fort eclatant ou d'un bel orient. Le sommet ou la *commissure* de la charniere est legerement crenelé, & dès la *charniere*, le long de chaque *coté*, court une *rangée* de plusieurs petits *tubercules* ou *dents*, le plus grand environ la grandeur de la tete d'une petite epingle; ils sont generalement plus distincts sur la valve *inferieure* que sur la *superieure*.

L'*exterieur* de la *valve superieure* ou *applatie* est d'une couleur brunatre sale, rudement chargée de lames, ou composée de ecailles transversales extremement minces, comme du papier ou des feuilles, posées en tuilée, ou en partie l'une sur l'autre. Ces *ecailles* ou *lames* sont serrées, conjointes, & compactement mises ensemble sur la partie superieure, jusques envers le milieu de la coquille, & de là au bas elles sont plus detachées, deviennent plus fines, minces, & etendues au delà de l'une l'autre, & comme elles approchent aux bords inferieurs, elles sont si laches, si detachées, & si etendues, que de *feuilleter* la coquille tres *elegamment*, & même fort au de-là des bords.

Outside of the *under valve* is very rugged, whitish, generally soil'd greenish. This is also of a plated or leaved structure, but seldom shews such fine foliations as the upper valve; they are chiefly apparent on the edges of the concentric and transverse wrinkles or furrows that cross the shell, which they furbelow or plait in a pretty manner, and the plaits are most commonly of a purplish colour. This valve is also wrought with several irregular prominent longitudinal *ribs*.

L'*exterieur* de la *valve inferieure* est fort raboteux, generalement blanchatre avec un teint verdatre. Elle est aussi d'une construction lamelleuse ou feuilletée, mais rarement montre de si beaux feuillages que la valve superieure; ils sont principalement apparents sur les bords de rides ou sillons concentriques & transversales qui traversent la coquille, qu'ils plissent & froncent d'une maniere mignonne ou en falbala, & les plissures sont communement de couleur pourprée. Cette valve est aussi chargée de plusieurs *cotes* elevées, irregulieres, & longitudinales.

Oysters are found in very great *plenty* on most of the coasts of *Great Britain* and *Ireland*.

In *England*, the counties of *Kent* and *Essex* are the most famous for oysters; they have *beds* or *stews*, and *fisheries* for them, and carry on a *great trade* in these shell fish.

Those of *Kent* are chiefly at *Faversham*, and throughout its manor, as also at *Rochester*.

Jacobs, in his History of Faversham, treats of the following particulars relative to this subject:—" Of the limits or boun-" daries of the oyster grounds; of the " company of free dredgers, which can be " traced by records from the reign of " Henry II. *viz.* 1154, for grant of which " they paid annually (and still continue " to pay) twenty-three shillings and four-" pence to the crown; of the two annual " courts, called Admiralty or Water " Courts, held for the fisheries; and of " the oyster trade, which, he says, is ge-" nerally about 3000 l. yearly from the " Dutch."

Les *huitres* se trouvent en tres grande *abondance* sur la pluspart des cotes de la *Grande Bretagne* & l'*Irlande*.

En *Angleterre*, les comtés de *Kent* & *Essex* sont les plus fameux pour les huitres; ils ont des *lits* ou *huitrieres*, & des *pescheries*, ils font même un *grand traffic* en huitres.

Celles de *Kent* se peschent principalement à *Faversham*, & dans tout son fief; aussi à *Rochester*.

Jacobs, dans son Histoire de Faversham, traite des circonstances suivantes relatives à cet sujet:—" Des limites ou bornes des " huitrieres; de la compagnie de pescheurs " francs, qu'on peut tracer par les archives " du regne de Henri II. *sçavoir* en 1154, " pour octroi de laquelle ils payoient an-" nuellement (qu'ils continuent de payer " même à present) vingt trois shelins & " quatre sous à la couronne; des deux " cours par an, nommées Cours d'Admi-" rauté ou de l'Eau, tenues pour les pes-" cheries; & du traffic en huitres, qu'il " dit, monte generalement à environ " 3000 l. par an, avec les Hollandois."

The

Le

The ſame author ſays, that from the mouths or entrances of *Sandwich Bay*, and the adjoining coaſt, or *Richborough* and *Reculver*, oyſters are got in great plenty, and are the moſt delicious that can be taken; but, as the beds do not afford native oyſters ſufficient for the demand, large quantities of ſmall ones, called *brood*, are annually collected from different parts of the ſurrounding ſea, even from the Land's End in Cornwall, from Scotland, and from France, in order to encreaſe, and be meliorated of their ſaltneſs, by the conſtant flow of freſh waters from the two great rivers, the Thames and Medway. It muſt therefore be *admitted*, that altho' *oyſters* are found round all the coaſt, yet thoſe of the bay of the *Roman Rutupiæ*, or *Richborough*, may juſtly claim the *preference* of all others.

This *preference* of the delicious Richborough oyſters is very *antient*, and was given them even by the *luxurious Romans*. The ſatyriſt *Juvenal* records the *Rutupian* or *Richborough oyſter*, for in Satyr IV. v. 140. ſatyrizing an epicure, he ſays,

———" *Circæis* nata forent, an
" *Lucrinum* ad ſaxum, *Rutupinove* edita
" fundo,
" Oſtrea, callebat primo deprendere morſu.

" He, whether *Circe*'s rock his oyſters bore,
" Or *Lucrine*'s lake, or diſtant *Richborough*'s
" ſhore,
" Knew at firſt taſte."

Rocheſter alſo carries on a great *trade* in oyſters, has its *beds*, *fiſheries*, and *company* of free *dredgers* of *Stroud*, ſubject to the authority of the mayor and citizens of this town. The company frequently buy *brood* or

Le même auteur nous dit, que des embouchures ou entrées de la *Baie* de *Sandwich*, & la cote voiſine, ou *Richborough* & *Reculver*, les huitres ſont peſchées en tres grande abondance, & ſont les plus delicieuſes qui puiſſent etre; mais, comme les lits ne produiſent les huitres en quantite ſuffiſante pour le debit, des grandes quantités de jeunes, appellées la *race*, ſont annuellement recueillies de differents lieux de la mer d'alentour, même du Land's End en Cornwall, de l'Ecoſſe, & de la France, pour les multiplier, & les ameliorer de leur ſalure, par les eaux fraiches qui courent continuellement des deux grandes rivieres, la Tamiſe & le Medway. Il faut donc etre *admis*, que quoique les *huitres* ſe trouvent ſur toute la cote, cependant celles de la baie du *Rutupiæ des Romains*, ou *Richborough*, peuvent juſtement pretendre la *preference* de toutes autres.

Cette *preference* des huitres delicieuſes de Richborough eſt fort *ancienne*, & leur fut donné même par les *Romains luxurieux*. L'ecrivain ſatyrique *Juvenal* rend temoignage de l'*huitre Rutupienne* ou de *Richborough*, car dans ſa Satyre IV. v. 140. ſatyriſant un epicurien, il dit,

" Lui, ſi les huitres etoient de *Circe*, de
" *Lucrine*, ou de *Rutupinum* ou *Richborough*,
" les connoiſſoit au premier gout."

Rocheſter auſſi à un grand *traffic* en huitres, & pareillement à ſes *lits*, *peſcheries*, & *compagnie* des *peſcheurs* francs de *Stroud*, ſujette à l'autorité du maire & des citoyens de cette ville. La compagnie achette ſouvent

or *ſpat*, which they lay on the oyſter grounds, where they ſoon grow to maturity.—Hiſt. of Rocheſter.

ſouvent des *jeunes huitres*, ou du *fray*, qu'ils mettent dans leurs huitrieres, ou elles deviennent adultes. — Hiſtoire de Rocheſter.

Colcheſter, in *Eſſex*, is another town famous for its oyſters: it alſo has its oyſter beds; fiſheries; company of free dredgers from Henry II.'s time; courts; and other privileges; and a great trade for them for all which particulars I ſhall refer my readers to Morant's hiſtory of that town.

Colcheſter, au comté de *Eſſex*, eſt une autre ville renommée pour ſes huitres: elle a auſſi ſes huitrieres; peſcheries; & compagnie des peſcheurs francs du tems de Henri II.; cours; & autres privileges; auſſi un grand traffic en huitres: mais pour toutes ces circonſtances je renvoye les lecteurs à l'hiſtoire de Colcheſter, par Morant.

Biſhop Sprat, in his Hiſtory of the Royal Society, p. 307, has given an account of the *generation* of *oyſters*, the *manner* of *forming* the *beds* or *ſtews* of them, and to make them *green*. Dr. Liſter republiſhed it (in Latin) in his Hiſt. Anim. Angl. It has alſo been copied by Dale, in his Hiſtory of Harwich: and laſtly, by Mr. Pennant, in his Britiſh Zoology. *Yet notwithſtanding*, as it is a *very curious memoir* relative to the *natural hiſtory* of the *oyſter*, and its *management*, I ſhall inſert it here; " for tho' (as Dr. Sprat obſerves) " the *Britiſh oyſters* have been famous in " the world ever ſince this iſland was diſ- " covered, yet the ſkill how to order them " aright has been ſo little conſidered " amongſt ourſelves, that we ſee at this " day it is confined to ſome few narrow " *creeks* of one ſingle county."

L'Eveque Sprat, dans ſon Hiſtoire de la Societé Royale, p. 307, nous à donné un recit de la *generation* des *huitres*, la *maniere* de *former* les *lits* ou *huitrieres*, & pour les faire *vertes*. Le Dr. Liſter le republia (en Latin) dans ſon Hiſt. An. Angl. Il à auſſi ete copié par Dale, dans ſon Hiſtoire de Harwich; & finalement par M. Pennant, dans ſa Zoologie Britannique. *Cependant nonobſtant*, comme elle eſt une *memoire tres curieuſe* relative a l'*hiſtoire naturelle* de l'*huitre*, & ſon *maniement*, je l'ajouterai ici; " car quoique (comme l'e- " veque remarque) les *huitres Britanniques* " ont eté toujours fameuſes dans le monde " depuis le tems que cette iſle à été de- " couverte, cependant l'addreſſe à les me- " nager comme il faut, à eté ſi peu conſi- " derée parmi nous mêmes, que nous voy- " ons juſques à cet jour qu'elle eſt bornée " à quelques petites *criques* d'un ſeul " comté."

The Hiſtory of the Generation *and Ordering of* Green Oyſters, *commonly called* Colcheſter Oyſters.

L'Hiſtoire de la Generation *& du Maniement des* Huitres Vertes, *communement appellées* Huitres de Colcheſter.

" IN the month of *May* the *oyſters* caſt " their ſpawn (which the dredgers call " their ſpat). It is like to a drop of " candle,

" DANS le moins de *May* les *huitres* " jettent leur fray (que les peſcheurs ap- " pellent leur ſpat). Il paroit comme " une

" candle, and about the bigneſs of an " halfpenny."

" The ſpat cleaves to ſtones, old oyſter " ſhells, pieces of wood, and ſuch-like " things, at the bottom of the ſea, which " they call *cultch*.

" 'Tis probably conjectured, that the " ſpat in twenty-four hours begins to " have a ſhell.

" In the month of *May*, the dredgers " (by the law of the Admiralty court) " have liberty to catch all manner of " *oyſters*, of what ſize ſoever

" When they have taken them, with a " knife they gently raiſe the ſmall brood " from the *cultch*, and then they throw " the *cultch* in again, to preſerve the " ground for the future, unleſs they be ſo " newly ſpat that they cannot be ſafely " ſevered from the *cultch*; in that caſe " they are permitted to take the ſtone or " ſhell, &c. that the *ſpat* is upon: one " ſhell having many times twenty *ſpats*.

" After the month of *May*, it is felony " to carry away the *cultch*, and puniſhable " to take any other *oyſters*, unleſs it be " thoſe of ſize (that is to ſay) about the " bigneſs of an half-crown piece, or when " the two ſhells being ſhut, a fair ſhilling " will rattle between them.

" The places where theſe *oyſters* are " chiefly catcht, are called the *Pont-Burn-* " *ham*, *Malden*, and *Colne* waters; the " latter taking its name from the river of " *Colne*, which paſſeth by *Colne-Cheſter*, " gives the name to that town, and runs " into a creek of the ſea at a place called " the *Hythe*, being the ſuburbs of the " town.

" This

" une goute de chandelle, & environ la " grandeur d'un demi-ſols Anglois.

" Cet fray s'attache aux pierres, vielles " coquilles, morceaux de bois, & telles " autres choſes, au fond de la mer, qu'ils " appellent *cultch*.

" On ſoupçon probablement, que le " fray commence à avoir une coquille en " vingt quatre heures.

" Dans le mois de *May*, les peſcheurs " (par les loix de la cour de l'Amiraute) " ont liberté de pecher toutes les *huitres*, " de quelle grandeur qu'elles ſoient.

" Quand il les ont priſes, ils ſeparent " tres legerement les petites jeunes huitres " avec un couteau du *fray*, & alors ils " jettent le *fray* derechef dans la mer, " pour conſerver la place à l'avenir, ex- " cepté que les huitres ſoient ſi re- " centes ou jeunes qu'elles ne puiſſent " etre bien ſeparées du *fray*; dans cet cas " il leur eſt permis de prendre la pierre ou " la coquille, &c. que le *fray* couvre: " une coquille etant ſouvent couverte de " vingt *frays*.

" Apres le mois de *May*, c'eſt un crime " capitale ou felonie d'enlever le *fray*, & " puniſſable de prendre autres *huitres*, que " celles qui ſont de juſte grandeur (c'eſt a " dire) environ la grandeur d'un demi-ecu, " ou que quand les deux battans ſont " fermés, un bon ſhelin Anglois ſera du " bruit entre eux.

" Les places ou ces *huitres* ſont princi- " palement peſchées, ſont a *Pont-Burnham*, " *Malden*, & *Colne Waters*; cet dernier lieu " prend ſon nom de la riviere de *Colne*, " qui paſſant par *Colne-Cheſter*, donne le " nom a cette ville, & court dans une " crique de la mer a une place nommée " *Hythe*, etant le fauxbourg de la ville.

" Cette

" This brood, and other *oysters*, they " carry to creeks of the sea at *Brickelsea*, " *Mersey*, *Langno*, *Fingrego*, *Wivenbo*, *Tolesbury*, and *Salt-coase*, and there throw " them into the channel, which they call " their beds or layers, where they grow " and fatten, and in two or three years " the smallest brood will be *oysters* of the " size aforesaid.

" Those *oysters* which they would have " green, they put into pits about three " foot deep, in the salt marshes, which are " overflowed only at spring tides, to " which they have sluices, and let out the " salt water until it is about a foot and a " half deep.

" These pits, from some quality in the " soil co-operating with the heat of the " sun, will become green, and communicate their colour to the *oysters* that are " put into them in four or five days, tho' " they commonly let them continue there " six weeks, or two months, in which time " they will be of a dark green.

" To prove that the sun operates in the " greening, *Tolesbury pits* will green only " in Summer; but that the earth hath " the greater power, *Bricklesea pits* green " both Winter and Summer: and, for a " further proof, a pit within a foot of a " greening-pit, will not green; and those " that did green very well, will in time " lose their quality.

" The *oysters*, when the tide comes in, " lie with their hollow shell downwards, " and when it goes out they turn on the " other

" Cette race, & autres *huitres*, ils portent aux criques de la mer, a *Brickelsea*, " *Mersey*, *Langno*, *Fringrego*, *Wivenbo*, *Tolesbury*, & *Salt-coase*, & là il les jettent " dans la mer, & les appellent les lits ou " couches, ou les huitres croissent & engraissent, & dans deux ou trois années " la plus petite race deviendra des *huitres* " de juste grandeur.

" Ces *huitres* qu'ils veulent faire vertes, " ils mettent dans des puits environ trois " pieds de profondeur, dans les marais salines, qui sont inondés seulement à " hautes marées, auxquels ils ont des " ecluses à faire l'eau sortir salée jusques " qu'elle soit environ un pied & demi en " profondeur.

" Ces puits, de quelque qualité dans le " terrein jointe a l'ardeur du soleil, deviendront vertes, & communiqueront " leur couleur aux *huitres* qu'on y mette " dans quatre ou cincq jours, quoique " communement on les laisse continuer là " six semaines, ou deux mois, dans lequel " tems elles seront d'une couleur verte " foncée.

" Pour prouver que le soleil opere en " les rendant vertes, les *puits de Tolesbury* " les feront vertes seulement en eté; mais " que le terrein a le plus grand pouvoir, " les *puits de Bricklesea* les feront vertes " tant en Hyver que en Eté: & pour une " preuve plus grande, un puits distant " seulement un pied d'un autre puits qui " les rend vertes, ne les rendront point " vertes; & les puits qui autrefois les ont " fort bien rendues vertes, en tems perdera cette qualité.

" Les *huitres*, quand la marée monte, sont " couchées la valve concave en bas, & " quand elle descend elles se tournent de " l'autre

" other side. They remove not from their " place, unless in cold weather, to cover " themselves in the ouse.

" The reason of the scarcity of *oysters*, " and consequently of their dearness, is, " because they are of late years bought up " by the *Dutch*.

" There are great penalties, by the Ad- " miralty court, laid upon those that fish " out of those grounds which the court " appoints, or that destroy the *cultch*, or " that take any *oysters* that are not of size, " or that do not tread under their feet, or " throw upon the shore, a fish which they " call a *five-finger*, resembling a spur rowel, " (the *star-fish*) because that fish gets into " the *oysters* when they gape, and sucks " them out.

" The reason why such a penalty is set " upon any that shall destroy the *cultch*, is, " because they find that, if that be taken " away, the ouse will increase, and then " *muscles* and *cockles* will breed there, and " destroy the *oysters*, they not having " whereon to stick their *spat*.

" The *oysters* are sick after they have " spat, but in *June* and *July* they begin " to mend, and in *August* they are perfectly " well. The *male oyster* is *black sick*, having " a black substance in the fin; the *female*, " *white sick* (as they term it) having a " milky substance in the fin. They are " salt in the pits, salter in the layers, but " saltest at sea."

" l'autre cote. Elles ne remuent point " de leur places, sinon en tems froid, pour " se couvrir dans le limon.

" La raison de la disette des *huitres*, & " consequemment leur haut prix, est que " les *Hollandois* les ont acheté ces années " passées.

" La cour d'Admirauté inflige de " grandes peines ou amendes sur ceux " qui peschent hors de ces huitrieres que la " cour fixe, ou qui detruisent le *fray*, ou " prennent les *huitres* qui ne sont de juste " grandeur, ou qui ne detruisent, ou jet- " tent au rivage, un poisson qu'ils appel- " lent *cincq doigts*, qui ressemble à la mol- " lette d'un eperon (l'*etoile de mer*) parce- " que cet poisson entre dans les *huitres* " quand elles s'ouvrent, & les mangent.

" La raison pourquoi une telle amende " est infligé sur ceux qui detruisent le *fray*, " est, à cause qu'ils trouvent que si on tire " le fray, le limon augmente, & alors les " *moules* & les *petoncles* engendreront là, " & detruiront les *huitres*, comme elles " n'ont aucune chose sur quoi deposer leur " *fray*.

" Les *huitres* sont malades apres qu'elles " ont deposé leur fray, mais en *Juin* & " *Juillet* elles commencent a se porter " mieux, & en *Aoust* elles sont parfaite- " ment gueris. L'*huitre male* est *malade* " *noire*, & a une substance noire dans sa " nageoire; la *femelle*, *malade blanche* " (comme ils s'expriment) ayant une sub- " stance laiteuse dans la nageoire. Elles " sont salées dans les puits, plus salées " dans les lits, & tres salées dans la mer."

The shores of *Scotland* yield *oysters* abundantly. The generality of the bays of all the *Western Islands* afford great plenty. In

Les cotes de l'*Ecosse* produisent des *huitres* en tres grande abondance. Presque toutes les baies des *Isles Occidentales* en four- nissent

In the *Isle of Wacksay* (says Martin) *they* are so *big*, that they must be cut into four to be eaten. Dr. Wallace also observes, that the largest oysters he had ever seen are got in some places in the *Orkneys*, and likewise mentions that they must be cut into two or three pieces.

The coasts of *Ireland* equally abound with this *species* of shell fish. There are several *oyster beds* near *Dublin*; a *rock oyster* at *Howth*; a bed of *oysters*, *large* as a horse-shoe, lies E. N. E. from *Ireland's Eye*; the bed of *Malahide* oysters, which are green finn'd, and very delicious; on the coasts of *Waterford*, and in most of the western harbours of *Cork*, &c.

The *anatomy* of this shell fish was first published by Dr. *Willis*, in his work *de Anima Brutorum*, p. 14. edit. 1676. Dr. *Lister* republish'd it in his *Hist. Anim. Angl.* but without the figures; and in his *Exercit. Anat.* 3. p. 62. with the figures tab. 5 & 6. with some observations; and also again republished the figures in his *History of Shells.*

Small *pearls*, but of no value, are often found in oysters.

nissent tres abondament. Dans l'*Isle de Wacksay* (Martin dit) *qu'elles* sont si *grandes* qu'il faut les couper en quatre pour les manger. Le Dr. Wallace observe pareillement, que les plus grandes huitres qu'il a jamais vu sont peschées dans quelques places des *Orcades*, & dit aussi qu'il faut les couper en deux ou trois pieces.

Les cotes de l'*Irlande* abondent egalement dans cette *espece* de coquille. Il y a plusieurs *lits* de *huitres* près de *Dublin*; une *huitre de rocher* à *Howth*; un lit de *huitres grandés* comme un fer de cheval, se trouve a E. N. E. de *Ireland's Eye*; le lit de huitres de *Malahide*, qui sont à nageoire verte, & fort delicieuses; sur les cotes de *Waterford*, & dans la plupart des havres occidentales de *Cork*, &c.

L'*anatomie* de cet coquillage fut premierement publiée par le Dr. *Willis*, dans son ouvrage *de Anima Brutorum*, p. 14. edit. 1676. Le Dr. *Lister* l'a republiée dans son *Hist. Anim. Angl.* mais sans les figures; & dans son *Exercit. Anat.* 3. p. 62. avec les figures pl. 5 & 6. & quelques observations. Il a derechef republié les figures dans son *Hist. Conchyliorum.*

Des petites *perles*, mais d'aucune valeur, se trouvent souvent dans les huitres.

IX.

O. STRIATUM. STRIATED.
Tab. XI. fig. 4. 4.

IX.

Pl. XI. fig. 4. 4.

O. mediæ magnitudinis veluti striatum intus virescente. Striatum. Tab. 11. fig. 4. 4.

Ostreum parvum veluti striatum, testa intus virescente, cardine utrinque canaliculato. List. H. An. Angl. p. 181. tit. 27 tab. 4. fig. 27.

Ostrea fere circinata, subviridis, leviter striata. List. H. Conch. tab. 202. 203. fig. 36. 37.

An

An *Ostreum vulgare, striatum, striis rotundis, crassioribus, interruptis radiatum, squamosum ex fusco viridescens.* Gualt. I. Conch. tab. 102. fig. B.?

This *shell* is thick, strong, and near opake, about one and a half inch in *diameter,* but most commonly much less. The *valves* are unequal; the *under* very concave; the *upper* flattish, of a round *shape,* but running into a peak or point in the center of the top. However it varies much in *shape,* as the common oyster does, some being variously sinuous, some very deep, and others shallow, &c.

Inside of a livid green colour, and glossy. The *hinge* is hidden within the shell, quite under the commissure; it is a broad, deep, and somewhat triangular gutter, transversely striated. About one quarter of an inch lower down in the shell, a large *oval mark* is seen, mostly of the colour of the shell, but at other times very *remarkable,* and looking as if *artificial,* being white or whitish, and exactly resembles a thick spot of white oil paint, or cerufs. This *mark* is always radiated with wrinkles from the center to the circumference, and is formed by the muscle of the fish, that, Lister says, is in the middle of the cavity, and whose impression is whitish. The *margins* on this side are plain.

The *outside* is a little uneven, but not rugged, nor of a leaved or flakey structure, as the common oyster, of a dirty rusty whitish *colour,* wrought with *longitudinal ridges.* These *ridges* are prominent, about the thickness of a coarse thread, very numerous, irregular, and run into one another; but towards the bottom, always furcate or divide.

Z The

Cette *coquille* est epaisse, forte, & presque opaque, environ un & demi pouce en *diametre,* mais generalement plus petite. Les *valves* sont inegales; l'*inferieure* fort concave; la *superieure* un peu platte, d'une *forme* ronde, mais etrecie en une pointe au centre du sommet. Cependant elle differe beaucoup en *forme,* comme l'huitre commune, quelqu'unes etant diversement sinueuses, d'autres fort concaves, & d'autres peu profondes, &c.

Le *dedans* d'une couleur verte livide, & lustré. La *charniere* est cachée dedans la coquille, tout à fait au dessous de la commissure; c'est une goutiere large, profonde, quelque chose triangulaire, & striée à travers. Environ un quart d'un pouce plus bas dans la coquille, une grande *marque ovale* se trouve, de même couleur que la coquille, mais tantot fort *remarquable,* & paroissant comme *artificielle,* etant blanche ou blanchatre, & ressemble exactement à une tache epaisse de fard blanc, ou ceruse en huile. Cette *marque* est toujours rayonnée des rides, qui partent du centre à la circonference, & est formée par le muscle de l'animal, que, Lister d'ecrit, etre dans le milieu de la cavité, & dont l'empreinte est blanchatre. Les *bords* de cet coté sont unis.

L'*exterieur* est un peu inegal, mais point raboteux, ni d'une structure feuilletée ou lamelleuse, comme l'huitre commune, d'une *couleur* sale rouille blanchatre, & à *cotes longitudinales.* Ces *cotes* sont saillantes, environ l'epaisseur d'un gros fil, tres nombreuses, fort irregulieres, & entrelassées, mais vers les bords toujours elles se divisent ou fourchent.

La

The *upper valve* is flattish, but varies very much in being more or less so, and has nearly the same colours, but the *outside work* is generally more obsolete.

This *oyster*, hitherto *only proposed* and *described* by Dr. *Lister*, is a very different *species* from the *common oyster*, but has been always overlooked as the same kind. Dr. Lister mentions it to be found in plenty at the mouth of the *Tees*, in Yorkshire, and says he first eat of it at Bourdeaux, in France, where it is greatly esteemed, and called *Rock Oyster*, for they get it among the rocks.

It is also found on many of our other shores, as *Kent*, *Sussex*, *Dorsetshire*, *Devonshire*, &c. and not in small quantities, but seldom perfect, for the valves are separated and greatly broken, rubb'd and worn. There are hundreds of the under or concave valve to one of the upper or flattish; they seem to have been torn from the rocks by the sea, as they are so greatly broken and shapeless, likewise so much worn that they are hardly recognizable, and have been *mistaken* for the shells of the following species, or *Anomia Cepa*, by the elegance and brightness of the colours which many wear; for being so worn as that their *outer coats* are rubb'd off, the *inner ones* only are seen, which are mostly of very fine *colours*, as *green*, *violet*, *pink*, *white* or *silvery*, *yellow*, and even *gold colour*, like a japan lacquer, very glossy and resplendent.

La *valve superieure* est applatie, mais varie beaucoup en etant plus ou moins platte, & elle a peu pres les mêmes couleurs, mais l'*ouvrage exterieur* est generalement plus foible.

Cette *huitre*, jusques à present *proposée* & *decrite seulement* par le Dr. *Lister*, est une *espece* fort differente de l'*huitre commune*, mais elle a eté toujours regardé ou meprise pour la même espece. Le Dr. Lister dit qu'elle se trouve en abondance à l'embouchure du *Tees*, au comté de York, & qu'il en a premierement mangé à Bourdeaux, en France, ou elle est beaucoup estimée, & ou on l'appelle *Huitre de Rocher*, parceque on les trouve parmi les rochers.

Elle se trouve aussi sur plusieurs de nos autres cotes, comme des comtés de *Kent*, *Sussex*, *Dorset*, *Devon*, &c. & pas en petite quantité, mais rarement parfaites, car les valves sont detachées, & tres brisées ou roulées. On trouve des centaines de la valve inferieure ou concave à une valve superieure ou applatie; elles paroissent etre arrachées des rochers par la mer, parcequ'elles sont si rompues & difformes, & tant roulées, que à peine sont elles reconnoissables, & ont eté *prises* pour des coquilles de l'espece suivante, ou *Anomia Cepa*, à cause de leurs couleurs vives & eclatantes que plusieurs portent; car etant tellement frottées qu'elles sont depouillées de leur *robes exterieures*, les *interieures* seulement s'offrent à la vue, qui sont generalement de fort belles *couleurs*, comme *verte*, *violette*, *d'oeillet*, *blanche* ou *argentée*, *jaune*, & même *couleur d'or*, comme du vernis de japon, fort lustrées & resplendissantes.

GENUS III.

ANOMIA. The ANOMIA.

Bivalves with *unequal valves*, that shut close, and the *beak* of *one valve* is always *perforated* with a *hole*. The animal *not yet perfectly known*, and is a *new genus*.

GENRE III.

ANOMIA. L'ANOMIE.

Bivalves à *valves inegales*, qui se ferment exactement, & le *bec* d'*une valve* est toujours *percé* d'un *trou*. L'animal *pas encore parfaitement connu*, & est un *nouveau genre*.

* MARINÆ. SEA.

* DE MER.

X.

A. TUNICA CEPÆ. ONION-PEEL.

Tab. XI. fig. 3.

X.

PELURE D'OIGNON.

Pl. XI. fig. 3.

A. *Subrotunda plicata pellucida levis, valva planiore perforata. Tunica cepæ.* Tab. 11. fig. 3.

Lift. H. Conch. tab. 204. fig. 38. muta.—*The perforated pearl oyster.* Petiv. Mus. p. 85. No. 823.—*A. Ephippium; larger.* Penn. Brit. Zool. No. 70. tab. 62.

Ostrea perlata capite foraminoso. Buon. Ricr. p. 163. fig. 56.

Huitre. Pelure d'oignon. Argenv. Conch. I. p. 316. tab. 22. fig. C. II. p. 277. tab. 19. fig. C.—D'Avila Cab. p. 281. No. 585. 586.

Concha subrotunda una valva perforata. Gualt. I. Conch. tab. 97. fig. B. B.

Ephipium. Anomia testa suborbiculata rugoso-plicata: planiore perforata. Lin. S. N. p. 1150. No. 218.

This *species* has the habit of an oyster. It affects a roundish *shape*, but always so sinuous and depress'd inwards, that it is never perfectly regular: the sides also are dissimilar, the right being round, and the left extended into a kind of wing or flap. It is of a transparent, extremely thin, fragil, pearly *substance*, like a film or flake of the *mica* called *Muscovy glass*. The *valves* are unequal, the *upper* flat, the *under* concave; and its most general *size* is about two inches in diameter.

Cette *espece* porte la ressemblance d'une huitre. Elle est d'une *forme* quasi arrondie, mais toujours si sinueuse & comprimée interieurement, qu'elle n'est jamais parfaitement reguliere: les cotés sont aussi inegaux, le droit etant arrondi, & le gauche allongé dans une sorte d'oreille ou aile. Elle est d'une *substance* nacrée, transparente, extremement mince, & fragile, comme une membrane ou un morceau de *mica* nommé le *verre de Moscovie*. Les *valves* sont inegales, la *superieure* platte, l'*inferieure* concave; & sa *grandeur* generale est environ deux pouces en diametre.

Outside. The *upper valve* is generally depress'd inwards, but slightly, from the upper part all round to near the bottom margins; on the left side or the flap, it slopes, or is much depress'd, like a gutter. At the *top* it is *perforated* by a large oval hole, the edges of which are much turned outwards, from its beginning just over the hinge to the other extreme, which is not closed, but lies incomplete or open. This *valve* is wrought with numerous slight transverse furrows. The *under valve* is uneven, with slight furrows, bumps, &c. but is no ways rugged.

Inside smooth, and like fine mother of pearl. The *hinge* lies hidden within the concave shell, quite under the commissure, and is an ovalish cavity. In the *upper* shell it is an oval claw or hook, placed under the perforation or hole.

These *shells*, especially the concave or under ones, have their pearly hue elegantly embellished with tints or glows of fine yellow, golden, red, rosy, bronze, and burnish'd brown colours; and often the *outsides* also of the *under* shell have a rosy hue for a great space about the middle.

This *species* is found in plenty on most of our shores, *Hampshire*, *Dorsetshire*, *Devonshire*, *Kent*, *Essex*, &c. They are frequently found sticking on the common oysters.

L'*exterieur.* La *valve superieure* est generalement enfoncée en dedans, mais legerement, dès la partie superieure tout autour à près des bords inferieurs; du cotè gauche ou l'aile, elle penche, ou est beaucoup enfoncée, comme une goutiere. Au *sommet* elle est *percée* d'un grand trou ovale, les bords duquel sont beaucoup retournés en dehors, dès son commencement justement au dessus de la charniere à l'autre bout, qui n'est point fermé, mais comme incomplet ou ouvert. Cette *valve* est à sillons nombreux legers & transversals. La *valve inferieure* est inegale, par de sillons legers, des bosses, &c. mais elle n'est point rude.

Le *dedans* lisse, & comme du beau nacre. La *charniere* est cachée en dedans la valve concave, tout à fait au dessous de la commissure, & c'est une cavité ovale. Dans la valve *superieure* c'est une petite patte ovale, ou crochet, situe au dessous du trou.

Ces *coquilles*, specialement les valves concaves ou inferieures, ont leur nacre si resplendissant & embelli par des teints d'une belle couleur jaune, d'or, rouge, de rose, bronzée, & brune brunies; & souvent l'*exterieur* aussi de la valve *inferieure* a une couleur de rose pour un grand espace au milieu.

Cette *espece* se trouve en abondance sur la pluspart de nos rivages, aux comtés de *Hants*, *Dorset*, *Devon*, *Kent*, *Essex*, &c. Elles sont frequemment adherentes aux huitres communes.

XI. | XI.

A. SQUAMULA. SCALE.

A. subrotunda albescens, piscium squamas æmulans. Squamula.

A. squammula; small. Penn. Brit. Zool. No. 71.

Squamula. Anomia testa orbiculata integerrima plana margine altero gibba lævi. Lin. S. N. p. 1151. No. 221.—F. Suec. II. No. 2151.—It. Wgot. 171.

A *small species*, like the scale of a fish, and about the *size* and *shape* of that of a carp or salmon, but sometimes much larger; generally very thin, like a film or membrane. The *valves* are unequal, one flat, the other slightly concave.

Outside. The *flat* valve whitish, with a slight tinge of brownish: the *beak* rises into a very sharp point: it is thickly wrought all over with very fine transverse *striæ*, and concentric *wrinkles*. The *concave* valve is little different.

Inside is smooth, glossy, and somewhat silvery, but brownish. The *hinge* lies hidden within the shell, quite under the commissure, and is a toothless cavity. The *margins* are plain and smooth.

This *species* adheres to lobsters, crabs, shells, and sea weeds, and is found, but not very frequent, on some of our shores. I have received it from *Yorkshire, Dorsetshire, Kent*, and *Essex*.

Une *petite espece*, comme l'ecaille d'un poisson, & environ la *grandeur* & la *forme* de celle d'une carpe ou d'un saumon, mais quelquefois plus grande; generalement tres mince, comme une peau ou membrane. Les *valves* sont inegales, une platte, & l'autre peu concave.

L'exterieur. La valve *platte* blanchatre, avec quelque teint leger de brunatre clair: le *bec* s'eleve en pointe tres aigue: elle est à *stries* transversales, extremement fines & serrées, & des *rides* concentriques. La valve *concave* est tres peu differente.

Le *dedans* lisse, lustré, & quelque peu argenté, mais brunatre. La *charniere*, qui est une cavité sans dent, se trouve tout à fait dans la coquille, au dessous de la commissure. Les *bords* sont unis.

Cette *espece* s'attache aux ecrevisses, cancres, coquilles, & plantes marines, & se trouve, mais point frequemment, sur quelques unes de nos cotes. Je les ai recu des comtés de *York, Dorset, Kent*, & *Essex*.

PART SE-

PART II.

Bivalves with *equal valves*, and ſhut cloſe.

SECONDE PARTIE.

Bivalves à battans egaux, & qui ſe ferment exactement.

DIVISION I.

Thoſe that have the *hinge* ſet with numerous teeth, or are *multarticulate*.

DIVISION I.

Celles a *charniere* compoſée de dents nombreuſes, ou *à pluſieurs dents*.

GENUS IV.

GLYCYMERIS.

Of a *round ſhape*, or approaching thereto. The animal a *Tethys*.

GENRE IV.

GLYCYMERIS.

D'une *forme ronde*, ou approchante. L'animal eſt un *Tethys*.

* MARINÆ. SEA.

XII.

G. ORBICULARIS. ORBICULAR.

Tab. XI. fig. 2. 2.

* DE MER.

XII.

LA CAME FURIE.

Pl. XI. fig. 2. 2.

G. *Orbicularis craſſa ſubalbida lineis rufulis ſagittæformibus variegata, intus obſuſcata margineque crenato. Orbicularis.* Tab. 11 fig. 2. 2.

The multarticulate oyſter with a bended baſe. Grew Muſ. p. 144. tab. 12. fig. 4.

Chama glycymeris Bellon. Pectunculus ingens variegatus ex rufo. Liſt. H.Conch. tab. 247. fig. 82.

Glycymeris cornubienſis craſſa marmorata. Muſ. Petiv. p. 84. No. 816.—*Baſtard* or *Dog's cockle.* Rutty Dublin, p. 379.—*Arca. Glycymeris. Orbicular.* Penn. Brit. Zool. No. 58. tab. 58. fig. 58.

Chama glycymeris Bellon. Aquat. 408.—Rondelet de teſtac. 31. Buon. Ricr. p. 165. fig. 60. a 62.—*Came furie.* D'Avila Cab. p. 330. No. 758.

Concha craſſa, levis, ſubalbida, luteis maculis radiata, ſignata, faſciata, & virgulata, intus macula fuſca obſcurata. Gualt. I. Conch. tab. 72. fig. G.

Glycymeris. Arca teſta ſuborbiculata gibba, ſubſtriata, natibus incurvis, margine crenato. Lin. S. N. p. 1143. No. 181.—Muſ. Reg. p. 521. No. 97.

This *ſhell* is exceeding thick, heavy, and ſtrong, of an orbicular *ſhape*, very concave,

Cette *coquille* eſt extremement epaiſſe, peſante, & forte, d'une *forme* ronde, tres concave,

cave, and of a large *size*, generally about two inches and a half in diameter.

Inside smooth and glossy, mostly with a large spot of dark chesnut brown on one side, or near wholly of that colour, with a little white round the margins and hinge. The *hinge* is semicircular, and on each side set with a curve row of transverse strong *teeth*, generally from five to ten on each side; but the center is quite smooth, and has none. The *margins* on this side are very finely notch'd.

Outside, with many slight concentric transverse wrinkles, and so very finely striated longitudinally as to be hardly perceivable. The *ground* sullied white, with concentric transverse girdles made up of fine angles or lines, like a chain, or range of pyramids or arrow-heads, close set, and of a *chesnut* colour, in a zig-zag and very pretty manner.

The *beaks* run to a point, and lie just below the *commissure*; for between *them* and the *commissure* a large *triangular space* intervenes, that slopes a little, and distances them from one another. This *slope* is wrought with two rows of oblique furrows, whose ends meet in an angle, the point whereof turns towards the beaks. The *like slope* obtains in most of the *multarticulate shells*, as in *this* and the *ark genus*.

This *species* is found about *Falmouth* and other places in *Cornwall*; also on the shores of the island of *Guernsey*; and on the coasts of *Ireland*, where it is called *dog cockle*. It is a frequent inhabitant of the Mediterranean and African seas.

XIII.

concave, & d'un grand *volume*, generalement environ deux pouces & demi de diametre.

Le *dedans* lisse & lustré, communement avec une grande tache couleur de chataigne brune foncée d'un coté, ou presque totalement de cette couleur, avec un peu de blanc sur la charniere & les bords. La *charniere* est en demi cercle, & de chaque cote a une rangée courbe de *dents* fortes transversales, communement de cinq à dix de chaque cote; mais le centre est uni, & n'a aucune. Les *bords* de cet coté sont finement dentelés.

L'*exterieur* à plusieurs rides legeres & concentriques, & à stries longitudinales si fines que à peine sont elles visibles. La *robe* est blanchatre sale, à zones concentriques transversales, composées des lignes ou angles, comme une chaine, ou rangée de pyramides ou fers à fleche, tres serrés, & d'une couleur *chataigne*, en zig-zag & d'une maniere mignonne.

Les *becs* sont pointus, & precisement au dessous de la *commissure*; car entre *eux* & la *commissure* se trouve un grand *espace triangulaire*, qui penche un peu, & les ecarte l'un de l'autre. Cet espace penchant qu'on nomme *la carenne*, à deux rangs de sillons obliques, dont les bouts aboutissent en un angle, & la pointe tourne vers les becs. Cette *carenne* se trouve dans la plupart de coquilles *à plusieurs dents*, comme dans *celle ci* & le *genre* des *arches*.

Cette *espece* se trouve à *Falmouth* & autres lieux en *Cornwall*; pareillement dans l'isle de *Guernsey*; & sur les cotes d'*Irlande*, ou on l'appelle *petoncle de chien*. Elle est une espece commune des mers Mediterranée & d'Afrique.

XIII.

XIII.

G. Argentea. Silvery.
Tab. XV fig. 6. Right-hand.

XIII.

Pl. XV. fig. 6. à droite.

G. parva ſubtriangularis, lævis, intus argentea. Argentea. Tab. 15. fig. 6. *right-hand.*

Pectunculus minimus lævis, intus argenteus, cardine ſerrato. Silver cockle. Muſ. Petiv. p. 87. No. 841. & Gaz. tab. 17. fig. 9.—*Arca nucleus. Silvery.* Penn. Brit. Zool. No. 59. *Male,* tab. 58. No. 59.

Buon. Ricr. p 160. fig. 134.

Tellina inæquilatera, margine interno minutiſſime dentato, ſed prope cardinem denticulis ſpiſſis, elatioribus, acutis conſpicua, oleagina, intus argentea. Gualt. I. Conch. tab. 88. fig. R.

Nucleus. Arca teſta oblique ovata læviuſcula, natibus incurvis, margine crenulato, cardine arcuato. Lin. S. N. p. 1143. No. 184.

A *ſmall ſpecies,* about the *ſize* of a filberd kernel, of a ſomewhat triangular *ſhape,* with diſſimilar ſides, and moderately concave. *When freſh and perfect,* the *outſide* is of an olive green colour, with ſome few tranſverſe wrinkles; but when *rubb'd* or *worn,* quite white, and almoſt ſmooth. It is thick for the ſize, and ſemitranſparent. The *beaks* are pointed and ſideways, or not central; and the *bottom margins* on this ſide are plain.

The *inſide* is of a fine ſilvery ſplendor, and ſmooth: the *bottom margins* are very finely notch'd; and the *hinge* is ſemicircular, and curiouſly ſet with numerous tranſverſe ſmall teeth, like plates.

This *ſpecies* is found in great plenty in *Kent,* as at *Margate, Shelneſs,* &c. I have alſo received it from *Scarborough,* &c. in *Yorkſhire,* and from ſome others of the *Engliſh coaſts.*

Une *petite eſpece,* environ la *grandeur* d'un noyau de noiſette, d'une *forme* quelque choſe triangulaire, à cotés inegaux, & moderement concave. *Quand vivante & parfaite,* l'*exterieur* eſt d'une couleur verte d'olive, avec un petit nombre de rides tranſverſales; mais quand *uſée* ou *frottée,* tout à fait blanche, & preſque liſſe. Elle eſt epaiſſe pour ſa grandeur, & demitranſparente. Les *becs* ſont pointus & de travers, ou point central; & les *bords inferieurs* de cet coté ſont unis.

Le *dedans* d'une belle ſplendeur argentée, & liſſe: les *bords inferieurs* de cet coté ſont crenelés tres finement; & la *charniere* eſt en demi cercle, & tres curieuſement garnie de petites dents, nombreuſes, tranſverſales, & comme des lames.

Cette *eſpece* ſe trouve en grande abondance au comté de *Kent,* comme à *Margate, Shellneſs,* &c. Je les ai reçu auſſi de *Scarborough,* &c. au comté de *York,* & de quelques autres *cotes Angloiſes.*

GENUS GENRE

GENUS V.

ARCA. ARKS, or BOATS.

Of a *squarish* or *oblong shape*. The animal a *Tethys*.

GENRE V.

ARCA. ARCHE.

D'une *forme* un peu *quarrée* ou *oblongue*. L'animal est un *Tethys*.

* MARINÆ. SEA.

XIV.

A. LACTEA. WHITE.

Tab. XI. fig. 5. 5.

* DE MER.

XIV.

Pl. XI. fig. 5. 5.

A. *Parva alba cancellata.* · *Lactea.* Tab. 11. fig. 5. 5.

Pectunculus exiguus albus, admodum tenuiter striatus. List. H. Conch. tab. 235. fig. 69.

Mytulus Garnseiæ albus, parvus, tenuiter cancellatus. Petiv. Gaz. tab. 73. fig. 1.

Lactea. A. testa subrhomboidea obsolete decussatim striata diaphana, natibus recurvis, margine crenulato. Lin. S. N. p. 1141. No. 173.

A *small species*, about the *size* of a horse bean, of a squarish *shape*, very thick, strong, semipellucid, and entirely milk *white*.

The *valves* are very deep or concave. The *hinge* is set with numerous teeth, and lies nearly on a parallel strait line with the bottom; but the *sides* are dissimilar, one *rounded*, the other *oblique* and *flatted*.

Inside smooth, but not glossy; and the *margins* are plain.

Outside thickly latticed or wrought with broad prominent transverse ridges and deep furrows, which are crossed by slighter longitudinal furrows, so as to appear like a net work. The *margins* on this side are notch'd very finely. The *beaks* are broad and obtuse, extremely prominent and twirl inwards, lie lower than the com-

Une *petite espece*, environ la *grandeur* d'une feve de cheval, d'une *forme* quelque peu quarrée, tres epaisse, forte, demi-transparente, & tout à fait *blanche*.

Les *valves* sont fort profondes ou concaves. La *charniere* à dents nombreuses, est située presque sur une ligne droite & parallele avec le bord inferieur; mais les *cotés* sont inegaux, un *arrondi*, l'autre *oblique* & *plat*.

Le *dedans* lisse, mais point lustré; & les *bords* sont unis.

L'*exterieur* à treillis tres serré ou travaillé à cotes saillantes & transversales & des sillons profonds, qui sont traversés par d'autres sillons longitudinaux legers, ainsi que de paroitre comme un ouvrage en reseau. Les *bords* sont tres finement crenelés de cet coté. Les *becs* larges, obtus, fort saillants, & recourbés en dedans; ils

missure, and have the *sloping space* between them, as *usual* in most *multarticulate bivalves*.

This *species* is found in great plenty on the shores of *Guernsey* island: It is also found on our western coasts, as *Dorsetshire*, *Devonshire*, and *Cornwall*.

sont placés au dessous de la commissure, & ecartés par la *carenne*, comme *usuelle* à la plupart de *bivalves à dents nombreuses*.

Cette *espece* se trouve en tres grande abondance sur les rivages de l'isle de *Guernsey*: Pareillement sur nos cotes occidentales, comme aux comtés de *Dorset*, *Devon*, & *Cornwall*.

DIVISION.

DIVISION.

DIVISION II.

Bivalves that have *few teeth* on the *hinge*.

DIVISION II.

Bivalves à *charniere* composée d'un *petit nombre de dents*.

GENUS VI.

CARDIUM. HEART COCKLE.

Bivalves with *equal valves*, and shut close; the *sides* nearly similar, and the *beaks* nearly central, very pointed, prominent, and twirled: the *hinge* is set with two teeth near the beaks, and another remote one on each side of the shell.—The animal a *Tethys*.

GENRE VI.

CARDIUM. COEUR.

Bivalves à *battans egaux*, qui se ferment exactement; les *cotés* à peu pres egaux, & les *becs* situès presque au centre, fort pointus, saillants, & recourbès: la *charniere* à deux dents près des becs, & une autre ecartée de chaque coté de la coquille.—L'animal est un *Tethys*.

* FLUVIATILES. RIVER.

XV.

C. NUX. NUT.

Tab. XIII. fig. 2. 2.

* FLUVIATILES.

XV.

CAME DES RUISSEAUX.

Pl. XIII. fig. 2. 2.

C. *Parvum globosum viride-fuscum. Nux.* Tab. 13. fig. 2. 2.

Musculus exiguus, pisi magnitudine, rotundus, subflavus, ipsis valvarum oris albidis. List. H. An. Angl. p. 150. tit. 31 tab. 2. fig. 31

Musculus exiguus, pisi magnitudine, subrotundus. Id. App. H. An. Angl. p. 14. tit. 34. tab. 1. fig. 5. & App. H. An. Angl. in Goedart. p. 22. tit. 34. tab. 1. fig. 5.

Pectunculus subviridis parvus subglobosus. Id. H. Conch. tab. 159. fig. 14.

Pectunculus fluviatilis nostras nuciformis. Petiv. Mus. p. 86. No. 831. — Morton Northampt. p. 417.—Wallis Northumb. p. 404. No. 45. — *Tellina cornea. Horny.* Penn. Brit. Zool. No. 36. tab. 49. fig. 36.

Musculus fluviatilis, æquilaterus, levis rotundus, pisiformis, ex rubro flavescens, ipsis valvarum oris albidis. Gualt. I. Conch. tab. 7. fig. C.

Came. Argenv. Conch. I. p. 376. tab. 31. fig. 9. II. p. 331. tab. 27. fig. 9. No. 4. & Zoom. p. 76. tab. 8. fig. 10.—*Chama, globosa, glabra, cornei coloris, sulco transverso. La came des ruisseaux.* Geoffroy Coquilles de Paris, p. 133. No. 1.

Concha testa subglobosa glabra cornei coloris: sulco transversali. Lin. F. Suec. I. p. 381. No. 1336. II. No. 2138.

Cornea. Tellina globosa glabra cornei coloris: sulco transversali. Id. S. N. p. 1120. No. 72.

This *species* is very *convex*, of an ovalish *shape*, and broad, or measuring more from side to side than from hinge to bottom; the sides are similar; from the *size* of a large pea to that of a hazel nut, thin, semitransparent and brittle.

Outside covered with a thin blackish *epidermis*, which, when taken off, shews it of a deep greenish brown or dark olive *colour*, & striated very finely in a transverse manner. The *beaks*, which are sharp pointed, project above the commissure.

Inside is of a livid bluish colour, smooth, but not glossy.

The *hinge* does not strictly answer to this *genus*, for the two *teeth* near the *beak* are not very apparent; but its *habit*, *shape*, *convexity*, &c. bring it nearer the *Cardium* than any *other kind*. *Linné* has rank'd it as a *Tellina*, but certainly it does not agree with his *own definition* of that *genus*.

This *species* is found in plenty in the rivers and stagnant waters of England. According to Mr. Geoffroy, it is a *viviparous* animal.

Morton, Northampt. p. 418, proposes a *pectunculus fluviatilis exiguus, figuræ subsphæricæ compressæ*, as a *non-descript species*, of several *sizes*, from that of a lettuce-seed to a hemp-seed, of a bluish yellowish white, and thin. He found it in plenty in the bogs in *Arthingworth* and *Oxendon* fields, and in the *river Ise*. However, I think it is only a *different growth*, or a *variety*, of this species.

* * MA-

Cette *espece* est fort *convexe*, d'une *forme* quasi ovale, & large, etant plus allongée de coté à coté que du sommet au bas; les cotés sont dès la *grandeur* d'un gros pois à celle d'une noisette, mince, demi-transparente & fragile.

L'*exterieur* est couvert d'une *epiderme* mince & noiratre, qui etant depouillée, la coquille est d'une *couleur* verdatre brune ou olive foncée, & striée fort finement à travers. Les *becs*, qui sont pointus, s'elevent au dessus de la commissure.

Le *dedans* de couleur livide bleuatre, lisse, mais point lustré.

La *charniere* ne repond pas scrupuleusement à cet *genre*, car les deux *dents* près du *bec* ne sont pas fort visibles; mais sa *ressemblance*, sa *forme*, & sa *convexité*, &c. approche cette coquille, au *genre* des *Cœurs*, plus que à aucun *autre*. *Linné* l'a rangé comme une *Telline*, mais asseurement elle ne repond pas à sa *propre definition* de cet *genre*.

Cette *espece* se trouve en abondance dans les rivieres & eaux croupissantes en Angleterre. Selon M. Geoffroy c'est un animal *vivipare*.

M. Morton, dans son Hist. du Comté de Northampton, a proposé un *pectunculus fluviatilis exiguus, figuræ subsphæricæ compressæ*, comme une *espece non-decrite*, de differentes *grandeurs*, dès la semence d'une laitue à celle de chanvre, d'une couleur bleuatre jaunatre blanche, & mince. Il l'a trouvé en abondance dans les lieux marecageux à *Arthingworth* & *Oxendon*, & dans la *riviere Ise*. Cependant, je pense qu'elle est seulement d'un *age different*, ou une *varieté*, de cette espece.

* * De

** MARINÆ. Sea. / ** De MER.

A. / A.

C. Aculeatum. Spiked. / Coeur de Boeuf Epineux.

Pectunculus maximus insigniter echinatus. Wallace Orkn. p. 44.

Cardium aculeatum. Aculeated. Penn. Brit. Zool. No. 37. tab. 50. fig. 37.

Buon. Ricr. p. 172. fig. 96. 97.—Boccone Obs. Nat. p. .—*Cœur de bœuf.* Argenv. Conch. I. p. 335. tab. 26. fig. B. II. p. 298. tab. 23. fig. B.—*Cœur de bœuf Epineux.* D'Avila Cab. p. 355. No. 817.

Concha cordiformis æquilatera, umbone cardinum unito, striata, striis latis canaliculatis muricata aculeis longis & acutis, aliquando recurvis in summitate striarum positis, albida, & parvis maculis luteis obscure fasciata. Gualt. I. Conch. tab. 72. fig. A.

Aculeatum. Cardium testa subcordata: sulcis convexis linea exaratis: exterius aculeato ciliatis. Lin. S. N. p. 1122. No. 78.—*Cardium muricatum.* Id. Mus. Reg. p. 485. No. 35.

This *species* is noted as a shell of the *Orkneys*, by Dr. Wallace; since which Mr. Pennant found it off those *islands*, and also off the *Hebrides*, or Western Islands.

Mr. Pennant describes it "as *large* as "a fist; the *marginal circumference* ten "inches and a half; the *ribs* longitudinal, "each with a furrow in the middle, and "near the circumference beset with large "and strong hollow'd processes. *One side* "projects further than the other, and "forms an angle. The *colour* is yellowish "brown."

Linné observes it is very like the *next* following *species*, and perhaps, *says he*, it is only a *variety* of it.

A *sort*, not so large, but congenerous to this shell, is not unfrequently found *fossil* in the *chalk-pits* of *Kent* and *Surrey*.

Cette *espece* est rapporté etre une coquille des *Orcades*, par le Dr. Wallace; & dernierement M. Pennant l'a trouve à ces *isles*, & aussi aux *Hebrides*, ou Isles Occidentales.

M. Pennant le decrit "comme de la "*grandeur* du poing; la *circonference marginale* dix pouces & demi; les *cotes* lon"gitudinales, chacune avec une canelure "au milieu, & près de la circonference à "epines fortes; grandes & creuses. *Un* "*coté* s'etend en longeur plus que l'autre, "& forme un angle. La *couleur* est jau"natre brune."

Linné observe qu'elle ressemble beaucoup à l'*espece suivante*, & peutetre, *dit il*, c'est seulement une *varieté* de celle là.

Une *sorte*, point si grande, mais de même genre que cette coquille, se trouve asses frequemment *fossile* dans les *puits à craie* de *Kent* & *Surrey*.

XVI. / XVI.

XVI.

C. Echinatum. Thorny.

Tab. XIV. fig. 2.

XVI.

Coeur Epineux.

Pl. XIV. fig. 2.

Cardium orbiculare, coſtis circiter viginti echinatis, ſpinis hamatis. Echinatum. Tab. 14. fig. 2.

Pectunculus echinatus, concha echinata. Rondelet, Geſn. & Aldrov. Liſt. H. An. Angl. p. 188. tit. 33. tab. 5. fig. 33.

Pectunculus orbicularis fuſcus, ſtriis mediis muricatis. Liſt. H. Conch. tab. 324. fig. 161.

Cardium echinatum. Penn. Brit. Zool. No. 38.

Geſner Aquat. 131. 132.—Rondelet de Teſtac. p. 22.—Buon. Ricr. p. 171. fig. 90.—Klein Oſtrac. tab. 10. fig. 40.—*Cœur epineux.* D'Avila Cab. p. 355. No. 817.

Concha cordiformis æquilatera, umbone cardinum unito, ſtriata & muricata, muricibus brevioribus, ſubalbida, & maculis fuſcis nigricantibus depicta. Gualt. I. Conch. tab. 72. fig. B.

Concha teſta ſubrotunda, ſulcis viginti longitudinalibus: dorſo antice aculeatis. Lin. F. Suec. I. p. 382. No. 1339. II. No. 2139. Muſ. Reg. p. 486. No. 36.—Et *Echinatum. Cardium teſta, ſubcordata: ſulcis exaratis linea ciliata aculeis inflexis plurimis.* S. N. p. 1122. No. 79.

This *ſpecies*, when *old*, is extremely thick, ſtrong, heavy, and opake: when *young*, it is conſiderably leſs thick, and ſometimes even thin. Of a roundiſh *ſhape*; the ſides not perfectly ſimilar, but nearly ſo; the *valves* extremely deep or concave; and the moſt general *ſize* is that of a middling peach, or about two inches and a half each way.

Inſide ſmooth, glazed, and white, perfectly ſo in the *old thick* ſhells; but in the *young* ones it is ridged and furrowed, anſwerable to the outſide: however, the *margins* are deeply indented by the outer ribs, in both old and young ſhells.

Outſide whitiſh, with a ſtrong browniſh or ruſty tinge, eſpecially on the upper part,

Cette *eſpece*, quand *vielle*, eſt extremement epaiſſe, forte, peſante, & opaque: quand *jeune*, conſiderablement moins epaiſſe, & quelquefois même mince. D'une *forme* quaſi ronde; les cotés pas tout à fait egaux, mais preſque ſemblables; les *valves* extremement profondes ou concaves; & ſa *grandeur* la plus generale eſt celle d'une peche moyenne, ou environ deux pouces & demi en longeur & en largeur.

Le *dedans* liſſe, luſtré, & blanc, parfaitement ainſi dans les coquilles *vielles & epaiſſes*; mais dans les *jeunes* il eſt a cotes & canelures, conforme à l'exterieur: cependant, les *bords* ſont profondement dentelès par les cotes exterieures, tant dans les vielles que les jeunes coquilles.

L'*exterieur* blanchatre, avec un teint brunatre ou de rouille foncé, ſpecialement ſur

part, or round the hinge, and oftentimes has fome tranfverfe *girdles*, of a fordid ruft *colour*. It is wrought with longitudinal *ribs*, moft commonly about twenty, large, prominent, tranfverfely ftriated, and along the *middle* of *each* runs a fine ftreak or *furrow*, on which a row of fmall, thick, flattifh, hooked *thorns* or *prickles* are fet; and thefe prickles grow larger and ftronger as they approach the bottom and fides. The *interftices* between the ribs form deep furrows, which are alfo ftriated in a tranfverfe manner.

fur la partie fuperieure, ou autour de la charniere, & fouvent à *zones* tranfverfales d'une *couleur* de rouille terne. Elle eft à *cotes* longitudinales, le plus generalement environ vingt, grandes, faillantes, friées à travers, & le long du *milieu* de *chacune* a une ligne fine ou *cannelure*, qui eft garnie d'une rangée *d'epines* ou *piquants*, petits, epais, plats & courbés; ces piquants deviennent plus forts & grands comme ils approchent au bas & aux cotés. Les *intervalles* entre les cotes forment des cannelures profondes, qui font pareillement friées à travers.

The *beaks* rife to the commiffure, are ftrong, pointed, and turn inwards.

This *kind* is found in quantity on many of our Britifh fhores, as *Kent*, *Dorfetfhire*, *Cornwall*, *Yorkfhire*, the *Orkney Iflands*, &c. They are moftly *dead* and *worn* fhells, with the *prickles* damaged or *broken off*.

Les *becs* s'elevent à la commiffure, ils font forts, pointus, & tournès en dedans.

Cette *efpece* fe trouve en abondance fur plufieurs de nos cotes Britanniques, comme de *Kent*, *Dorfet*, *Cornwall*, *York*, les *Ifles Orcades*, &c. La plupart font des coquilles *mortes* & *roulées*, avec les *piquants* endommages ou *detruits*.

XVII. XVII.

C. Parvum. Small.

Cardium parvum tenue, coftis triquetris aculeatis. Parvum.

Pectunculus albus exiguus, muricibus infigniter exafperatus. Wallace Orkn. p. 44.

Pectunculus minimus triquetrus Effexienfis. Petiv. Gaz. tab. 93. fig. 11.

C. Ciliare. Fringed. Penn. Brit. Zool. No. 39. tab. 50. fig. 39.

Concha cordiformis æquilatera, umbone cardinum unito, ftriata, ftriis latis angularibus, in quarum extremitate prope peripheriam aculei totidem producuntur, candida, lineis luteis circumdata. Gualt. I. Conch. tab. 72. fig. C.

Ciliare. Cardium tefta fubcordata: fulcis elevatis triquetris: extimis aculeato-ciliatis. Lin. S. N. p. 1122. No. 80.

A *fmall fpecies*, with the habit of the above *C. Echinatum*, but I have never feen

Une *petite efpece*, avec l'apparence du *Cœur Epineux* fufdit, mais je ne l'ay jamais

seen it larger than a nutmeg; of a roundish *shape*; the *valves* very concave or deep; and the *shell* is thin, brittle, and semitransparent.

Inside whitish, sometimes with a glance of rust, hardly glossy, and wrought with ridges and furrows answerable to the outside ones.

Outside wrought with about fifteen longitudinal *ribs*, broad and prominent, transversely striated, and somewhat triangular, rising sharp or into a ridge in the middle, which is set with a row of small prickles. The *furrows* or interstices between the ribs are also transversely striated. The *colour* whitish, but at the top and sides generally tinged faint brown, and also faint brown regular transverse *girdles* are oftentimes seen on it.

This *kind* is found on several of our coasts. I have received it from *Cornwall*, *Dorsetshire*, and *Devonshire*. Mr. Petiver notes it about *Maldon*, and in the *Hundreds of Essex*; and Dr. Wallace as a shell of the *Orkneys*.

jamais vu plus grande q'une noix muscade; d'une *forme* quasi ronde; les *valves* sont fort profondes ou concaves; & la *coquille* est mince, fragile, & demi-transparente.

Le *dedans* blanchatre, quelquefois teint de rouille, à peine lustré, & à cotes & canelures conformes à celles de l'exterieur.

L'*exterieur* à environ quinze *cotes* longitudinales, larges & saillantes, striées à travers, & quelque chose triangulaires, car elles s'elevent aigue ou en vive arrete au milieu, qui est garni d'une rangée de petits piquants. Les *canelures* ou intervalles entre les cotes sont pareillement striés à travers. La *couleur* est blanchatre, mais au sommet & aux cotes communement teinte de brun pale, & aussi souvent à *zones* transversales regulieres brun pale.

Cette *espece* se trouve sur plusieurs de nos cotes. Je les ai reçu dès comtés de *Cornwall*, *Dorset*, & *Devon*. Mr. Petiver dit aux environs de *Maldon*, & dans les *Cantons de Essex*; & le Dr. Wallace les rapporte comme une coquille des *Orcades*.

XVIII.

C. Lævigatum. Smooth.

Tab. XIII. fig. 6. 6.

XVIII.

Pl. XIII. fig. 6. 6.

Cardium obovatum striis obsoletis longitudinalibus. Levigatum. Tab. 13. fig. 6. 6.

Pectunculus maximus, et minus concavus; pluribus minutioribus & parum eminentibus striis donatus, rostro acuto, minusque incurvato. List. H.An.Angl. p. 187. tit. 32. tab. 5. fig. 32.

Pectunculus subfuscus striis leviter tantum incisis. List. H. Conch. tab. 332. fig. 169.

Pectunculus major striis angustis. Petiv. Gaz. tab. 93. fig. 10. Wallace Orkn. p. 44.—*Large high-beaked cockle.* Wallis Northumb. p. 395.—*C. Lævigatum. Smooth.* Penn. Brit. Zool. No. 40. tab. 51. fig. 40.

Levigatum.

Lævigatum. Cardium testa obovata, striis obsoletis longitudinalibus. Lin. S. N. p. 1123. No. 88.—Mus. Reg. p. 490. No. 44.

The *shell* is thick, heavy, and strong, of a suboval *shape*, about the *size* of a large pippin, or two inches and a half in diameter; the *valves* deep or concave, and the sides dissimilar.

Inside smooth, glossy, and pure white, but sometimes with a very faint reddish glance; and the *margins* are finely dentated.

Outside. The *epidermis* thin and blackish, under it, the *shell* is very sleek, and whitish, with a pretty strong glance of reddish. It is wrought with numerous longitudinal *striæ*, close set, but so slightly prominent as to be near obsolete, for only towards the bottom they shew themselves more strong, and greatly dentate the *margins.* The *interstices* between the striæ are furrows that seem transversely striated, and several *concentric wrinkles* run across the shell from top to bottom.

The *beaks* are pointed, turn inwards, and just over-top the commissure.

This *species* is found on several of our shores, as *Yorkshire, Northumberland, Dorsetshire, Cornwall, Carnarvenshire,* &c. also on the shores of the *Orkneys*; but, however, it is not a *common* shell.

La *coquille* est epaisse, pesante, & forte, d'une *forme* quasi ovale, environ *le volume* d'une grande pomme renette, ou deux pouces & demi en diametre; les *valves* profondes ou concaves, & les cotés inegaux.

Le *dedans* lisse, lustré, & blanc comme niege, mais quelquefois avec un teint roussatre leger; & les *bords* sont finement dentelès.

L'exterieur. *L'epiderme* mince & noiratre; au dessous d'elle, la *coquille* est fort unie, & blanchatre, avec un teint asses fort de roussatre. Elle est à *stries* longitudinales nombreuses, tres serrées, mais si peu marquées ou elevées que d'etre presque effacées, car seulement vers le bas elles se montrent plus vives, & dentelent tres fortement les *bords.* Les *intervalles* entre les stries sont des canelures qui paroissent striés à travers, & plusieurs *rides concentriques* passent à travers la coquille du sommet au bas.

Les *becs* sont pointus, tournès en dedans, & dejettent à poinc nommé la commissure.

Cette *espece* se trouve sur plusieurs de nos cotes, comme aux comtés de *York, Northumberland, Dorset, Cornwall, Carnarvon,* &c. aussi sur les rivages des *Orcades*; mais, cependant, elle n'est pas une coquille *commune.*

XIX.

C. VULGARE. COMMON.

Tab. XI. fig. 1. 1.

XIX.

PETONCLE COMMUN.

Pl. XI. fig. 1. 1.

Pectunculus vulgaris, albidus, subrotundus, circiter viginti-sex striis majusculis at planioribus donatus. Vulgaris. Tab. 11. fig. 1. 1.

Pectunculus vulgaris, albidus, rotundus, circiter viginti-sex striis majusculis at planioribus donatus. The Cockle. List. H. An. Angl. p. 189. tit. 34. tab. 5. fig. 34.

Pectunculus striatus vulgaris. Cockles Anglice dictus. List. Exerc. Anat. 3. p. 20. tab. 3. fig. 1. 2. 3.

Pectunculus capite minore, rotundiore, & magis æquali margine. List. H. Conch. tab. 334. fig. 171.

Cockles. Leigh. Lancash. p. 134.—Dale Harw. p. 387. 1.—Borlase Cornw. p. 274.—Wallis Northumb. p. 394.—Wallace Orkn. p. 45.—Martin W. Isles, p. 343.—Smith Cork, p. 318.—State of Down, p. 239.—Rutty Dublin, p. 378.

Pectunculus maritimus nostras edulis vulgatissimus. The common cockle. Mus. Petiv. p. 86. No. 835.—*Cardium edule. Edible.* Penn. Brit. Zool. No. 41. tab. 50. fig. 41.

Concha cordiformis æquilatera, umbone cardinum unito, striata striis crassis, elatis, subrotundis, quatuor lineis fuscis circumdata. Gualt. I. Conch. tab. 71. fig. F.

Concha testa subrotunda; sulcis viginti-sex longitudinalibus, tribus transversalibus. Lin. F. Suec. I. p. 383. No. 1340. II. No. 2141.—*Edule. Cardium testa antiquata, sulcis viginti-sex obsolete recurvato-imbricatis.* Id. S. N. p. 1124. No. 90.

This *shell* is thick and strong; the most general *size* is that of a walnut, but sometimes double as large; of a roundish *shape*; the *valves* very concave, or deep; and the *sides* dissimilar, one being rounded, the other lengthened to almost an angle.

Inside, pure or milk white, except that the lengthened side has a large space of violet livid deep brown, which extends over the hinge on that side; quite smooth and glossy, for it has no marks of the outside ribs, but the *margins* are deeply toothed.

Outside has frequently a thin blackish *epidermis*. The shell is of a whitish *colour*, gene-

Cette *coquille* est epaisse & forte; la *grandeur* la plus generale est celle d'une noix, mais quelquefois du double cet volume; d'une *forme* quasi ronde; les *valves* fort concaves, ou profondes; & les *cotés* inegaux, un etant arrondi, l'autre etendu à presque un angle.

Le *dedans* blanc pur, ou comme niege, excepté que le coté allongé a un grand espace de couleur violette livide brune foncée, qui s'etend sur la charniere de cet coté; tout à fait lisse & lustré, car il ne porte aucunes marques des cotes exterieures, mais les *bords* sont fort grandement dentelés.

L'*exterieur* se trouve frequemment couvert d'une *epiderme*, mince & noiratre. La coquille

generally with a rusty hue, and is wrought with about twenty-six strong longitudinal *ribs*, close set, broad, flatted, and striated transversely, by which striæ they are notched or divided into knobs. The *interstices* between them are slight furrows.

coquille est d'une *couleur* blanchatre, generalement avec un teint de rouille, & à environ vingt-six *cotes* fortes & longitudinales, tres serrées, larges, applaties, & striées à travers, & par les stries elles sont crenelées ou decoupées en petites tubercules. Les *intervalles* entre les cotes sont des canelures legeres.

The *margins* are deeply toothed by the ribs, and the shell is set with several strong concentric transverse wrinkles. The *beaks* are pointed, turn inwards, and over-top the commissure.

Les *bords* sont profondement dentelés par les cotes, & la coquille est traversée de plusieurs rides concentriques. Les *becs* sont pointus, tournés en dedans, & dejettent la commissure.

The *cockle* is found in great plenty on all the *coasts* of *Great Britain* and *Ireland*, more especially on the *sandy shores*. They are sold in great quantities in London and other towns, and are much esteemed as a wholesome and palatable food. The *season* for them is from Autumn to Spring. The cockles from *Selsea*, near *Chichester*, in *Sussex*, are accounted the most delicious in England.

Le *petoncle commun* se trouve en tres grande abondance sur toutes les *cotes* de la *Grande Bretagne* & de l'*Irlande*, plus specialement sur les *rivages sablonneux*. Ils sont vendus en grande quantité à Londres & les autres villes, & tres estimés comme une nourriture salutaire & agreable. Leur *saison* est dès l'Automne au Printems. Les petoncles de *Selsea*, près de *Chichester*, en *Sussex*, sont reputès les plus delicieux de l'Angleterre.

XX. XX.

C. Carneosum. Flesh-coloured.

Cardium parvum subrotundum oblique striatum colore carneoso. Carneosum.

Concha parva subrotunda, ex parte interna rubens. List. H. An. Angl. p. 175. tit. 25. tab. 4. fig. 25.

Tellina æquilatera levis, tenuis subrubra. Gualt. I. Conch. tab. 77. fig. 1.

Carnaria. Tellina testa suborbiculata lævi utrinque incarnata oblique striata: striis [illegible] reflexis. Lin. S. N. p. 1119. No. 66.

A *small species*, about the *size* of an apricot stone, or one inch both ways; of a roundish *shape*, and compress'd or *flattish*, for the *valves* are rather shallow; the *sides* nearly

Une *petite espece*, environ la *grandeur* du noyau d'un abricot, ou un pouce & demi en longueur & largeur; d'une *forme* quasi ronde, & comprimée ou *applatie*, car les valves,

nearly ſimilar, and the *beaks* ſharp pointed and central.

The *ſhell* is pretty thick for its ſize, and ſemitranſparent.

Outſide moſt generally white, with a fine tinge of pale roſe or fleſh *colour* in the middle; others, the ſmalleſt or *young* ſhells, are entirely of a fine purple roſe or fleſh *colour*; and ſome others have a bright *yellow liſt* round the margins. It is wrought all over with extreme fine delicate *ſtriæ*, like the fineſt ſtrokes of engraving. Theſe *ſtriæ* are numerous, cloſe ſet, and run the length of the ſhell, not in ſtrait, but very *oblique lines*, and diverge ſo much on the ſides as almoſt to be *tranſverſe* to thoſe on the body. On one ſide it has a tendency to a ſlope or *flexure*, like the *tellens*, but it is ſcarcely perceivable without much attention.

Inſide ſmooth and very gloſſy, of the ſame *colours*, but in much ſtronger ſhades.

Dr. Liſter acquaints us it is a *litoral* or *ſhore ſhell*, and found very frequently in the ſhallows of *Lancaſhire*, and near *Filey*, &c. in *Yorkſhire*; I have alſo received it from *Scarborough*, and other places in that county, and likewiſe in plenty from *Devonſhire* and *Cornwall*.

GENUS

valves ſont plutot peu profondes; les *cotés* à peu pres egaux, & les *becs* fort pointus & centrals.

La *coquille* eſt aſſes epaiſſe pour ſa grandeur, & demitranſparente.

L'*exterieur* plus generalement blanc, avec un beau teint de *couleur* roſe ou de chair au milieu; d'autres, les plus petites ou les *jeunes* coquilles, ſont totalement d'une belle *couleur* pourpre roſe ou de chair; & quelqu'unes ont une *bordure jaune* eclatante, qui environne les bords. Elle eſt entierement à *ſtries* fines & belles, comme les plus beaux traits de graveure. Ces *ſtries* ſont nombreuſes, ſerrées, & courent la longeur de la coquille, pas en *lignes* droites, mais fort *obliques*, & s'ecartent tant vers les cotés à preſque devenir *tranſverſales* à celles du corps. D'un coté elle a une diſpoſition à un pli ou *ſinuoſité*, comme les *tellines*, mais il eſt à peine apperceu ſans beaucoup d'attention.

Le *dedans* liſſe & fort luſtré, de mêmes *couleurs*, mais plus vives ou foncées.

Le Dr. Liſter nous apprend qu'elle eſt une *coquille* des *rivages*, & ſe trouve tres frequemment dans les bas fonds au comté de *Lancaſter*, & près de *Filey*, &c. au comté de *York*; je les ai auſſi reçu de *Scarborough*, & autres lieux du dit comté, & pareillement des comtés de *Devon* & *Cornwall* en abondance.

GENRE

GENUS VII.

PECTUNCULUS. COCKLE.

Bivalves with *equal valves*, and shut close; the *sides* very dissimilar; the *beaks* not central, but greatly turned towards one side: the *hinge* is set with three teeth, two near to each other; the third is divergent from the beaks.—The animal a *Tethys*.

GENRE VII.

Bivalves à *battans egaux*, & se ferment exactement; les *cotés* sont fort inegaux; les *becs* point central, mais beaucoup tournès vers un coté: la *charniere* à trois dents, deux proche de l'une l'autre; la troisieme s'ecarte des becs.—L'animal est un *Tethys*.

* MARINÆ. SEA.

XXI.

P. CRASSUS. THICK.

Tab. XIV. fig. 5.

* DE MER.

XXI.

Pl. XIV. fig. 5.

PECTUNCULUS *major crassus, albo castaneus. Crassus.* Tab. 14. fig. 5.

Concha é maximis, admodum crassa, rotunda, ex nigro rufescens. List. H. An. Angl, p. 173. tit. 22. tab. 4. fig. 22.

Pectunculus maximus, subfuscus, valde gravis. List. H. Conch. tab. 272. fig. 108.

Wallace Orkn. p. 42. 44.—*Pectunculus maximus crassus nostras nigricans.* Mus. Petiv. p. 86. No. 832.—*Large, round, thick and tawney conch.* Wallis Northumb. p. 396. No. 12.

Venus mercenaria. Commercial. Penn. Brit. Zool. No. 47. tab. 53. fig. 47.

Chama inæquilatera, levis, crassa, subalbida. Gualt. I. Conch. tab. 85. fig. B.

A very thick, strong, heavy and opake *shell*; *size* of a very large orange, or three inches and a half long, and somewhat broader; of a roundish *shape*; the *valves* very concave or deep, and their *sides* dissimilar, one being considerably lengthened, while the other is foreshortened and rounded.

Outside covered with a strong brownish black and very stringy or fibrous *epidermis*, under which the *shell* is whitish, with large shades of a chesnut or russet *colour*, and wrought

Une *coquille* tres epaisse, forte, pesante & opaque; de la *grandeur* d'une tres grande orange, ou trois & demi pouces en longeur, & quelque peu plus large; d'une *forme* quasi ronde; les *valves* fort concaves ou profondes, & leurs *cotés* inegaux, un considerablement allongé, pendant que l'autre est raccourci & arrondi.

L'*exterieur* couvert d'un *epiderme* fort brunatre noir & tres fibreux ou cordé, au dessous de laquelle la *coquille* est blanchatre, avec des grandes nuances de *couleur* de chataigne

wrought all over with numerous ſtrong tranſverſe *ſtriæ* and *wrinkles*, cloſe ſet. The *beaks* are ſtrong and prominent, but not higher than the commiſſure, pointed, and turn ſideways. The *margins* are plain.

Inſide pure white, but ſeldom gloſſy. The *margins* plain.

This *ſpecies* is common on ſeveral of the Britiſh coaſts, as *Northumberland, Yorkſhire, Lancaſhire, Dorſetſhire, Carnarvonſhire* and other ſhores of Wales, *Aberdeenſhire* and the iſlands of *Orkney*, &c. in Scotland.

OBS.—*Linné S. N. p.* 1131, *No.* 123, has quoted this *ſhell* of Liſter H. An. Angl. for his *Venus Mercenaria*, which is an *American ſhell* of a quite *diſtinct ſpecies:* he indeed ſeems not to have this *kind* in his *Syſtem*, unleſs perhaps it be his *V. Iſlandica. Ib. No.* 124; but that is doubtful. Mr. *Pennant* has alſo followed this *Linnean error*, by propoſing the *North-American clam* or *wampum ſhell* as the ſame *ſpecies* with *this*.

chataigne ou brunatre, à *ſtries* & *rides* nombreuſes, fortes, tranſverſales, & ſerrées. Les *becs* forts & bombés, mais ne dejettent la commiſſure, pointus, & tournès de coté. Les *bords* ſont unis.

L'*interieur* blanc pur, mais rarement luſtré. Les *bords* unis.

Cette *eſpece* eſt commune ſur pluſieurs des cotes Britanniques, comme aux comtés de *Northumberland, York, Lancaſter, Dorſet, Carnarvon* & autres cotes de la principauté de Galles, au comté de *Aberdeen* & les iſles *Orcades*, &c. en Ecoſſe.

OBS.—*Linné S. N. p.* 1131, *No.* 123, a cité cette coquille de Liſter H. An. Angl. pour ſon *Venus Mercenaria*, qui eſt une *coquille Americaine* d'une *eſpece* totalement *diſtincte:* il paroit, en verité, n'avoir point cette *ſorte* dans ſon *Syſteme*, ſinon peutetre qu'elle ſoit ſon *V. Iſlandica. Ib. No.* 124; mais c'eſt douteux. M. *Pennant* a auſſi ſuivi cette *erreur Linnéene*, en propoſant la *coquille clam* ou *wampum* de l'*Amerique Septentrionale* comme la méme *eſpece* que *celle-ci*.

XXII.

P. GLABER. SMOOTH.

Tab, XIV. fig. 7.

XXII.

Pl. XIV. fig. 7.

Pectunculus major craſſus, politus, caſtaneus, lucide radiatus. Glaber. Tab. 14. fig. 7.

Pectunculus maximus craſſus, lævis fere radiatus. Muſ. Petiv. p. 86. No. 833.—*Curviroſtrum.* Leigh Lancaſh. tab. 3. fig. 5.

Chama inæquilatera, rugis tranſverſis circumdata, ex rubro & albo tenuiter, & lucide lineata, & radiata, intus candida. Gualt. I. Conch. tab. 86. fig. A.

Chione. Venus teſta cordata tranſverſe ſubrugoſa lævi, cardinis dente poſteriori lanceolato. Lin. S. N. p. 1131. No. 125.—Muſ. Reg. p. 500. No. 58.

The

La

The *shell* is thick, strong, and heavy; *size* of an orange, or from two and a half to three inches long; of a roundish *shape*; rather *compressed*, for the *valves* are not very concave; and the *sides* are dissimilar.

Outside extremely smooth, glossy, and as if polished, with numerous fine concentric transverse wrinkles; of a light chesnut *colour*, with many broad and narrow longitudinal rays of a darker shade; the *margins* plain; the *beaks* pointed, and turned sideways; the *slopes*, one with a strong cartilage, the other with a long pointed oval depression.

Inside milk white finely glazed, and the *margins* plain.

This *species* is rare in England. I found it at *Mount's Bay*, in *Cornwall*, where the fishermen told me they called it *queen fish*; it is also found near *Fowey* and other shores of that county. I have seen some from *Weymouth*, and Mr. Petiver received it from the *island* of *Purbeck*, in *Dorsetshire*. Dr. Leigh mentions that it is got on the coasts of *Cheshire*.

Linné says, it is an Asiatic shell, and perhaps (adds he) an European one.

La *coquille* est epaisse, forte & pesante; de la *grandeur* d'une orange, ou de deux & demi à trois pouces en longeur; d'une *forme* quasi ronde; plutot *comprimée*, car les *valves* ne sont point fort concaves; & leurs *cotés* sont inegaux.

L'*exterieur* extremement uni, lustré, & comme poli, à *rides* fines nombreuses, concentriques & transversales; d'une *couleur* de chataigne, à plusieurs rayés longitudinales larges & etroites, d'une nuance plus foncée; les *bords* unis; les *becs* pointus, & tournès de coté; les *pentes*, une munie d'un cartilage fort, l'autre avec un enfoncement ovale, long & pointu.

Le *dedans* blanc de lait parfaitement bien lustré, & les *bords* sont unis.

Cette *espece* est rare en Angleterre. Je l'ai trouve à *Mount's Bay*, en *Cornwall*, ou les pescheurs m'ont dit qu'ils l'appelloit le *queen fish*, ou *coquille reine*; elle se trouve pareillement près de *Fowey*, & autres rivages de cet comté. Je les ai vu de *Weymouth*, & M. Petiver rapporte l'avoir reçu de l'*isle* dé *Purbeck*, au comté de *Dorset*. Le Dr. Leigh dit qu'elle se trouve sur les cotes du comté de *Chester*.

Linné dit, c'est une coquille Asiatque, & peutetre (ajoute il) aussi une Europeenne.

XXIII. P. Strigatus. Ridged.

Tab. XII. fig. 1. 1.

XXIII. Clonisse.

Pl. XII. fig. 1. 1.

Pectunculus crassissimus strigatus, strigis ex latere bullatis. Strigatus. Tab. 12. fig. 1. 1.

Pectunculus omnium crassissimus, fasciis ex latere bullatis donatus. List. H. Concli. tab. 284. fig. 122.

Cornwall heart cockle with rugged girdles. Petiv. Gaz. tab. 93. fig. 17.

Concha

Concha cinerea densa, margine dentato, striis rugosis & è lateribus undose tuberculosis. The wrinkled, notched, and high-beaked concha or cockle. Borlase Cornw. p. 278. tab. 28. fig. 32.

Venus Erycina. Sicilian Penn. Brit. Zool. No. 48. tab. 54. fig. 48.

Concha marina valvis æqualibus æquilatera, notabiliter umbonata, & oblique incurvata, subrotunda, vulgaris, striis circularibus profundis, elatis, bullatis exasperata, & circumdata, crassa, subalbida. Gualt. I. Conch. tab. 75 fig. H.

Clonisse de la Mediterranée. D'Avila Cab. p. 333. No. 762.

Concha testa subrotunda; sulcis profundis viginti, margine crenato. Lin. F. Suec. I. p. 383. No. 1341.

Verrucosa. Venus testa subcordata: sulcis membranaceis striatis reflexis, antice imprimis, verrucosis, margine crenulato. Lin. S. N. p. 1130. No. 116.

The *shell* is extremely thick, strong, heavy and opake; about the *size* of a large apricot, or one and three quarters of an inch long; of a roundish *shape*; the *valves* moderately concave; and the *sides* dissimilar, but not greatly so.

Outside whitish, but always very dirtied or soiled, sometimes with a *reddish* or rusty glance. It is wrought with very regular concentric transverse *ridges*. These *ridges* are numerous and close set, thick, prominent, rugged, strong, chapped or notched, and scooped or hollowed on the fore part or towards the hinge, but towards the sides of the shell are so greatly cut as to form rows of large thick separate knobs. The *interstices* between them are narrow deep furrows. The *margins* are plain.

The *beaks* are pointed, and turn sideways. The *cartilage slope* is a long slanting hollow, and has a narrow short cartilage on its upper part; the other *slope* has a strong convex heart-like mark.

Inside smooth, glazed, and pure *white*, but sometimes on the edge of the longer side

La *coquille* est extremement epaisse, forte, pesante & opaque; environ le *volume* d'un grand abricot, ou un & trois quarts d'un pouce en longueur; d'une *forme* quasi ronde; les *valves* moderement profondes; & les *cotés* inegaux, mais pas beaucoup.

L'*exterieur blanchatre*, mais toujours fort sale ou terne, quelquefois avec un teint *rougeatre* ou de rouille. Il est à *sillons* reguliers, concentriques & transversals. Ces *sillons* sont nombreux & serrés, epais, en vive arrete, raboteux, forts, crenelés, & creux en devant ou envers la charniere, mais vers les cotés de la coquille ils sont si grandement decoupés que de former des rangées de tubercules, separés, gros & epais. Les *intervalles* entre eux sont des cancelures etroites & profondes. Les *bords* sont unis.

Les *becs* pointus, & tournés de coté. La *pente* du *cartilage* est un creux oblique & long, avec un cartilage, etroit & court, sur sa partie superieure; l'autre *pente* à une marque convexe & forte, en forme de cœur.

Le *dedans* lisse, lustré, & *blanc* pur, mais quelqufois sur le bord du coté allongé à un

side it has a space of a violet *colour*. The *margins* are finely or delicately crenated.

This *species* is rare in our seas. The shores of *Cornwall* afford them, and they have been got in *Devonshire* and *Dorsetshire*. I am also informed that they are found on the *eastern coast* of *Sussex*, but not frequently.

This species is likewise a native of the Mediterranean sea.

un espace de *couleur* violette. Les *bords* sont finement & delicatement crenelés.

Cette *espece* est rare dans nos mers. Les cotes de *Cornwall* nous en fournit, & elles ont eté peschées aux comtés de *Devon* & *Dorset*. Je suis aussi informé qu'elles se trouvent sur la *cote orientale* du comté de *Sussex*, mais point frequemment.

Cette espece est pareillement native de la mer Mediterranée.

XXIV.

P. Capillaceus. Hair-Streaked.
Tab. XII. fig. 5. 5.

XXIV.

Pl. XII. fig. 5. 5.

Pectunculus planus, crassus, striis capillaceis dense striatus. Capillaceus. Tab. 12. fig. 5. 5.

Pectunculus rostro productiore, capillaceis fasciis donatus. List. H. Conch. tab. 290. fig. 126.—*P. dense fasciatus, ex rubro variegatus & undatus.* Tab. 291. fig. 127.—*P. crassus, dense fasciatus, leviter ex rufo variegatus.* Tab. 292. fig. 128.—*P. subfuscus tenuiter admodum fasciatus.* Tab. 293. fig. 129.—*P. planus, crassus, ex rufo radiatus.* Tab. 299. fig. 136.

Lunata alba rotunda, dense fasciata. Petiv. Gaz. tab. 93. fig. 18.—*L. fasciata undis rubris.* Tab. 93. fig. 15.—Et *L. Orcadensis alba, fasciis capillaceis.* Tab. 76. fig. 1.

Venus exoleta. Antiquated. Penn. Brit. Zool. No. 49. tab. 54 & 56. fig. 49 A. & 49.

Concha marina valvis æqualibus æquilatera, notabiliter umbonata & oblique incurvata, subrotunda, vulgaris, striis densissimis & profundis transversim striata & exasperata, candida, leviter ex fusco variegata, & radiata. Gualt. I. Conch. tab. 75. fig. F.

Exoleta. Venus testa lentiformi transversim striata pallida, obsolete radiata, ano cordato. Lin. S. N. p. 1134. No. 142.—*Concha testa subrotunda: striis transversis innumeris, margine levi.* Id. F. Suec. I. p. 383. No. 1342. II. No. 2145.—Id. Mus. Reg. p. 506. No. 70.

The *shell* is very thick, strong and heavy; from half to the full *size* of an apricot, or about one and a half inch in length; of a round *shape*; sometimes much flatted, sometimes near globose, by the shallowness or depth of the *valves*, whose *sides* are nearly similar.

La *coquille* est tres epaisse, forte & pesante; de moitie, au juste *grandeur* d'un abricot, ou environ un & demi pouce en longeur; d'une *forme* ronde; quelquefois asses comprimée, quelquefois bombée, par le peu de profondeur ou la concavite des *valves*, dont les *cotés* sont presque egaux.

Outside whitish; sometimes sullied ash colour, at other times slightly russety; and many are also very prettily variegated with regular broad rays of a russet *colour*, that run from top to bottom. It is wrought all over with regular concentric transverse *striæ*, very numerous and close set, fine as threads or hairs, and sharp or thin. The *interstices* between them are linear furrows. The *beaks* very sharp pointed, and turned sideways. The *cartilage slope* and its cartilage so small as to be hardly observable, but the other *slope* is strongly impressed with a cordiform mark. The *margins* are plain.

Inside milk white, smooth and glazed. In numbers of shells it is remarkably thick, as if a new coat was superadded, and has a tongue-like thick prominent space, that stretches from the margin to the middle, as *Linné* has observed in his *Fauna Suecica*. The *margins* are quite plain.

This *species* is found, and in some plenty, on several of the shores of *Great Britain*, as *Cornwall*, *Dorsetshire*, *Devonshire*, *Yorkshire*, &c. in plenty in the *isle of Guernsey*, and also in the *Orkney islands*.

L'exterieur blanchatre; terni de couleur de cendre, tantot quelque peu roussatre; & plusieurs sont joliment variées par des rayes larges & regulieres, *brunes*, qui courent du haut en bas. Elle est à *stries* regulieres concentriques transversales, fort nombreuses, & serrées, fines comme des fils ou cheveux, & aigues ou minces. Les *intervalles* entre elles sont des canelures comme des lignes. Les *becs* sont fort pointus, & tournès de coté. La *pente du cartilage* & son cartilage si petit que a peine est il visible, mais l'autre *pente* est fortement empreinte d'une marque en forme de cœur. Les *bords* sont unis.

Le *dedans* blanc de lait, lisse & lustré, & dans un grand nombre de coquilles il est tres remarquable en epaisseur, comme si un surcroit d'epaisseur y avoit eté ajouté, & avec un espace epais & elevé, comme une languette, qui s'etend du bord au milieu, comme *Linné* a remarqué dans son *Fauna Suecica*. Les *bords* sont tout à fait unis.

Cette *espece* se trouve, & en quelque abondance, sur plusieurs des cotes de la *Grande Bretagne*, comme aux comtés de *Cornwall*, *Dorset*, *Devon*, *York*, &c. en quantité dans l'*isle de Guernsey*, & pareillement dans les *Orcades*.

XXV.
P. FASCIATUS. FASCIATED.
Tab. XIII. fig. 3.

XXV.
Pl. XIII. fig. 3.

Pectunculus parvus, planior, crassus, dense fasciatus. Fasciated. Tab. 13. fig. 3.

Pectunculus orbicularis planior, rugosus. Lift. H. Conch. tab. 281. fig. 119.—Et *Pectunculus fuscus dense fasciatus, elegant quadam pictura undulata insignitus.* Tab. 282. fig. 120.

A Une

A *small species*, about the *size* of a filberd, or three quarters of an inch long, of a roundish *shape*, pointed towards the top. The *shell* is thick and strong, semiperspicuous, and compressed or *flattish*, for the *valves* are shallow, and their *sides* very dissimilar.

Outside of a whitish or reddish *ground*, variegated in a very pretty manner with broadish *rays* of purple red from top to bottom, and also oftentimes with long *streaks* of purple, irregularly placed, all over the shell. It is wrought with about a dozen concentric and regular transverse *girdles*, not much raised, but smooth and flatted, broad, and augment in breadth from the top to the bottom, where the latter four become remarkably broad. The *interstices* between them are deep furrows. The *beaks* very pointed, and turned sideways. The *cartilage slope* small; the *other* also small, with a suboval depression. The *margins* are quite plain.

Inside smooth, glossy and white. The *margins* plain.

This *kind* is found in *Cornwall*, and the other *western coasts*. I am informed it is also found near *Bangor*, among the rocks, from *Bangor Ferry* to *Anglesea*, in *Wales*.

Obs.—In habit it has some resemblance to the next, or *Vetula*, and congenerous to it, but nevertheless is a very distinct species.

Une *petite espece*, environ la *grandeur* d'une noisette, ou trois quarts d'un pouce en longueur, d'une *forme* quasi ronde, pointue vers le haut. La *coquille* est epaisse & forte, demitransparente, & comprimée ou *applatie*, car les *valves* sont peu profondes, & leurs *cotés* sont fort inegaux.

L'exterieur. La *robe* blanchatre ou rougeatre, variée d'une tres jolie maniere par des *rayes* asses larges de pourpre rouge du haut en bas, & aussi souvent avec des *lignes* pourpres, irregulierement placées, par tout la coquille. Il est travaillé avec environ douze *zones* regulieres concentriques & transversales, point tres elevées, mais lisses & applaties, larges, & deviennent plus larges du haut en bas, ou les quatre dernieres ont une largeur remarquable. Les *intervalles* entre elles sont en canelures fort profondes. Les *becs* tres pointus, & tournès de coté. La *pente du cartilage* petite; l'*autre* aussi petite, avec un enfoncement quasi ovale. Les *bords* sont tout à fait unis.

Le *dedans* lisse, lustré & blanc. Les *bords* unis.

Cette *sorte* se trouve en *Cornwall*, & autres *cotes occidentales*. Je suis informé qu'elle se trouve près de *Bangor*, entre les rochers, du *Bac de Bangor* à *Anglesea*, dans la principauté de *Galles*.

Obs.—En apparence elle a quelque rapport à la prochaine, ou *Vielle ridée*, & est d'affinité, mais cependant c'est une espece fort differente.

XXVI.

P. Vetula. Old Wife.

Tab. XIII. fig. 5. 5.

XXVI.

Vielle Ridee.

Pl. XIII. fig. 5. 5.

Pectunculus variegatus gravis, fasciis latis & pulvinatis conspicuus. Vetula. Tab. 13. fig. 5. 5.—List. H. Conch. tab. 279. fig. 116.

Rumph. Mus. tab. 48. fig. 5.—*Vielle ridée. Vetula.* Argenv. Conch. I. p. 324. tab. 24. fig. B. II. p. 286. tab. 21. fig. B. Gualt. I. Conch. tab. 85. fig. A.—*Paphia. Venus testa subcordata, rugis incrassatis, pube rugis attenuatis, labris complicatis.* Lin. S. N. p. 1129. No. 113.

The *shell* is extremely thick and strong, *size* of a small walnut, of a roundish *shape*, and rather *compressed*, for the *valves* are not very deep or concave, and their *sides* are dissimilar.

Outside. The *ground* white, thickly and prettily *mottled* with small *spots* and *angled lines*, like arrowheads, of a *bay brown* colour. It is wrought with thick or swell'd and broad concentric transverse *ridges*, which strongly dentate the edge of the cartilage slope. The *interstices* between them are broad furrows. The *beaks* sharp pointed, and turn greatly sideways. The *cartilage slope* large and concave; the other *slope* is marked with a cordiform depression.

Inside glazed, smooth and white, except under the cartilage slope, where is a large *space* of a *bay brown colour*.

This *species*, which is an inhabitant of the Mediterranean, has been sometimes fish'd up on our *western shores*. The *Patronage* of Dr. *Fothergill* granted me the inspection of these shells, from his *Elegant* and *Scientific Collection*, and also of an *extraordinary lusus* of this *kind*, likewise found on *our coast*, whereon the *ridges* had *run together*,

Cette *coquille* est extremement epaisse & forte, de la *grandeur* d'une petite noix, d'une *forme* quasi ronde, & plutot *comprimée*, car les *valves* ne sont point fort profondes ou concaves, & leurs *cotés* sont inegaux.

L'*exterieur* à *robe* blanche, *tachetée* d'une maniere mignonne & tres serrée de petites *taches* & *lignes angulaires*, comme des fers de fleche, de *brun chataigne*: à *sillons* epais ou enflés, larges, transversals & concentriques, qui dentelent fortement le bord de la pente du cartilage. Les *intervalles* entre eux sont des canelures larges. Les *becs* tres pointus, & beaucoup tournés de coté. La *pente du cartilage* est grande & concave; l'autre *pente* est marquée d'un enfoncement en forme de cœur.

L'*interieur* lustré, lisse & blanc, exceptè au dessous de la pente du cartilage, ou il se trouve un grand *espace* de *couleur rouge brune*.

Cette *espece*, qui est habitante de la mer Mediterranée, à eté quelquefois pesché sur nos *cotes occidentales*. La *Protection* du Dr. *Fothergill* m'a accordé la connoissance de ces coquilles, de sa *Collection Elegante & Scientifique*, comme aussi d'un *jeu de la nature extraordinaire* de cette *espece*, pareillement trouvé sur *notre cote*, dans lequel les *sillons* se sont

together, or were *confluent* into only four lower ones, by which they were so extremely *swollen* and *broad* that the shell appeared *disfigured* and *quite monstrous.*

sont *unis*, ou *concourus* ensemble en seulement quatre inferieurs, de sorte qu'ils etoient si extremement *larges* & *enflés* que la coquille paroissoit *difforme* & *tout à fait monstreuse.*

XXVII.

P. Striatulus. Striated.

Tab. XII. fig. 2. 2.

XXVII.

Pl. XII. fig. 2. 2.

Pectunculus parvus transversim striatus fusco radiatus. Striatulus. Tab. 12. fig. 2. 2.

An *Gallina. Venus testa subcordata radiata: striis transversis obtusis, cardinis dente postico minimo, margine crenulato.* Lin. S. N. p. 1130. No. 119.—F. Suec. II. No. 2143?

A *small species*, about the *size* of a peach-stone, or an inch long, but generally less; the *shell* thick and strong, of a roundish *shape*, and rather *compressed*, the *valves* not being very concave; and the *sides* are dissimilar.

Outside, most generally of a white *ground*, with three or four longitudinal *rays*, that are narrow at top, and widen greatly as they approach the bottom, of a light *chesnut colour*; but this *colour* is not an uniform *field*, for it is made up of very numerous fine small longitudinal *streaks*, extremely delicate and curious: these *streaks* also appear dispersed all over the shell. Some shells are of a washy flesh-coloured *ground*, with like rays of a darker shade, and others of same *ground* and also *white*, without any rays; but all have the small streaks. The whole *shell* is wrought with numerous concentric transverse *striæ*, like fine threads, close set, and as *Linné observes* seem crenulated, tho' really not so, for they are smooth; and it is the delicate longitudinal streaks of brown that imposes on

Une *petite espece*, environ la *grandeur* d'un noyau de peche, ou un pouce en longeur, mais generalement moins; la *coquille* est epaisse & forte, d'une *forme* quasi ronde, & plutot *comprimée*, les *valves* n'etant point fort concaves; & leur *cotés* sont inegaux.

L'*exterieur* plus generalement à *robe* blanche, avec trois ou quatre *raies* longitudinales, qui sont etroites au haut, & s'elargissent beaucoup comme elles approchent du bas, d'une *couleur chataigne* clair; mais cette *couleur* ne forme pas un *corps* uni, car elle est composée de petites *lignes* fines longitudinales, tres nombreuses, curieuses & extremement mignonnes: ces *lignes* paroissent aussi etre repandues par tout la coquille. Quelques coquilles ont la *robe* couleur de chair foible, avec des semblables raies d'une nuance plus foncée, & d'autres la même *robe* & aussi *blanche*, sans raies; mais toutes ont les petites lignes. Toute la *coquille* est à *stries* nombreuses, tres serrées, concentriques & transversales, comme des fils deliés, & comme *Linné remarque* paroissent

on the fight, and makes them appear as if rough and cut with fuch very minute notches. The *beaks* are pointed, and turn greatly fideways. The *cartilage flope* is a ftrong depreffion that runs the length of the fide, but the cartilage is very fmall and fhort. The other *flope* has a ftrong cordiform depreffion. The *margins* on this fide are plain.

Infide milk white, fmooth and gloffy, and the *margins* are delicately notched.

This *pretty fpecies* is found on feveral of our coafts. I have received them from *Dorfetfhire, Cornwall,* the *ifles* of *Scilly, Chefhire, Yorkfhire,* and *Carnarvonfhire* and *Flintfhire* in *Wales.*

fent crenelées, quoique vraiment point, car elles font liffes; & c'eft les lignes longitudinales mignonnes brunes qui dupe la veue, & les font paroitre comme apres & decoupées avec des entailleures fi extrement petites. Les *becs* font pointus, & beaucoup tournès de coté. La *pente du cartilage* eft un grand enfoncement qui s'etend le long du coté même, mais le cartilage eft fort petit & court. L'autre *pente* à un enfoncement en forme de cœur. Les *bords* de cet coté font unis.

Le *dedans* blanc, liffe & luftré, & les *bords* font tres delicatement entaillés.

Cette *jolie efpece* fe trouve fur plufieurs de nos cotes. Je les ai reçu des comtés de *Dorfet, Cornwall,* les *ifles* de *Scilly, Chefter, York,* & *Carnarvon* & *Flint* dans la principauté de *Galles.*

XXVIII.

P. Sulcatus. Sulcated.

Pectunculus parvus denfe fulcatus. Sulcatus.
Venus rugofa. Wrinkled. Penn. Brit. Zool. No. 50. tab. 56. fig. 50.

A *fmall fpecies,* about the *fize* of a nut, of a roundifh *fhape;* the *fhell* very thick and ftrong; the *valves* with diffimilar fides, and not very concave.

Outfide. Generally, when recently fifh'd, is very filthy, with vermiculi, dirt, coral dregs, &c. and is covered with a ftrong tho' thin blackifh *epidermis,* which being cleared off, the *fhell* is *whitifh* or *reddifh,* with numerous concentric tranfverfe *ftriæ,* very prominent, regular, and clofe fet. The *interftices* between them are broad furrows.

XXVIII.

Une *petite efpece,* environ la *grandeur* d'une noifette, d'une *forme* quafi ronde; la *coquille* eft tres epaiffe & forte; les *valves* ont les cotés inegaux, & ne font point tres profondes.

L'exterieur. Generalement, quand tirée de la mer, eft couvert de faletés, comme vermiffeaux, fange, petits coraux, &c. & auffi couvert d'un *epiderme* fort, quoique mince & noiratre, qui etant depouillé, la *coquille* eft *blanchatre* ou *rougeatre,* à *ftries* nombreufes concentriques & tranfverfales, tres elevées, regulieres, & ferrées. Les *intervalles* entre elles

furrows. The *beaks* pointed, and much turned sideways. The *margins* on this side are plain. The *cartilage slope*, a narrow and pointed depression; the other *slope* has a very deep cordiform cavity.

Inside white, smooth and glossy. The *margins* are delicately notched.

I have received this species from *Scarborough* in *Yorkshire*.

elles sont des canelures larges. Les *becs* pointus, & beaucoup tournés de coté. Les *bords* de cet coté sont unis. La *pente du cartilage* est un enfoncement pointu & etroit; l'autre *pente* à une cavité fort profonde, & en forme de cœur.

Le *dedans* blanc, lisse & lustré. Les *bords* sont delicatement crenelés.

J'ai reçu cette espece de *Scarborough* au comté de *York*.

XXIX.

P. Membranaceus. Membranous.
Tab. XIII. fig. 4. left-hand.

XXIX.

Pl. XIII. fig. 4. à gauche.

Pectunculus strigis transversis remotis, acutis, membranaceis, donatus. Tab. 13. fig. 4. left-hand.

An *Cancellata. Venus subcordata, striis transversis membranaceis remotis, ano cordato.* Lin. S. N. p. 1130. No. 118.—*V. Ziczac.* Id. Mus. Reg. p. 506. No. 71?

This *shell* is thick; *size* of an apricot, or about one inch and a half long; of a roundish *shape*, and somewhat *flattish*, for the *valves* are but moderately concave, and their *sides* dissimilar.

Outside of a whitish *colour*, with a strong cast of brown or pale carnation, with concentric transverse *ridges*: these *ridges* are set equidistant, but remote, and are thin and sharp, like plates or blades. The *beaks* are pointed, and turn very much sideways. The *slope* under the beaks has a strong cordiform depression.

Inside white, smooth and glazed. The *margins* finely crenated.

Mr.

Cette *coquille* est epaisse; de la *grandeur* d'un abricot, ou environ un pouce & demi en longeur; d'une *forme* quasi ronde, & quelque chose *applatie*, car les *valves* ne sont que moderement concaves, & leurs *cotés* inegaux.

L'*exterieur* de *couleur* blanchatre, avec un teint fort de brun ou incarnat foible, à *sillons* concentriques & transversales: ces *sillons* sont à même distance, mais eloignés, & minces & aigus, comme des plaques ou lames. Les *becs* pointus, & tournès beaucoup de coté. La *pente* au dessous des becs à un enfoncement fort, en forme de cœur.

Le *dedans* blanc, lisse & lustré. Les *bords* sont finement crenelés.

M.

Mr. Pennant seems to have *figured* this *shell*, tab. 54. fig. 48. A. as a *worn shell* of his *Venus Erycina*.

From the *Western Shore*.—Dr. Fothergill's Collection.

M. Pennant paroit avoir *figuré* cette *coquille*, pl. 54. fig. 48. A. comme une *coquille roulée* de son *Venus Erycina*.

De la *Cote Occidentale*.—Du Cabinet de Dr. Fothergill.

XXX.

P. Depressior. Flatted.
Tab. XIII. fig. 4. right-hand.

XXX.

Pl. XIII. fig. 4. à droite.

Pectunculus depressior subrotundus, dense & transversim strigatus. Depressior. Tab. 13. fig. 4. right-hand.

A thick, strong, heavy *shell*, yet semitransparent; of a round *shape*, the *sides* being nearly similar, and the *beaks* almost central; *flatted*, or much depressed, for the *valves* are very shallow; and *size* of a small apricot, or about one inch and a quarter long.

Outside white, with a cast of yellowish round the upper or hinge part, with fine concentric transverse *ridges*, little prominent, and rounded like threads, extremely close set, with intervening slight *furrows*. The *beaks* are very small, pointed, and hardly turned aside. The *cartilage slope* small, narrow, pointed, and fill'd with a like cartilage; the other *slope* has a very small, narrow, pointed depression. The *margins* plain.

Inside smooth, glazed, and pure white, except in the middle, which is tinged yellowish. The *margins* are plain.

I received this *species* from the coast of *Cornwall*.

Une *coquille* epaisse, forte & pesante, quoique demitransparente; d'une *forme* ronde, les *cotés* etant quasi egaux, & les *becs* presque centrals; *applatie*, ou beaucoup comprimée, car les *valves* sont peu profondes; & de la *grandeur* d'un petit abricot, ou environ un & un quart de pouce en longueur.

L'*exterieur* blanc, avec un petit teint jaunatre autour de la partie superieure ou la charniere, à *sillons* fins concentriques & transversales, peu saillants, & arrondis comme des fils, extremement serrés, avec les intervalles en *canelures* legeres. Les *becs* sont fort petits, pointus, & à peine tournès de coté. La *pente du cartilage* petite, etroite, pointue, & ramplie d'un cartilage semblable; l'autre *pente* à un fort petit enfoncement, etroit & pointu. Les *bords* sont unis.

L'*interieur* lisse, lustré, & blanc pur, excepté au milieu, qui est teint jaunatre. Les *bords* sont unis.

J'ai reçu cette *espece* de la cote de *Cornwall*.

XXXI. XXXI.

P. TRUNCATUS. TRUNCATED.

Pectunculus parvus, crassus, truncatus, strigis eminentibus. Truncatus.

Concha in vertice leviter purpurascens striis eminentibus, margine parallelis distincta. The light purple tellina, &c. Borlase Cornw. p. 278. tab. 28. fig. 23.—*Tellina Cornubiensis. Cornish.* Penn. Brit. Zool. No. 35.

A *small species*, very thick and strong, of a somewhat triangular *shape*; the *valves* pretty concave; the *sides* very dissimilar, one being rounded, the other near perpendicular, flattish, or *truncated.*

Outside. Whitish, except towards the upper part, or round the beaks, which is of a light *purple* colour. It is wrought with concentric transverse broad and very thick *ridges*, close set, with large intermediate furrows. On the *turn* or *edge* of the *truncated side* the ridges generally divide or fork, and on the *truncated side* they tend obliquely upward from those of the body: this *truncated* part is rather in a slope than quite perpendicular. The *beaks* are strong, pointed, and turn sideways. The *margins* are plain.

Inside white, smooth and glossy. The *margins* delicately notched.

I received this *species* from the coast of *Cornwall.*

Une *petite espece*, tres epaisse & forte, d'une *forme* un peu triangulaire; les *valves* asses profondes; les *cotés* fort inegaux, l'un etant arrondi, l'autre presque perpendiculaire, applati, ou *tronqué.*

L'*exterieur. Blanchatre*, excepté vers la partie superieure, ou autour des becs, qui est d'une couleur *pourpre* claire. Il est à *sillons* concentriques, transversals, larges, fort epais, & tres serrés, avec des grandes canelures intermediates. Sur le *tour* ou *bord* du *coté tronqué* les sillons generalement se divisent ou deviennent fourchus, & sur le *coté tronqué* ils tendent par haut obliquement à ceux du corps: cet *coté tronqué* est plutot en pente que tout à fait perpendiculaire. Les *becs* sont forts, pointus, & tournès de coté. Les *bords* sont unis.

Le *dedans* blanc, lisse & lustré. Les *bords* mignonnement crenelès.

J'ai reçu cette *espece* de la cote de *Cornwall.*

GENUS VIII.
TRIGONELLA.

Bivalves with *equal valves*, and ſhut cloſe; of a ſubtriangular *ſhape*. The *hinge* is ſet with a ſmall middle complicated tooth, aſide of which lies a broad oblique ſubtrigonal cavity. The lateral teeth large and ſtrong.—The animal a *Tethys*.

GENRE VIII.

Bivalves à *battans egaux*, & ſe ferment exactement; d'une *forme* quaſi triangulaire. La *charniere* avec une petite dent compliquée au milieu, à coté de laquelle ſe trouve une cavité large, oblique, & un peu triangulaire. Les dents laterales ſont grandes & fortes.—L'animal eſt un *Tethys*.

* MARINÆ. SEA.
XXXII.
T. RADIATA. RAYED.
Tab. XII. fig. 3. 3.

* DE MER.
XXXII.
Pl. XII. fig. 3. 3.

TRIGONELLA tenuis admodum concava ferrugineo-cinerea radiata. Radiata. Tab. 12. fig. 3. 3.

Pectunculus triquetrus ex flavo radiatus. Liſt. H. Conch. tab. 251. fig. 85.—*Mactra ſtultorum. Simpleton.* Penn. Brit. Zool. No. 42. tab. 52. fig. 42.

Stultorum. Mactra teſta ſubdiaphana lævi obſolete radiata, intus purpuraſcente, &c. Lin. S. N. p. 1126. No. 99.

This *ſhell* is light, brittle, thin, and ſemitranſparent; of a ſubtriangular *ſhape*; the *ſize* of a large plumb, or about one and a half inch *long*, and near the ſame *breadth*. The *valves* are very concave or deep, and the *ſides* nearly equal.

Outſide, near ſmooth, and gloſſy, being only tranſverſely and thickly ſtriated with ſuch extreme delicate or fine ſtriæ as ſcarcely to be perceptible. Some ſhells are of a light uniform aſh *colour*, but moſt generally of a reddiſh or ferruginous aſh *colour*, with numerous pale longitudinal regular

Cette *coquille* eſt legere, fragile, mince & demitranſparente; d'une *forme* quaſi triangulaire; de la *grandeur* d'une grande prune, ou environ un & demi pouce en *longueur*, & près de la même *largeur*. Les *valves* ſont fort concaves ou profondes, & les *cotés* à peu près egaux.

L'*exterieur*, preſque liſſe, & luſtré, etant ſtrié ſeulement à ſtries ſerrées & tranſverſales, ſi delicates & fines qu'a peine elles ſont viſibles. Quelques coquilles ſont d'une *couleur* unie cendre claire, mais plus generalement d'une *couleur* rouſſâtre ou ferrugineuſe cendrée, à *raies* longitudinales regulieres,

regular *rays* of reddish and of whitish. The *beaks* are near central, small, very pointed, and level with the commissure. The *sides* are somewhat flatted, and the *margins* plain. Sometimes remains of a silky slight brown *epidermis* are found on it.

Inside. In the *thinner* or *young ones* whitish and glazed, with a strong tint or glance of reddish or brownish; but the *old* or *thick* ones are entirely of a fine deep violet *colour*. The *margins* are plain.

This *species* is found on the coasts of *Kent*, *Dorsetshire*, *Cornwall*, and the other western shores; at *Highlake* in *Cheshire*, ten miles from *Liverpool*; at the mouth of the river *Mersey*; and on the coast of *Aberdeenshire*, and other *shores* of *Scotland*.

regulieres, nombreuses & pales, rousfatres & blanchatres. Les *becs* presque centrals, petits, fort pointus, & de niveau avec la commissure. Les *cotés* sont un peu applatis, & les *bords* unis. Quelquefois des restes d'une *epiderme* legere, soyeuse & brune, se trouve.

Le *dedans*, des coquilles *minces* ou *jeunes*, est blanchatre & lustrè, avec un grand teint de rougeatre ou brunatre; mais les coquilles *epaisses* ou *vielles* sont tout à fait d'une belle *couleur* violette foncée. Les *bords* sont unis.

Cette *espece* se trouve sur les cotes des comtés de *Kent*, *Dorset*, *Cornwall*, & les autres cotes occidentales; à *Highlake* au comté de *Chester*, dix miles de *Liverpool*; à l'embouchure de la riviere *Mersey*; & sur la cote du comte de *Aberdeen*, & autres *cotes* de l'*Ecosse*.

XXXIII.
T. Zonaria. Girdled.
Tab. XV. fig. 1. 1.

XXXIII.
Pl. XV. fig. 1. 1.

Trigonella crassa transversim fasciata. Zonaria. Tab. 15. fig. 1. 1.

Concha crassa, ex altera parte compressa, ex altera subrotunda. Lift. H. An. Angl. p. 174. tit. 24. tab. 4. fig. 24.—*Pectunculus crassiusculus albidus.* Lift. H. Conch. tab. 253. fig. 87.

Wallace Orkn. p. 43.—*Chama media fasciata crassa.* Petiv. Gaz. tab. 94. fig. 7.—*Chama minor plurimis fasciis.* Id. ib. fig. 6.—*A pectunculus with azurine circular lines interpolated.* Leigh Lancash. tab. 3. fig. 6.—*Thick white striated chama.* Wallis Northumb. p. 395.—*Mactra solida; strong.* Penn. Brit. Zool. No. 43. tab. 51. fig. 43. A. & tab. 52. fig. 43.

Buon. Ricr. p. 163. fig. 51. & 54.—Klein Ostrac. p. 141. § 365. No. 4. tab. x. fig. 42.

Cardium Solidum. Lin. F. Suec. II. No. 2140.—*Solida. Mactra testa opaca læviuscula subantiquata.* Id. S. N. p. 1126. No. 100.

 This Cette

This *shell* is extremely thick, ſtrong, opake and heavy; of a ſubtriangular ſhape; rather depreſſed, for the *valves* are but moderately deep. The moſt general *ſize* is that of a common plumb, or about one and a quarter inch *long*, and one and three quarters of an inch in *breadth*.

Cette *coquille* eſt extremement epaiſſe, forte, opaque & peſante; d'une *forme* quaſi triangulaire; plutot comprimée, car les *valves* ne ſont que moderement profondes. La *grandeur* la plus generale eſt celle d'une prune commune, ou environ un pouce & un quart en *longueur*, & un pouce & trois quarts en *largeur*.

Outſide, generally without any gloſs, but of a rough appearance; of various *colours*, moſtly whitiſh, or pale yellowiſh brown, with ſeveral concentric tranſverſe *girdles* or *zones*, blackiſh, ſlate-colour'd, brown, &c. Theſe *girdles* are thick, broad, and ſtrongly marked, in the *live ſhells*; but in the *dead ſhells* are very raiſed and prominent, like ribs, the ſurface between them being often greatly eroded or worn down. The *margins* are plain; the *beaks* pointed and central.

L'*exterieur*, generalement ſans luſtre, mais d'une apparence rude; de diverſes *couleurs*, la pluſpart blanchatre, ou pale jaunatre brun, à pluſieurs *bandes* ou *zones* concentriques & tranſverſales, noiratres, couleur d'ardoiſe, brunes, &c. Ces *bandes* ſont epaiſſes, larges, & fortement marquées, ſur les *coquilles vivantes*; mais ſur celles qui ſont *mortes* elles ſont tres hauſſées & ſaillantes, comme des cotes, la ſurface entre elles etant ſouvent tres frottée ou uſée. Les *bords* ſont unis; les *becs* pointus & centrals.

Inſide, ſmooth, but without any gloſſineſs; whitiſh, brown, or livid, &c. moſt generally paler than the outſide colours.

Le *dedans*, liſſe, mais ſans aucun luſtre; blanchatre, brun, ou livide, &c. generalement plus foible en couleur que l'exterieur.

This *ſpecies* is found in plenty on many of our ſhores, as *Kent*, *Dorſetſhire*, *Lancaſhire*, *Yorkſhire*, *Northumberland*, &c.

Cette *eſpece* ſe trouve en abondance ſur pluſieurs de nos cotes, comme aux comtés de *Kent*, *Dorſet*, *Lancaſter*, *York*, *Northumberland*, &c.

XXXIV.

XXXIV.

T. Subtruncata. Subtruncated.

Trigonella albeſcens lævis, lateribus ſubtruncatis. Subtruncata.

This *ſpecies*, like the laſt, is thick, ſtrong, heavy and opake, but about half its *ſize*; of a triangular *ſhape*, for the *ſides* are much flatted, or almoſt truncated,

Cette *eſpece*, comme la derniere, eſt epaiſſe, forte, peſante & opaque, mais environ la moitie en *grandeur*; d'une *forme* triangulaire, car les *cotés* ſont tres

cated, and, when the ſhell is ſhut, have a ſtrong heart-like impreſſion. The *valves* very deep or concave; and the *beaks* ſtrong, pointed, turned inwards, and juſt overtop the commiſſure.

tres applatis, ou preſque tronqués, &, quand la coquille eſt fermée, repreſentent une empreinte forte en forme de cœur. Les *valves* ſont tres profondes ou concaves; & les *becs* forts, pointus, tournés en dedans, & dejettent à point nommé la commiſſure.

Outſide, ſmooth and gloſſy, of a pale whitiſh colour, and is thick ſet with very fine concentric tranſverſe *ſtriæ*, which, towards the upper part and bottom, are much ſtronger. The *margins* are plain.

L'*exterieur*, liſſe & luſtré, d'une couleur pale blanchatre, & à *ſtries* fort fines, concentriques & tranſverſales, qui, vers la partie ſuperieure & les bords, ſont beaucoup plus marquées. Les *bords* ſont unis.

Inſide, white and ſmooth in the middle, but pale; whitiſh and gloſſy on the borders. The *margins* plain.

Le *dedans* blanc & liſſe au milieu, mais pale; blanchatre & luſtré ſur les bords. Les *bords* ſont unis.

This is not a *common ſpecies*. I have received it from *Hampſhire* and *Devonſhire*.

Cette *eſpece* n'eſt pas *commun*. Je l'ai reçu des comtés de *Hants* & *Devon*.

Obs.—*Incredible quantities* of this ſpecies are found *foſſil* in the *ſand-pits* at *Woolwich*, in *Kent*.

Obs.—Des *quantités incroyables* de cette eſpece ſe trouvent *foſſiles* dans les *ſablonnieres* à *Woolwich*, au comté de *Kent*.

XXXV. T. Gallina. Hens. Tab. XIV. fig. 6. 6.

XXXV. Pl. XIV. fig. 6. 6.

Trigonella ex flavo maculata. Gallina. Tab. 14. fig. 6. 6.

An *Pectunculus ex flavo variegatus, quadrangularis, rhomboeides, ſive margine ima quaſi in angulum exeunte.*—Liſt. H. Conch. tab. 250. fig. 84?

This *ſpecies* is ſomewhat *leſs* than the *Zonaria*, No. 33, but equally thick, ſtrong, opake and heavy; of a ſubtriangular *ſhape*; rather depreſſed, for the *valves* are but moderately concave; the *ſides* nearly ſimilar, and ſlightly flatted; and the *beaks* are ſtrong, very pointed, turn inwards, and juſt overtop the commiſſure.

Cette *eſpece* eſt quelque peu plus *petite* que la *Zonaria*, No. 33, mais egalement epaiſſe, forte, opaque & peſante; d'une *forme* quaſi triangulaire; plutot comprimée, car les *valves* ſont que moderement concaves; les *cotés* ſont preſque egaux, & legerement applatis; & les *becs* forts, tres pointus, tournès en dedans, & dejettent à point nommé la commiſſure.

Outſide,

L'ex-

Outside, very sleek and glossy, with numerous fine transverse concentric *striæ*, and a few strong *wrinkles*; of a cream white *colour* with a tint of yellowish, round the top; a large long *spot*, like an irregular transverse streak, near the middle, and under it another long transverse *streak*, both of a fine light chesnut *colour*: these *spots* or *streaks* lie only on one side of the shells. The *margins* are plain.

Inside, white, smooth and glossy. The *margins* plain.

This *kind* is found in great plenty on the coast of *Cornwall*, and are much eaten, like the common cockle; they vulgarly call it *Hens*. I have also seen it from the coast of *Dorsetshire*.

L'*exterieur*, extremement uni & lustré, à *stries* nombreuses fines, transversales & concentriques, & quelques *rides* fortes; d'une *couleur* blanche de creme, avec un teint jaunatre, autour du haut; une *tache* grande & longue, comme une raie irreguliere transversale, près du milieu, & en dessous d'elle une autre *raie* longue & transversale, toutes les deux d'une belle *couleur* chataigne claire: ces *taches* ou *raies* se trouvent seulement sur un coté des coquilles. Les *bords* sont unis.

Le *dedans*, blanc, lisse & lustré. Les *bords* unis.

Cette *espece* se trouve en grande abondance sur la cote de *Cornwall*, & est beaucoup mangée, comme le petoncle commun; on les appelle vulgairement *Hens*, ou *Poules*. Je les ai aussi vu de la cote du comté de *Dorset*.

XXXVI.

T. Plana. Flat.

Tab. XIII. fig. 1. 1.

XXXVI.

Pl. XIII. fig. 1. 1.

Trigonella tenuis subrotunda plana. Plana. Tab. 13. fig. 1. 1.

Concha tenuis, subrotunda, omnium minime cava, cardinis medio sinu & amplo & pyriformi. List. H. An. Angl. p. 174. tit. 23. tab. 4. fig. 23.

Pectunculus latus, admodum planus, tenuis, albidus. List. H. Conch. tab. 253. fig. 88.

Wallace Orkn. p. 42.—*Slender smooth chama.* Wallis Northumb. p. 395.—*Venus borealis. Northern.* Penn. Brit. Zool. No. 52. and also his *Tellina crassa. Flat.* No. 28. fig. 28. 28.

Buon Ricr. p. 163. fig. 52. 55.

This *species* is very thin, semitransparent and brittle; of a somewhat triangular *shape*; greatly depressed, or flat, for the *valves* are extremely shallow; of the *size* of an apricot, or about one and a half inch *long*,

Cette *espece* est fort mince, demitransparente & fragile; d'une *forme* quelque peu triangulaire; beaucoup comprimée, ou applatie, car les *valves* ont tres peu de profondeur ou concavité; de la *grandeur* d'un

long, and near two inches *broad*. The *ſides* are nearly equal.

Outſide, of a pale browniſh *colour*, with ſome blackiſh ſtreaks, and a little gloſſy when *alive*, but *whitiſh* when *dead*. It is thick ſet with concentric tranſverſe fine capillary *ſtriæ*, intermixed with ſeveral ſmall *wrinkles*. The *beaks* are level with the commiſſure, very ſmall, pointed, and central; and the *margins* are plain.

Inſide, white, ſmooth, and gloſſy. The *margins* plain.

The *hinge* of this *kind* is of a different ſtructure from the *Trigonellæ*, for it conſiſts of *two* minute thin plate-like parallel *teeth*, aſide of which is a large *triangular cavity*, and has no *lateral teeth*.

This *ſhell* is common on ſeveral of our coaſts, as *Kent*, *Dorſetſhire*, *Cheſhire*, *Yorkſhire*, *Northumberland*, &c.

Obs.—Though *Linné* quotes this *ſhell* of *Liſter H. An. Angl.* for his *Venus Borealis*, *S. N. p.* 1134, *No.* 143, it is an *error*, for they are *different ſpecies*; and I think this very *ſpecies* is *undeſcribed* in his *Syſtem*.

GENUS

d'un abricot, ou environ un pouce & demi en *longueur*, & près de deux pouces en *largeur*. Les *cotés* ſont preſque egaux.

L'*exterieur* d'une *couleur* pale brunatre, avec quelques raies noiratres, & un peu luſtré quand *vivantes*, mais *blanchatres* quand *mortes*; à *ſtries* fines comme des cheveux, tres ſerrées, tranſverſales & concentriques, entremelées de pluſieurs petites *rides*. Les *becs* ſont de niveau avec la commiſſure, fort petits, pointus, & centrals; & les *bords* ſont unis.

Le *dedans*, blanc, liſſe & luſtré. Les *bords* unis.

La *charniere* de cette *eſpece* eſt d'une ſtructure differente des *Trigonellæ*, car elle eſt compoſée de *deux* tres petites *dents*, minces, comme des lames, & paralleles, à cote deſquelles ſe trouve une grande *cavité triangulaire*, & ſans *dents laterales*.

Cette *coquille* eſt commune ſur pluſieurs de nos cotes, comme des comtés de *Kent*, *Dorſet*, *Cheſter*, *York*, *Northumberland*, &c.

Obs.—Quoique *Linné* cite cette *coquille* de *Liſt. H. An. Angl.* pour ſon *Venus Borealis*, *S. N. p.* 1134, *No.* 143, c'eſt une *erreur*, car elles ſont des *eſpeces differentes*; & je crois que cette *eſpece preciſe* n'*eſt pas decrite* dans ſon *Syſteme*.

GENRE

GENUS IX.

CUNEUS. PURR.

Bivalves with *equal valves*, and shut close; of greater *breadth* than *length*; the *sides* very dissimilar, for the *beaks* are placed near to or at one extremity; the *hinge* various, both in regard to *structure* and the number of *teeth*.—The animal a *Tethys*.

GENRE IX.

Bivalves à *battans egaux*, qui se ferment exactement; d'une plus grande *largeur* que *longueur*; les *cotés* tres inegaux, car les *becs* sont situés près de ou à une extremité; la *charniere* diverse, tant à l'egard de la *structure* que du nombre des *dents*.—L'animal est un *Tethys*.

* MARINÆ. SEA.

XXXVII.

C. RETICULATUS. RETICULATED PURR.

Tab. XIV. fig. 4. 4.

* DE MER.

XXXVII.

Pl. XIV fig. 4. 4.

CUNEUS reticulatus, longitudinaliter & transversim vel decussatim striatus, subrufus, intus ex parte violaceus. Reticulatus. Tab. 14. fig. 4. 4.

Concha quasi rhomboides, in medio cardine utrinque circiter tribus exiguis denticulis donata. List. H. An. Angl. p. 171. tit. 20. tab. 4. fig. 20.

Chama fusca striis tenuissimis donata. List. H. Conch. tab. 423. fig. 271.

Chama purrs Anglice dicta, et *Tellina fasciata compactilis radiata intus ex parte subaurea, interdum subpurpurea.* List. Exercit. Anat. 3. p. 25. 27. tab. 3.

Wallace Orkn. p. 42.—*Chama nostras striis capillaceis.* Mus. Petiv. p. 83. No. 811.—*Purra fasciata & radiata. Cornwall purr.* Petiv. Gaz. tab. 95. fig. 8.—*Chama, purrs.* Dale Harw. p. 387. No. 5. Smith Cork, p. 318.—*Venus Litterata. Lettered.* Penn. Brit. Zool. No. 53. tab. 57. fig. 53.

Buon. Ricr. p. 166. fig. 68.—*Chama inæquilatera, minutissime striata, nonnullis lineis, sive rugis super impositis raro, & gradatim circumdata, subflava, punctis cyaneis, & fuscis aliquando densissime aspersa, & signata.* Gualt. I. Conch. tab. 85. fig. L.—Et *Chama inæquilatera, striis minimis, & aliquibus lineis fasciata, subrufa, aliquando fulvida.* Id. tab. 85. fig. C. E. & tab. 86. fig. B.

An *Decussata. Venus testa ovata antice angulata decussatim striata.* Lin. S. N. p. 1133. No. 149. Mus. Reg. p. 509. No. 77.?

This *species* is a *broad* shell, and measures from near an inch to above two inches

Cette *espece* est une coquille *large*, & porte de près d'un pouce à plus de deux pouces

202 Da Costa is very faulty in his Synonyms to this Shell. There are in fact three species of this shell, and he supposes there are but two. All the three are found at Sandwich. His Cuneus fasciatus is another of the three species.

inches in *breadth*, or from ſide to ſide, and in *length*, or from the hinge to the bottom, only from half an inch to an inch and a half, of an oval *ſhape*, and thick, ſtrong and heavy.

Outſide, generally *whitiſh*, and often of a *ruſt colour*. The *former* are the *ſmaller* or *young* ſhells; the latter the *large* or *full grown*, and theſe have ſeldom any lines or characters; but the whitiſh ones are ſometimes, not frequent, eſpecially the ſmalleſt, prettily variegated with ſtraggling ſpots and ſtreaks, like characters, of a browniſh, aſh, or livid *colour*. The *whole ſhell* is *reticulated*, or wrought with longitudinal and tranſverſe *ſtriæ*, thick ſet. In the *larger*, the *ſtriæ* are very ſtrong and bold, and on the ſides are cut or notched into ſmall bumps; but in the *ſmaller*, the *ſtriæ* are delicate and fine, and only run a little ſtrong on the ſides. The *beaks* prominent, ſmall, pointed, and turn outwards, are level with the commiſſure, and ſituated almoſt at one extreme of the ſhell; the *ſides* therefore are very diſſimilar, one rounded and ſhort, the other lengthened and ſubangulated. The *margins* are plain.

Inſide, ſmooth, gloſſy, and white; but in *moſt*, even the *larger* and *old* ſhells, the border of the hinge on the longer ſide, from the teeth (not on them, nor on the other ſide beyond them,) is of a very fine violet *colour*; and *generally*, the *ſmaller* or *young* ones have a large ſpace of the like fine violet *colour* all down the ſame ſide of the ſhell. The *margins* are plain.

The *hinge* is ſet with three ſtrong parallel teeth; the *middle one* bifid.

pouces en *largeur*, ou d'un coté a l'autre, & en *longueur*, ou dès la charniere au bas, ſeulement de un demi pouce à un pouce & demi, d'une *forme* ovale, & epaiſſe, forte & peſante.

L'*exterieur*, generalement *blanchatre*, & ſouvent d'une *couleur* de *rouille*. Les *premieres* ſont les *petites* ou *jeunes* coquilles; les dernieres les *grandes* ou *adultes*, & celles ci ont rarement des lignes ou caracteres; mais les blanchatres ſont quelquefois, pas frequemment, ſpecialement les plus petites, joliment variées de taches & lignes ecartées, comme des caracteres, d'une *couleur* brunatre, cendre, ou livide. *Toute la coquille* eſt *reticulée*, ou à *ſtries* longitudinales & tranſverſales, tres ſerrées. Dans les *grandes*, les *ſtries* ſont tres fortes & bien marquées, & ſur les cotés ſont coupées ou entaillées en des petits tubercules; mais dans les *petites* coquilles, les *ſtries* ſont fines & delicates, & ſeulement ſe trouvent un peu fortes ſur les cotés. Les *becs* ſaillants, petits, pointus, & tournès en dehors, de niveau avec la commiſſure, & ſituès preſque à une extremité de la coquille; c'eſt pourquoi les *cotés* ſont tant inegaux, un arrondi & court, l'autre allongé & un peu angulaire. Les *bords* ſont unis.

Le *dedans*, liſſe, luſtré, & blanc; mais dans la *pluſpart*, même les *grandes* ou *vielles* coquilles, la bordure de la charniere ſur le coté allongé, dès les dents (non pas ſur elles, ni de l'autre coté au dela d'elles,) eſt d'une fort belle *couleur* violette; & *generalement*, les *petites* ou *jeunes* ont un grand eſpace de la même belle *couleur* violette le long du même coté de la coquille. Les *bords* ſont unis.

La *charniere* eſt à trois dents fortes & paralleles; *celle* du *milieu* eſt fourchue.

This *species* is found in plenty on most of our shores, as *Kent*, *Essex*, *Sussex*, *Hampshire*, *Dorsetshire*, *Cornwall*, *Yorkshire*, on the coasts of *Wales*, &c.

Cette *espece* se trouve en abondance sur la pluspart de nos cotes, comme aux comtés de *Kent*, *Essex*, *Sussex*, *Hants*, *Dorset*, *Cornwall*, *York*, & aussi sur les cotes de la principauté de *Galles*, &c.

XXXVIII.

XXXVIII.

C. Fasciatus. Fasciated.

Cuneus fasciatus sive transversim sulcatus. Fasciatus.

Pectunculus fasciatus. The clam. Dale Harw. p. 386. No. 3.—*Venus rhomboides. Rhomboid.* Penn. Brit. Zool. No. 55.—Et *Tellina depressa. Depressed.* Id. No. 27. tab. 47. fig. 27.

An *Rotundata. Venus testa ovata antice subangulata: striis transversis, cardinis dente intermedio bifido.* Lin. S. N. p. 1135. No. 148.—Mus. Reg. p. 509. No. 76.?

This *species*, in most particulars, is like the last, but differs in being of a smaller *size*, more smooth, and is *fasciated* or wrought with regular transverse concentric *striæ*, of a pale brown or russet ash *colour*, and *sometimes*, but very *rarely*, is prettily *variegated* with spots and characters like the last.

Inside, livid or whitish, and glossy, without any of the violet colour. The *hinge*, and other particulars, the same.

I have received this kind from *Yorkshire*, *Dorsetshire*, and *Cornwall*.

Cette *espece*, en plusieurs circonstances, ressemble a la derniere, mais differe en *grandeur* etant plus petite, plus lisse, & à *stries* regulieres concentriques & transversales, d'une *couleur* pale brune ou rougeatre cendrée, & *quelquefois*, mais tres *rarement*, est joliment *variée* par des taches & caracteres comme la derniere.

Le *dedans*, livide ou blanchatre, & lustré, sans aucune couleur violette. En *charniere*, & autres particularités, de même.

J'ai reçu cette sorte des comtés de *York*, *Dorset*, & *Cornwall*.

XXXIX.

XXXIX.

C. Foliatus. Foliated.

Tab. XV. fig. 6. left-hand.

Pl. XV. fig. 6. à gauche.

Cuneus parvus albescens, rugis foliaceis & membranaceis erectis transversim cinctus. Foliatus. Tab. 15. fig. 6. left-hand.

Concha trisidos, striata, rugosa, ex albido cinerea, in uno extremo ad angulos rectos truncata. Gualt. I. Conch. tab. 95. fig. A. *Irus.*

Irus. Donax testa ovali, rugis membranaceis erectis striatis cincta. Lin. S. N. p. 1128. No. 111.

A very small *species*, of the *size* of a kidney-bean, or about half an inch *broad* and a quarter of an inch *long*; of an oval *shape*, rather depressed, and very thin and brittle.

Outside, sullied *white*, rugged or uneven. It is set with many regular concentric transverse and strong *wrinkles*, that *rise* into flat, sharp, curled or waved *plates*, and thin, or like membranes, in a very remarkable manner, and are so very raised towards the bottom margins as to fringe them very prettily. The *sides* are dissimilar, one short and rounded, the other greatly lengthened. The *beaks* slightly prominent, pointed, level with the commissure, and turn towards the short side. The *margins* plain.

Inside, white, smooth and glossy. The *hinge* set with two small teeth, *one* of which is *bifid*, or cloven. The *margins* plain.

This *species* is found abundantly in *Cornwall*, buried in the sands; and it is not uncommon on the shores of *Dorsetshire*.

Une tres petite *espece*, de la *grandeur* d'une faseole, ou environ un demi pouce en *largeur*, & un quart d'un pouce en *longueur*; d'une *forme* ovale, plutot comprimée, & fort mince ou fragile.

L'*exterieur*, *blanc* sale, rude ou inegale, à plusieurs *rides* fortes, regulieres, concentriques & transversales, qui s'*elevent* en *lames* plattes, en vive arrete, frisées ou onduleuses, & minces, ou comme des membranes, d'une maniere tres remarquable, & sont si saillantes vers les bords inferieures que de les franger tres joliment. Les *cotés* sont inegaux, un court & arrondi, l'autre beaucoup allongé. Les *becs* dejettent peu, sont pointus, de niveau avec la commissure, & tournès vers le coté court. Les *bords* unis.

Le *dedans*, blanc, lisse & lustré. La *charniere* à deux petites dents, dont *une* est *fourchue*, ou fendue. Les *bords* unis.

Cette *espece* se trouve en abondance au comté de *Cornwall*, ensevelie dans les sables, & elle n'est pas rare sur les cotes du comté de *Dorset*.

XL.

C. Truncatus. Truncated.

XL.

Cuneus ex albo & violaceo radiatus, intus vero violaceus, latere altero gibbo & truncato. Truncatus.

Tellina intus ex viola purpurascens, in ambitu serrata. List. H. An. Angl. p. 190. tit. 35. tab. 5. fig. 35.

Tellina crassa, admodum leviter striata, intus violacea. List. H. Conch. tab. 375. fig. 216 & 376. fig. 218. 219.

Wallace

Wallace Orkn. p. 45.—*The tooth'd broad muſcle.* Grew Muſ. p. 147. Smith Cork, p. 318.—*Donax denticulata. Purple.* Penn. Brit. Zool. No. 46

Buon. Ricr. p. 160. fig. 37.—Klein Oſtrac. p. 159. § 397. No. 3. tab. 11. fig. 61.—Argenv. Conch. I. p. 331. tab. 25. fig. L. II. p. 293. tab. 22. fig. L.

Tellina inæquilatera, altero latere truncato, anguſto, & ſtriato, margine interno dentato, candida, intus purpuraſcens. Gualt. I. Conch. tab. 89. fig. D.

Rugoſa. Donax teſta antice rugoſa gibba, marginibus crenatis Lin. S. N. p. 1127. No. 104. Muſ. Reg. p. 494. No. 50.

A ſmall *ſpecies*, about the *ſize* of a hazel nut, or from half to above an inch in *breadth*, and from half to three quarters of an inch long; thick, ſtrong, and moſtly ſemitranſparent; of a triangular *ſhape*, and depreſſed. The *ſides* diſſimilar, one ſtretching out and narrow, the other blunt or truncated.

Outſide, gloſſy, and as if poliſhed, thickly ſtriated with very fine and delicate longitudinal *ſtriæ*; of different *colours*, pale brown, often with a violet tinge, aſh, and more generally milk white. The latter are often ſet with narrow and very broad regular longitudinal *rays*, of a fine violet *colour*; but ſometimes browniſh or pale. The *narrow ſide* is rounded; the *truncated ſide* near flat, and riſes into a very prominent *ridge* at its juncture with the body. Another *prominent line* runs alſo its length, near the *middle*, by which it is divided into two *compartments* or *areas*: the *inner* one, or that next the body, is finely ſtriated its length; the *outer* one, or that near the opening of the ſhell, is delicately wrought with very fine tranſverſe zigzag or undulated ſtreaks. This *truncated part*, when the valves are cloſed, forms a pointed heart-like figure. The *beaks* are ſmall, pointed, turn

Une petite *eſpece*, environ la *grandeur* d'une noiſette, ou dès un demi à plus d'un pouce en *largeur*, & de un demi à trois quarts d'un pouce en *longueur*; epaiſſe, forte, & pour l'ordinaire demitranſparente; d'une *forme* triangulaire, & comprimée. Les *cotés* inegaux, un allongé & etroit, l'autre obtus ou tronqué.

L'*exterieur*, luſtré, & comme poli, ſerrement ſtrié à *ſtries* longitudinales, tres fines & delicates; de differentes *couleurs*, brune pale, ſouvent teinte de violette, cendre, & plus generalement blanche de lait. Les dernieres ſont ſouvent à *raies* longitudinales, regulieres, etroites, & tres larges, d'une belle *couleur* violette; mais quelquefois brunatres ou pales. Le *coté etroit* eſt arrondi; le *coté tronqué* preſque plat, & s'eleve en un *bord* tres ſaillant ou il eſt joint au corps. Une autre *ligne ſaillante* court auſſi ſon long, près du *milieu*, par leſquels il eſt diviſé en deux *compartiments* ou *aires*: l'*interieur*, ou celui proche du corps, eſt finement ſtrié ſa longueur; l'*exterieur*, ou celui près de l'ouverture de la coquille, eſt à lignes tranſverſales, en zigzac ou onduleuſes, tres delicates. Cette *partie tronquée*, quand la coquille eſt fermée, repreſente la forme d'un cœur pointu. Les *becs* ſont petits, pointus,

turn inwards, and just over-top the commissure. The *margins* delicately notched.

Inside, smooth, very glossy, and in the *larger* or *old* shells of a deep violet *colour*; in the *young* shells, as the small and white, sometimes quite white, and sometimes with a large space of a pale violet tinge. The *margins* finely notched. The *hinge* with two teeth.

This *species* is frequent on most of our shores. I have received it from *Essex*, *Sussex*, *Cornwall* and other *western coasts*, the *isles* of *Scilly*, *Yorkshire*, and from *Wales*; also from *Aberdeenshire*, and the *Orkney islands*, in *Scotland*; and from the shores of the county of *Cork*, in *Ireland*.

pointus, tournès en dedans, & dejettent à point nommé la commissure. Les *bords* sont delicatement crenelés.

Le *dedans*, lisse, tres lustré, & dans les *grandes* ou *vielles* coquilles d'une *couleur* violette foncée; dans les *jeunes* coquilles, comme les petites & les blanches, quelquefois tout à fait blanc, & tantot avec un espace large teint pale violet. Les *bords* finement crenelés. La *charniere* à deux dents.

Cette *espece* est frequent sur la pluspart de nos cotes. Je l'ai reçu des comtés de *Essex*, *Sussex*, *Cornwall* & autres *cotes occidentales*, des *isles* de *Scilly*, de *York*, & de la principauté de *Galles*; aussi du comté de *Aberdeen*, & des *isles Orcades*, en *Ecosse*; & des cotes du comté de *Cork* en *Irlande*.

XLI.

C. Vittatus. Ribband.
Tab. XIV. fig. 3.

XLI.

Pl. XIV. fig. 3.

Cuneus angustior lævis subfuscus vittis purpurascentibus fasciatus. Vittatus. Tab. 14. fig. 3.

Tellina subfusca angustior, intus purpurascens. List. H. Conch. tab. 376. fig. 217.

Tellina vittis albo luteis & purpurascentibus, leviter striata, margine serrato. Borlase Cornw. p. 278. tab. 28. fig. 25.—*Donax trunculus. Yellow.* Penn. Brit. Zool. No. 45. tab. 55. fig. 45.

Tellina inæquilatera lævis, margine interno minutissime dentato, ex albido, & violacco fasciata, & ex fulvido maculata & radiata. Gualt. I. Conch. tab. 88. fig. Q.

Trunculus. Donax testa antice lævi, intus violacea, marginibus crenatis. Linn. S. N. p. 1127. No. 105. F. Suec. II. No. 2142. Mus. Reg. p. 494. No. 51.

A small, short and broad *species*, about the *size* of an almond; above half an inch *long* from the hinge to the margin, and one inch and a quarter *broad*, or from side to side; sleek, and glossy as if polished; of a narrow or oblong *shape*, depressed or shallow,

Une petite *espece*, large & courte, environ la *grandeur* d'une amande; plus d'un demi pouce en *longueur* dès la charniere au bas, & un pouce & un quart en *largeur*, ou de coté à coté; lisse, & lustrée comme si polie; d'une *forme* etroite ou oblongue, com-

ſhallow, and pretty thick, ſtrong, and ſemitranſparent.

Outſide, longitudinally ſtriated; the *ſtriæ* ſo extremely fine as not to be very apparent, by which the ſhell appears ſleek or ſmooth; of a whitiſh or yellowiſh brown *colour*, with many regular concentric tranſverſe narrow *zones* or *belts* of a pale violet *colour*. The *ſides* very diſſimilar; one extended near three quarters of the ſhell, narrow and rounded; the other with a ſmall ſlope. The *beaks* lie on the ſloping ſide, and turn inwards to it, ſmall, pointed, and juſt over-top the commiſſure. The *margins* are finely notched.

Inſide, ſmooth, gloſſy, white, and ſometimes with a tinge of yellow, but moſt generally of a fine ſtrong violet *colour*. The *margins* are finely toothed.

This *ſpecies* is commonly found, and in plenty, on the ſame ſhores as the laſt.

Obs.—*The miſ-quotations of authors by Linné, relative to this and the laſt ſpecies, is a capital error. Such capital errors are very frequent in that celebrated naturaliſt's Syſtem of Shells.*

comprimée ou peu profonde, & aſſes epaiſſe, forte, & demitranſparente.

L'*exterieur* ſtrié longitudinalement; les *ſtries* ſi extremement fines que à peine elles ſont viſibles, c'eſt pourquoi la coquille paroit liſſe ou unie; d'une *couleur* blanchatre ou jaunatre brune, avec pluſieurs *zones* ou *bandes* regulieres, concentriques, tranſverſales, & etroites, d'une *couleur* pale violette. Les *cotés* fort inegaux; un allongé près de trois quarts de la coquille, etroit & arrondi; l'autre à une petite pente. Les *becs* ſont du coté de la pente, & tournès en dedans envers elle, petits, pointus, & dejettent à point nommé la commiſſure. Les *bords* ſont finement entaillés.

Le *dedans*, liſſe, luſtré, blanc, & quelquefois avec un teint jaunatre, mais plus generalement d'une belle *couleur* violette foncée. Les *bords* ſont finement dentelés.

Cette *eſpece* ſe trouve communement, & en abondance, ſur les mêmes rivages que la derniere.

Obs.—*Les fauſſes citations des auteurs par Linné, relatives a cette & la derniere eſpece, eſt une erreur capitale. Des telles erreurs capitales ſont frequent dans le Syſteme de Teſtacées de cet celebre naturaliſte.*

GENUS

GENRE

GENUS X.

TELLINA. TELLEN.

Bivalves with *equal valves*, generally *short* and *broad* shells, of equal *sides*, and the *beaks* central. At one *extreme* commonly *sloping* down, or with a *flexure*, its length; and on that *side* several species do not *shut close*, but *gape* a little. The *hinge* mostly with three teeth.—The animal a *Tethys*.

GENRE X.

TELLINA. TELLINE.

Bivalves à *battans egaux*, generalement des coquilles *courtes* & *larges*, à *cotés* egaux, & à *becs* centrals. Elles ont communement, à une *extremité*, une *pente*, *pli*, ou *sinuosité*, qui court sa longueur; & de cet *coté* plusieurs especes ne se *ferment exactement*, mais sont un peu *beantes*. La *charniere* generalement à trois dents. —L'animal est un *Tethys*.

* MARINÆ. SEA.

XLII.

T. RADIATA. RAYED.

Tab. XIV. fig. 1.

* DE MER.

XLII.

Pl. XIV. fig. 1.

TELLINA *depressa transversim striata albescens e rubro radiata. Radiata.* Tab. 14. fig. 1.

Concha rugosa, tellinæ formis, lineola quadam paululum eminente ab ipso cardine ad imum ambitum donata. List. App. Hist. An. Angl. p. 19. tit. 36. tab. 1. fig. 8. & App. Hist. An. Angl. in Goedart. p. 32. tit. 36. tab. 1. fig. 8.

Tellina ex rufo maculata, fasciis exasperata. List. H. Conch. tab. 394. fig. 241.

Tellina cuneata compressa, e rubro radiata. Red waved bastard tellen. Petiv. Gaz. tab. 94. fig. 9.—*Tellina incarnata. Carnation.* Penn. Brit. Zool. No. 31. tab. 47. fig. 31.

Concha testa ovata: altero latere angulo plano à cardine ad ambitum. Lin. F. Suec. I. p. 381. No. 1337.

An *Trifasciata. Tellina testa ovata læviuscula sanguineo triradiata, pube rugosa.* Lin. S. N. p. 1118. No. 57. F. Suec. II. No. 2133.

This *species* is a *broad* and *short* shell, thin, brittle, and semitransparent; double the *size* of an almond, about one inch and a half in *breadth*, or from side to side, and only about three quarters of an inch *long*, or from the hinge to the bottom; of a

Cette *espece* est une coquille *large* & *courte*, mince, fragile, & demitransparente; du double la *grandeur* d'une amande, environ un pouce & demi en *largeur*, ou de coté à coté, & porte en *longueur*, ou dès la charniere au bas, seulement

a narrow oval *shape*, and very depressed or shallow.

Outside, thickly and transversely striated; the *striæ* strong and well wrought. Of a pale rosy white *colour*; sometimes of a very rosy hue, and with several broad and narrow interrupted longitudinal *rays* of *red*. It is often covered with a thin blackish fibrous *epidermis*. The *beaks* central, very small, and sharp. The *sides* equal. The *cartilage side* has a strong *flexure* its length, marked on the back by a crenated *ridge*, and projects like a neck, somewhat sharp or angular. This *flexure* is pretty *sloping* or *oblique*, and its striæ run into an angle with those of the body, are stronger and more rough. The *margins* plain.

Inside, smooth, glossy, and white, oftentimes with a rosy hue. The *margins* are plain.

This *pretty species* is rather *uncommon* on our coasts. I have received it from *Scarborough* in *Yorkshire*. It is *scarce* on the coasts of *Cornwall*, but of a larger size.

ment environ trois quarts d'un pouce; d'une *forme* ovale etroite, & fort comprimée ou peu profonde.

L'*exterieur*, serrement & transversalement strié; les *stries* fortes & bien marquées. D'une *couleur* blanche teinte de rose pale; quelquefois d'une vermeille plus foncée, & à plusieurs *raies* larges & etroites, longitudinales & interrompues, de *rouge*. Il est souvent couvert d'une *epiderme* mince, cordée & noiratre. Les *becs* centrals, tres petits, & pointus. Les *cotés* egaux. Le *coté* du *cartilage* à un *pli* fort son long, marqué sur le dos par un *sillon* crenelé, & s'etend comme une languette, un peu aigue ou angulaire. Cet *pli* est asses *penchant* ou *oblique*, & ses stries forment un angle avec celles du corps, & sont plus fortes & plus aspres. Les *bords* unis.

Le *dedans*, lisse, lustré, & blanc, souvent avec un teint vermeil. Les *bords* sont unis.

Cette *jolie espece* est plutot *rare* sur nos rivages. Je l'ai reçu de *Scarborough*, au comté de *York*. Elle est aussi *rare* sur la cote de *Cornwall*, mais plus grande.

XLIII.

XLIII.

T. Tenuis. Thin.

Tellina valde tenuis, parva, subrotunda, plerumque rubra.

Tellina parva, intus rubra, ad alterum latus sinuosa. List. H. Conch. tab. 405. fig. 250. Et *Tellina lævis intus & extra rubra, ad latus sinuosa.* Ib. fig. 251.

Tellina parva radiata, intus omnino purpurascens. Petiv. Gaz. tab. 18. fig. 4.—*Tellina planata. Plain.* Penn. Brit. Zool. No. 29. tab. 48. fig. 29.

Tellina æquilatera lævis, parva subrubra. Gualt. I. Conch. tab. 77. fig. M.

Concha testa subrotunda glabra incarnata. Lin. F. Suec. I. p. 381. No. 1335.—*Planata.* *Tellina testa ovata compressa, transversim substriata lævi: marginibus acutis, &c.* S. N. p. 1117. No. 52. & Mus. Reg. p. 480. No. 25.

A *small* shell, about the *size* of a finger nail, very delicate, thin, brittle, and semitransparent; near smooth, being set only with a few hardly perceptible *striæ*, and very glossy; of a roundish *shape*, or less broad than tellens usually are, and depressed or shallow.

Outside, most generally of a bright *carnation red,* with regular concentric transverse *streaks,* of a paler *colour,* and *whitish.* *Many* shells are *whitish, some yellowish,* and a *few* of a deep *violet,* but all have the transverse *streaks* of *paler colours.* The *beaks* almost central, very small, and pointed. The *cartilage side* has a *flexure,* but less sloping or remarkable than common. The *margins* are plain, and very sharp.

Inside, smooth and glossy; of the same *colours* as the outside, but not so vivid. The *margins* plain.

This *species* is found in plenty on many of our shores, as *Kent, Essex, Dorsetshire, Cornwall, Yorkshire,* &c.

Une *petite* coquille, environ la *grandeur* de l'ongle, tres delicate, mince, fragile, & demitransparente; presque lisse, ou seulement à quelques *stries* à peine visibles, & tres lustré; d'une *forme* quasi ronde, ou moins large que les tellines sont generalement, & comprimée ou peu profonde.

L'*exterieur,* plus communement d'une belle couleur *rouge incarnatte,* à *raies* concentriques & transversales, plus *pales,* & *blanchatres. Plusieurs* coquilles sont *blanchatres, quelqu'unes jaunatres,* & un *petit nombre violet* foncé, mais toutes ont les *raies* transversales de *couleurs plus pales.* Les *becs* presque centrals, tres petits, & pointus. Le *coté* du *cartilage* à un petit *pli,* mais moins en pente ou remarquable que à l'ordinaire. Les *bords* sont unis, & fort tranchants.

Le *dedans,* lisse & lustré; de mêmes *couleurs* que a l'exterieur, mais point si eclatantes. Les *bords* unis.

Cette *espece* se trouve en abondance sur plusieurs de nos cotes, comme aux comtés de *Kent, Essex, Dorset, Cornwall, York,* &c.

XLIV.

T. RUBRA. RED.

Tab. XII. fig. 4. 4. 4.

XLIV.

Pl. XII. fig. 4. 4. 4.

T. parva subrotunda fasciata plerumque rubra. Rubra. Tab. 12. fig. 4. 4. 4.

Concha parva subrotunda, ex parte interna rubens. List. H. An. Angl. p. 175. tit. 25. tab. 4. fig. 25.

Chama minor fasciis paucioribus & fasciis plurimis. Spoon eggs. Petiv. Gaz. tab. 94. fig. 5 & 6. Wallace Orkn. p. 43. Dale Harw. p. 387. No. 1. Smith Cork, p. 318.

Ff —*Small*

—*Small smooth conchs with red and white fillets*; *with bright yellow and white fillets*; *white tinged with red*; also the *rose-coloured Tellina with white fillets*; and the *white Tellina with pearl-coloured fillets.* Wallis Northumb. p. 397. No. 15. 16. 17. 19. & 20.—*Tellina carnaria. Flesh-coloured.* Penn. Brit. Zool. No. 32. tab. 49. fig. 32. 32.

Buon. Ricr. p. 161. No. 44.

This *shell* is thick and strong for its bigness, yet semitransparent; of a rounded *shape*; the *size* of a small or black cherry, or about three quarters of an inch *either way*, and pretty concave or deep.

Cette *coquille* est epaisse & forte pour son volume, cependant demitransparente; d'une *forme* quasi ronde; de la *grandeur* d'une petite cerise ou cerise noire, ou environ trois quarts d'un pouce tant en *largeur* que *longueur*, & asses concave ou profonde.

Outside, smooth; of different *ground colours*, chiefly *red*, *yellow* of several shades, and some *milk white* and *whitish*, marked with regular concentric transverse *streaks* and *belts*, of same colours, but deeper. The *belts* are of one *colour* on *each* particular *shell*, as *white* on the *yellow* or *red grounds*, &c. and form *pretty variegations*. These *belts* seem to rise a little prominent from the surface of the shell, and also of a more compact or solid nature. The *beaks* near central, small and pointed. The *sides* almost equal, one rounded, the other somewhat angular by the *flexure* on it, which is narrow, and not strongly marked. The *margins* are plain and sharp.

L'*exterieur*, lisse, ; de differentes *robes*, principalement *rouge*, *jaune* de plusieurs nuances, & quelqu'unes *blanches* comme le *lait* & *blanchatres*, marquées par des *raies* & *zones* regulieres, concentriques & transversales, de mêmes couleurs, mais plus foncées. Les *zones* sont d'une même *couleur* sur *chaque coquille* particuliere, comme *blanches* sur les *robes jaunes* ou *rouges*, &c. & forment de *jolies bigarrures*. Ces *zones* paroissent un peu elevées ou saillantes au dessus de la surface de la coquille, & aussi d'une substance plus solide ou compacte. Les *becs* presque centrals, petits, & pointus. Les *cotès* à peu pres egaux, un arrondi, l'autre quelque peu angulaire par le *pli* qu'il a, qui est etroit, & pas tres marqué. Les *bords* sont unis & tranchants.

Inside, smooth, glossy, and most generally of a *deeper red*, *sometimes* pale *yellow* or citron, and very *seldom white*. The *margins* plain and sharp.

Le *dedans*, lisse, lustré, & le plus generalement d'une couleur *rouge* plus *foncée quelquefois jaune* pale ou *citron*, & *tres rarement blanc*. Les *bords* unis & tranchants.

This *pretty species* is very common on many of the *British* shores.

Cette *jolie espece* est tres commune sur plusieurs des cotes *Britanniques*.

XLV. XLV.

XLV. XLV.

T. Bimaculata. Double Spot.

T. minima lævis alba, intus maculis duabus ſanguineis oblongis notata. Bimaculata.

Bimaculata. Tellina teſta triangulo-ſubrotunda latiore lævi albida: intus maculis duabus ſanguineis oblongis. Lin. S. N. p. 1120. No. 67.—F. Suec. II. No. 2135.

A very *ſmall ſpecies*, about the *ſize* of a large pea; delicate, thin, brittle, and tranſparent; of a ſubtriangular rounded *ſhape*, and moderately concave.

Outſide, milk white, ſleek, and gloſſy, with a few very ſlight or obſolete tranſverſe *wrinkles*. The *beaks* central, ſmall, and pointed. The *ſides* near equal; the *flexure* is hardly perceivable. The *margins* plain. The two inſide *ſpots*, by the deepneſs of their colour, and by the thinneſs or tranſparency of the ſhell, are viſible on this ſide.

Inſide, milk white, ſmooth, and gloſſy, with two long and broad parallel *ſpots* or *rays* of a deep and fine red *colour*, that riſe from each ſide of the hinge, and run ſtrait till near the bottom, in a pretty and remarkable manner. The *margins* plain.

I have received this *remarkable ſpecies* from the ſhores of *Lancaſhire* and *Hampſhire*.

Une tres *petite eſpece*, environ la *grandeur* d'un gros pois; delicate, mince, fragile, & tranſparente; d'une *forme* quaſi triangulaire arrondie, & moderement concave.

L'*exterieur*, blanc de lait, uni, & luſtré, avec quelques *rides* tres legeres ou peu marquées & tranſverſales. Les *becs* centrals, petits, & pointus. Les *cotés* preſque egaux; le *pli* eſt à peine viſible. Les *bords* unis. Les deux *taches* en dedans, par leur couleur foncée, & par la tenuité & tranſparence de la coquille, ſont viſibles de cet coté.

Le *dedans*, blanc de lait, liſſe, & luſtré, à deux *taches* ou *raies* paralleles, longues & larges, d'une *couleur* rouge belle & foncée, qui s'elevent des cotés de la charniere, & courent directement preſque aux bords, d'une maniere tres remarquable & jolie. Les *bords* ſont unis.

J'ai reçu cette *eſpece remarquable* des cotes des comtés de *Lancaſter* & *Hants*.

GENUS XI.
MYTILUS. MUSCLE.

Bivalves with *equal valves*, and ſhut cloſe. The *hinge* is toothleſs, and conſiſts of a longitudinal and linear furrow.—The animal, an *Aſcidia?*

GENRE XI.
MYTILUS. MOULE.

Bivalves à *battans egaux*, qui ſe ferment exactement. La *charniere* eſt ſans dents, & composée d'une rainure longitudinale & en ligne. — L'animal eſt un *Aſcidia?*

* FLUVIATILES. RIVER.
XLVI.
M. CYGNEUS. GREAT HORSE MUSCLE.

* FLUVIATILES.
XLVI.

MYTILUS *fluviatilis maximus, admodum tenuis ex fuſco virideſcens. Cygneus.*

Muſculus latus maximus, teſta admodum tenui, ex fuſco virideſcens, paluſtris. Liſt. App. H. An. Angl. p. 8. tit. 30. tab. 1. fig. 3. & App. H. An. Angl. in Goedart. p. 9. tit. 30. tab. 1. fig. 3.—*Muſculus latus maximus & tenuiſſimus & cæruleo virideſcens, fere paluſtris.* H. Conch. tab. 156. fig. 11.

Mytilus fluminum maximus ſubviridis. Plott Oxfordſh. ch. 7. § 32.—*M. Cygneus. Swan.* Penn. Brit. Zool. No. 78. tab. 67. fig. 78. *and in many of our Engliſh writers this and the following kind are indiſcriminately put together.*

Muſculus fluviatilis maximus, profunde ſtriatus, intus; teſta admodum tenui, ex fuſco virideſcens, interdum ruſeſcens, intus argenteus. Gualt. I. Conch. tab. 7. fig. F.

Concha teſta oblonga ovata longitudinaliter ſubrugoſa, poſtice compreſſo-prominula. Lin. F. Suec. I. p. 380. No. 1332.—*Cygneus. Mytilus teſta ovata antice compreſſiuſcula fragiliſſima cardine laterali.* S. N. p. 1158. No. 257.

This is a large and *broad ſpecies*, from five to ſix inches *broad*, or from ſide to ſide, and only about two inches and a half to three inches *long*, or from the hinge to the oppoſite or bottom margin; of an oval *ſhape*; thin, ſemitranſparent, brittle, and light; and the *valves* are very deep or concave.

Outſide, wrought with numerous ſtrong concentric tranſverſe *wrinkles*, and has often

Une *eſpece* grande & *large*, de cinq à ſix pouces en *largeur*, ou de coté à coté, & ſeulement environ deux pouces & demi à trois pouces en *longueur*, ou dès la charniere au bord oppoſé; d'une *forme* ovale; mince, demitranſparente, fragile, & legere; & les *valves* ſont fort profondes ou concaves.

L'*exterieur*, à *rides* nombreuſes, fortes, tranſverſales & concentriques, & porte ſouvent

often great remains of a thin, ſtringy, ſtrong *epidermis*, of a browniſh dull green *colour*. The *ſides* very diſſimilar or unequal; one quite rounded and ſhort, the other very extended and narrow, but alſo rounded at its end: this is the *cartilage ſide*, and has a ſtrong *ſlope* or declivity from near the middle of the ſhell. The *beaks* are placed near the ſhort and rounded extreme, ſmall, quite level with the ſhell, or no wiſe prominent, pointed, even with the commiſſure, and generally have the outer coat worn off, ſo as to be very pearly. The *margins* are plain.

Inſide, ſmooth, very pearly, and has often ſmall rugged knobs or *pearls*. The *beaks* on this ſide are even with the reſt of the ſhell, and no wiſe apparent. The *margins* plain.

This *kind* is common (tho' leſs ſo than the next ſpecies) in the Britiſh *rivers*, *ponds*, and other waters.

ſouvent des grands reſtes d'une *epiderme* mince, forte, & cordée, d'une *couleur* terne brunatre verte. Les *cotés* fort inegaux, un tout à fait arrondi & court, l'autre fort etendu & etroit, mais pareillement arrondi à ſon extremité: cet *coté* eſt celui du *cartilage*, & à un *pente* forte dès le milieu de la coquille. Les *becs* ſont ſitués près du coté arrondi & etroit, petits, tout à fait applatis à la ſurface de la coquille, ou nullement ſaillants, pointus, de niveau avec la commiſſure, & ſont generalement tant depouillés ou frottés de la robe exterieure, que d'etre tres nacrès. Les *bords* ſont unis.

Le *dedans*, liſſe, fort nacrè, & ſouvent à petits boutons rudes ou *perles*. Les *becs* de cet coté de niveau avec le reſte de la coquille, & nullement apparents. Les *bords* ſont unis.

Cette *eſpece* eſt commune (quoique moins frequente que la ſuivante) dans les *rivieres*, *etangs*, & autres eaux de la Grande Bretagne.

XLVII.

M. Anatinus. Small Horse Muscle.

Tab. XV. fig. 2.

XLVII.

La Grande Moule des Etangs.

Pl. XV. fig. 2.

Mytilus fluviatilis minor. Anatinus. Tab. 15. fig. 2.

Muſculus latus, teſta admodum tenui, ex fuſco viridſcens, interdum rufeſcens, &c. Liſt. H. An. Angl. p. 146. tit. 29. tab. 2. fig. 29.—*Muſculus tenuis minor latiuſculus.* App. H. An. Angl. p. 10. tit. 30. tab. 1. fig. 2. App. H. An. Angl. in Goedart, p. 13. tit. 30. tab. 1. fig. 2. Exercit. Anat. 3. tab. 2.—*Muſculus tenuis, minor, ſubfuſcus, latiuſculus.* H. Conch. tab. 153. fig. 8.

Mytuli majores à noſtratibus. Horſe muſcles. Merret Pin. p. 193.—*Broad thin horſe muſcle.* Petiv. Gaz. tab. 93. fig. 8. Morton Northampt. p. 418. Leigh Lancaſh. p. 138. Wallis Northumb. p. 402.—*M. Anatinus. Duck.* Penn. Brit. Zool. No. 79. tab. 68. fig. 79.

Conch.

Concha lunga. Buon. Ricr. p. 161. fig. 40. Klein Ostrac. p. 129. No. 2. tab. 9. fig. 26. Argenv Conch. I. tab. 27. fig. 10. No. 5. 6. 7. II. tab. 8. fig. 12.

Musculus fluviatilis, striatus, angustior, umbonibus acutis, valvarum cardinibus veluti pinnis donatis, sinuosis, ex flavo viridescens intus argenteus. Gualt. I. Conch. tab. 7. fig. E.

La grande moule des etangs. Mytulus, testa tenui, è fusco viridescente, umbone non prominulo. Geoffroy Coquilles de Paris, p. 139. No. 1.

Anatinus. Mytilus testa ovali compressiuscula, fragilissima margine membranaceo, natibus decorticatis. Lin. S. N. p. 1158. No. 258. F. Suec. II. No. 2158.

This *species*, in most *particulars*, is very *like* the *last*, but differs in being only of about half the *size*, rather more *compressed* and *oblong*, of a clearer green *colour*, and that the *cartilage* or *lengthened side* extends or runs on a strait line to an acute *angle*, like a *fin*, and thence continues down the side in an oblique line, till near the bottom, where it is rounded.

This shell is *extremely common* in our rivers and stagnant waters.

Cette *espece*, dans plusieurs *circonstances*, *ressemble fort* à la *derniere*, mais elle differe en etant seulement la moitie si *grande*, plutot plus *comprimée* & *oblongue*, & d'une *couleur* verte plus claire. Le *coté allongé* ou du *cartilage* s'etend sur une ligne droite à un *angle* aigu, comme une *nageoire*, & de la continue le long du coté sur une ligne oblique, jusques près du bas, ou il est arrondi.

Cette coquille est *extremement commune* dans nos rivieres & eaux croupissantes.

* MARINÆ. SEA.

XLVIII.

M. VULGARIS. COMMON MUSCLE.

Tab. XV. fig. 5. left-hand.

* DE MER.

XLVIII.

LA MOULE COMMUNE.

Pl. XV. fig. 5. à gauche.

Musculus vulgaris sublævis ex cæruleo niger. Vulgaris. Tab. 15. fig. 5. left-hand.

Musculus ex cæruleo niger. List. H. An. Angl. p. 182. tit. 28. tab. 4. fig. 28.—*Musculus subcæruleus, fere virgatus.* H. Conch. tab. 362. fig. 200.—*Musculus vulgaris marinus subcæruleus fere virgatus, aut nigricans.* Exercit. Anat. 3. p. 30. tab. 4.

Mytuli. Common sea muscles. Merret Pin. p. 193. Wallace Ork. p. 45. Martin W. Isles, p. 6. 343. &c.—*Musculus maritimus vulgatissimus edulis.* Petiv. Mus. p. 84. No. 820. Leigh Lancash. p. 134. tab. 3. fig. 1. a 3. Dale Harw. p. 388. Borlase Cornw. p. 274. Smith Waterford & Cork, p. 272. & p. 318. Rutty Dublin, p. 385. Wallis Northumb. p. 395.—*Mytulis edulis. Edible.* Penn. Brit. Zool. No. 73 tab. 63. fig. 73.

Mytilus. Gesn. Rondelet, Bellon & Aldrovand.—Anth. de Heide Anatome Mytuli, Amst. 1634, cum fig.—Buon. Ricr. p. 158. fig. 30.—Mem. de l'Acad. de Paris, 1711. —Argenv. Conch. II. p. 52. tab. 5. fig. D. E. F. H.

Mytilus

Mytilus latus, aliquando lævis, aliquando rugosus, ex albido nitide violaceus, seu obscure hyanthinus. Gualt. I. Conch. tab. 91. fig. E.

Concha testa oblonga lævi subviolacea. Lin. F. Suec. I. p. 381. No. 1333. II. No. 2156. —*Edulis. Mytilus testa læviuscula violacea, valvulis antice subcarinatis, postice retusis, natibus acuminatis.* S. N. p. 1157. No. 253.—*Edulis. Mytilus testa læviuscula violacea, valvulis obliquis postice acuminatis.* Muf. Reg. p. 541. No. 135.

This *species* is pretty thick, strong, and somewhat semipellucid, of an oblong or suboval *shape*, broad and rounded at the bottom, narrow and pointed at the top. Properly speaking, it is a *long* and narrow *shell*, generally from two and a half to three inches *long*, and about one inch across the middle in *breadth*; but in some places muscles are found of a much larger *size*. The *valves* are very deep or concave.

Outside, near smooth in the *young* or *small shells*, and thickly wrought with numerous concentric fine striæ, and some strong wrinkles, in the *large* or *old* ones. It is covered by a thin or filmy brownish black *epidermis*, which being taken off, the *shell* is of a deep or blackish blue *colour*, but when *uncoated* or *polished*, it is then of a fine deep violet *colour*, sometimes *wholly* so, but at other times *longitudinally striped* with brown or whitish *rays*, from near the middle to the bottom. The *sides* are very dissimilar, one running on a near strait line, and slopes, the other is convex, and extends into a rounded contour, and the whole shell has a somewhat incurvated appearance. The *beaks* are pointed, narrow, prominent, and even with the commissure, and in this part the shell is exceeding thick; but the bottom and side *margins* run to a fine delicate sharp edge, like a knife, and are plain.

Inside,

Cette *espece* est asses epaisse, forte, & un peu demitransparente, d'une *forme* oblongue ou ovale, large & arrondie au bas, etroite & pointue au sommet: à proprement parler, c'est une *coquille longue* & etroite, generalement de deux & demi à trois pouces en *longueur*, & environ un pouce à travers le milieu en *largeur*; mais dans quelques parages les moules se trouvent beaucoup plus *grandes*. Les *valves* sont tres profondes ou concaves.

L'*exterieur*, presque lisse dans les *jeunes* ou *petites coquilles*, & à stries fines, nombreuses, serrées & concentriques, & quelques rides fortes, dans les *grandes* ou *vielles* coquilles. Il est couvert d'une *epiderme* brunatre noire, mince ou membraneuse, qui etant depouillée, la *coquille* est d'une *couleur* noiratre bleue foncée, mais quand cette robe est *depouillée* ou *polie*, alors elle est d'une belle *couleur* violette foncée, tantot *tout à fait*, mais tantot aussi *raiée longitudinalement* des *raies* brunes & blanchatres, de près du milieu au bas. Les *cotés* sont fort inegaux, un presque en ligne droite & penchant, l'autre est convexe, & s'etend dans un contour arrondi, & toute la coquille paroit etre un peu courbée. Les *becs* sont pointus, etroits, bossus, & de niveau avec la commissure, & dans cette partie la coquille est extremement epaisse; mais les *bords* des cotés & du bas deviennent fins, delicats & tranchants, comme un couteau, & sont unis.

Le

Inside, smooth, glossy, and somewhat pearly; the center white; the margins all round deep blue; very often with rugged pearl bumps and small *seed pearl*. The *beaks* run concave under the summits; and the *margins*, as on the outside, are delicate, thin, sharp, and plain.

This *species* abounds on all the British and Irish coasts.

It is a *rich food* and *much eaten*, but often occasions *disorders* to some constitutions, the *symptoms* whereof are great swellings, eruptions of blotches or pimples, shortness of breath, convulsive motions, and even sometimes delirium. *Sudorifics*, *vomits*, *oils*, *&c.* are the *remedies*. The *remedy* used by the Dutch is, two spoons-full of oil and one of lemon-juice, or, in defect of this, a little more of vinegar, well shaken together, and swallowed immediately.

This *venom* has been attributed to the *little crab* sometimes found in the *muscles*; *however*, it seems not to have its seat in any thing essential to the *muscle*, for it has been observed that, when *accidents* of this kind have *happened*, they have happened to *particular persons*, while *others*, who at the same time have eaten a *larger quantity*, and perhaps of the same parcels of *muscles*, have been quite *free* of complaints.

XLIX.

Le *dedans* lisse, lustré, & un peu nacré; le centre blanc; tout autour des bords bleu foncé; tres souvent avec des bosses raboteuses nacrées ou perles baroques, & des *tres petites perles*. Les *becs* sont tres concaves au dessous des sommets; & les *bords*, comme à l'exterieur, sont delicats, minces, tranchants, & unis.

Cette *espece* se trouve en abondance sur toutes les cotes de Bretagne & de Irlande.

C'est un *aliment delicieux*, & *beaucoup mangé*, mais souvent il cause des *maladies* à quelques temperements, les *symptomes* desquelles sont des grandes enflures, des pustules ou boutons, courte haleine, des mouvements convulsifs, & même quelquefois le delire. Des *sudorifiques*, *vomissements*, *huiles*, *&c.* sont les *remedes*. Le *remede* usé par les Hollandois est, de deux cuillierées d'huile, & une de jus de citron, ou, en cas de besoin, un peu plus de vinaigre, bien remuês ensemble, & pris immediatement.

On a attribuè cet *venin* au *petit cancre* qui se trouve quelquefois dans les *moules*; *cependant*, il paroit n'avoir point son origine dans aucune chose essentielle à la *moule*, car il a ete remarqué que, quand des tels *accidents* ont *arrivé*, ils ont arrivé à des *gens particuliers*, pendant que d'*autres*, qui au même tems ont mangé une plus *grande quantité*, & peutetre des mêmes amas de *moules*, se sont trouvés tout à fait *libres* de douleurs.

XLIX.

XLIX.

M. Modiolus. Great.

Tab. XV. fig. 5. right-hand.

XLIX.

Pl. XV. fig. 5. à droite.

Mytilus magnus nigrescens. Modiolus. Tab. 15. fig. 5. right-hand.

Lift. H. Conch. tab. 359. fig. 198.—*M. Modiolus. Great.* Penn. Brit. Zool. No. 77. tab. 66. fig. 77.

Musculus papuanus authorum. Rumph. Muf. tab. 46. fig. B.—Klein Oftrac. tab. 11. fig. 67.—Gualt. I. Conch. tab. 91. fig. H.—Argenv. Conch. I. tab. 25. fig. C. II. tab. 22. fig. C.

Modiolus. Mytilus testa lævi, margine anteriore carinato, natibus gibbis, cardine fublaterali. Lin. S. N. p. 1158. No. 256. Muf. Reg. p. 542. No. 157.

A large, thick, ftrong and heavy *fhell*, from fix to fix and a half inches in *length*, and from two and a half to three inches in *breadth*; of a fuboval *fhape*, narrow and blunted at the top, dilated and rounded at the bottom. The *valves* very concave or deep.

Outfide, black, and covered with a flight filmy *epidermis*, and under it is thickly fet with tranfverfe concentric *ftriæ*, that run ftronger, or like wrinkles, on the fides and bottom. From the beaks to near the bottom it is diagonally gibbous or convex, and thence flopes on all fides. The *beaks* very ftrong, large, and convex or fwell'd, as is alfo the part afide them; and the body under them likewife fwells into a roundifh convexity. The *fides* are nearly fimilar, but one extends on a more ftrait line than the other, which is dilated. All the upper or top part of the fhell is extremely thick; the lower part thinner and patulous. The *margins* are plain, and fomewhat fharp.

Infide, fmooth and pearly; fometimes very orient with fine glows of flame, purple,

Une grande *coquille*, epaiffe, forte & pefante, de fix à fix & demi pouces en *longueur*, & de deux & demi à trois pouces en *largeur*; d'une *forme* quafi ovale, etroite & obtufe au fommet, etendue & arrondie au bas. Les *valves* font concaves ou profondes.

L'*exterieur*, noir, & couvert d'une *epiderme* membraneufe legere, & au deffous eft à *ftries* concentriques tranfverfales, qui deviennent plus fortes, ou comme des rides, fur les cotés & au bas. Dès les becs prefque au bas il eft diagonalement boffu ou convexe, & de la penche de touts cotés. Les *becs* tres forts, grands, & convexes ou enflès, comme auffi la partie à leur coté; & le corps au deffous eft pareillement enflé en convexité arrondie. Les *cotés* font à peu pres egaux, mais un court en une ligne plus droite que l'autre, qui eft etendu. Toute la partie fuperieure ou le fommet de la coquille eft extremement epaiffe; la partie inferieure plus mince & etendue ou ouverte. Les *bords* font unis, & un peu tranchants.

Le *dedans*, liffe & nacré; quelquefois d'un bel orient à flammes de feu, de pourpre,

ple, rosy, and green. The *margins* as on the outside.

This *species* is the *largest* of the *British muscles*, and is found on several of our coasts; as very fine and large at *Scarborough* in *Yorkshire*, where they are only got in deep waters, take hold of the fishermen's baits, and are hawled up with the lines, but are never found cast on shore. They are likewise found in other places on the coast of that county. I have also received them of large size from the coasts of *Wales*, the *Orkneys* and other coasts of *Scotland*; and of a smaller size from *Margate Flats* in *Kent*, from *Cornwall*, and other English shores.

pre, de rose, & de verd. Les *bords* comme à l'exterieur.

Cette *espece* est la plus *grande* de *moules Britanniques*, & se trouve sur plusieurs de nos cotes; comme de tres grandes & belles à *Scarborough*, au comté de *York*, ou elles habitent seulement les eaux profondes, avalent les amorces des pescheurs, & sont prises par leurs lignes, mais ne se trouvent jamais jettées sur le rivage. Elles se trouvent pareillement en d'autres lieux de la cote de cet comté. Je les ai aussi reçu, & tres grandes, dès cotes de la principauté de *Galles*, dès isles *Orcades* & autres cotes de l'*Ecosse*; & de moindre grandeur dès *Basses de Margate* en *Kent*, de *Cornwall*, & autres cotes Angloises.

L.

L.

M. Curtirostratus. Wry-Beak.

Mytilus nigrescens apice tortuoso. Curvirostratus.
M. umbilicatus. Umbilicated. Penn. Brit. Zool. No. 76. tab. 65. fig. 76.

This *species* is somewhat above half the *size* of the foregoing, or *Modiolus*, and is much thicker, but in most other respects resembles it.

The *principal* and *distinguishing character* lies in the top and hinge; for the space opposite to the hinge is depressed and bent inwards, in a crooked or winding manner, into a deep rugged cavity, which, when the shells are closed, forms a long deep hollow, or *umbilicus*, as if beat or bruised in. On one *valve*, this *depression* is more inflected inwards,

Cette *espece* est quelque peu au dessus de *mi-grandeur* de la precedente, ou *Modiolus*, & beaucoup plus epaisse, mais dans la pluspart d'autres circonstances la ressemble.

Son *caractere principal* & *distingué* se trouve dans le sommet & la charniere; car l'espace opposé à la charniere est enfoncé & plié en dedans, d'une maniere tortue ou sinueuse, en une cavité profonde & raboteuse, qui, quand la coquille est fermée, forme un creux ou *umbilic* long & profond, comme si il etoit froissé ou poussé en dedans. Sur une *valve* cet

wards, beat down and winding, than on the other.

Mr. *Pennant* is the only *author* who has *proposed* this *rare* and *new species*; he says, it is sometimes dredged up off *Priestholme*, in the island of *Anglesea*. My specimen was from that place.

cet *enfoncement* est plus tourné ou poussé en dedans, & plus sinueux, que dans l'autre.

M. *Pennant* est le seul *auteur* qui à *proposé* cette *espece rare* & *nouvelle*; il note qu'elle se trouve quelquefois pesché à *Preistholme*, dans l'isle de *Anglesea*, en *Galles*. Notre coquille est de cette place.

LI.
M. Discors. Divided.
Tab. XVII. fig. 1.

LI.
Pl. XVII. fig. 1.

Mytilus minor tenuis areis tribus distinctus. Discors. Tab. 17. fig. 1.
Discors. Mytilus testa ovali cornea subdiaphana, antice longitudinaliter, postice transversaliter striata. Lin. S. N. p. 1159. No. 261.

This is a very *small species*, of the *size* of a kidney-bean, extremely thin, semitransparent or horny, and brittle, of a light greenish *colour*, with a faint tint of brown or rosy; it is a *broad shell*, of a squarish *shape*, and the *valves* are extremely deep or concave.

Outside, longitudinally divided into *three areas* or *compartments*, well defined; the *middle* one broad and smooth, or at most with such exceeding fine transverse striæ as hardly to be perceptible; the *other two* exterior ones, one of which is as broad again as the other, are finely *striated* their length. The *margins* are delicately crenated. The *sides* dissimilar, one greatly extended, the other very short, but both rounded. The *beaks* strong and prominent, but obtuse, level with the commissure, posited near the short side, like to the *Cunei* or *Purrs*, and turn towards it. The

Celle-ci est une tres *petite espece*, de la *grandeur* d'une faseole, extremement mince, demitransparente ou comme de la corne, & fragile, d'une *couleur* verdatre claire, avec un teint leger de brun ou de rose; elle est une *coquille large*, d'une *forme* quasi quarrée; & les *valves* sont extremement profondes ou concaves.

L'*exterieur*, divisé longitudinalement en *trois aires* ou *compartiments*, bien marqués; celui du *milieu* large & lisse, ou au plus dire avec des stries transversales, si fines que à peine sont elles visibles; les *deux autres* exterieurs, un desquels est autant plus large que l'autre, sont delicatement *striés* leur longueur. Les *bords* sont mignonnement crenelés. Les *cotés* inegaux, un beaucoup allongé, l'autre tres court, mais touts deux arrondis. Les *becs* forts & elevés, mais obtus, de niveau avec la commissure, situés près du coté court, comme dans les *Cunei*, & tournès envers ce coté.

The lower *margins*, or those opposite to the hinge, run on a near strait line, which gives the subsquarish *figure* to the shell.

Inside, smooth and glossy, sometimes pearly, at other times same as the outside, with a more rosy blush. The *margins* finely crenated; and, by the thinness of the shell, the outside striæ are seen.

Linné is the *first author* who has described this *pretty species*, found by one of his curious pupils in Norway and Iceland. His *description* is excellent, bating only the *mistake* he makes in saying the *third area* is transversely *striated*, whereas it is longitudinally striated, exactly similar to the first area or compartment.

The *discovery* of this pretty and rare species on our *coasts* is owing to the unwearied pursuits of Dr. *Richard Pulteney*, *F. R. S.* of *Blandford* in *Dorsetshire*, who obligingly communicated to me this, with many other *curious observations*, set forth in *my work*. He found it on an *ascidia*, at *Weymouth* in that county, and I have received several since from that coast.

All that *Linné* had *seen*, as well as all those found on our *coasts*, are very small, thin, and delicate; but a kind no wise different, except in *size* and *colour*, being larger than a great walnut, and quite brown, was brought from the southern hemisphere by that great and national honour Capt. Cook, the circumnavigator, in the late expeditions for the discovery of new countries. These also were intirely *unknown to* all our collectors; and, as they only differ in *size*, *thickness*, and *colour*,

coté. Les *bords* du bas, ou ceux opposés à la charniere, courent sur un ligne presque droite, ce qui cause la *forme* quasi quarrée de la coquille.

Le *dedans*, lisse & lustré, quelquefois nacré, & tantot de même que l'exterieur, avec un teint plus foncé de rose. Les *bords* delicatement crenelés; &, par la tenuité de la coquille, les stries exterieures sont visibles.

Linné est le *premier auteur* qui a decrit cette *jolie espece*, trouvée par un de ses disciples curieux en la Norwege & l'Iceland. Sa *description* est excellente, moyennant seulement l'*erreur* qu'il fait en disant que la *troisieme aire* est *striée* transversalement, d'autant qu'elle est striée longitudinalement, tout à fait semblable au premier aire ou compartiment.

La *decouverte* de cette jolie & rare espece sur nos *cotes* est due aux poursuites assidues du Dr. *Richard Pulteney*, *Membre de la Societé Royale*, de *Blandford* au comté de *Dorset*, qui me la communiqua d'une maniere tres obligeante, avec plusieurs autres *observations curieuses*, inserrées dans *mon ouvrage*. Il l'a trouvée sur un *ascidia*, à *Weymouth* dans ce comté, & j'en ai reçu plusieurs depuis de cette cote.

Toutes celles que *Linné* a *vu*, comme aussi toutes celles trouvées sur nos *cotes*, sont tres petites, minces, & delicates; mais une sorte point differente, excepté en *grandeur* & *couleur*, etant de volume d'une grande noix, & tout à fait brune, nous fut apporté de l'hemisphere meridionale, par le fameux navigateur, l'honneur de la nation Angloise, le Cap. Cook, dans ses dernieres expeditions faites à la decouverte des pais inconnus. Celles-ci furent entierement *nouvelles à* touts nos connoisseurs,

lour, but are exactly the same in structure, way of life, and other particulars, as these of our coasts, is it a *distinct species*, or *variety* only?

seurs; &, comme elles different seulement en *grandeur*, *epaisseur*, & *couleur*, mais sont exactement la même en structure, maniere de vivre, & autres circonstances, que celles de nos cotes, est elle une *espece distincte*, ou seulement une *varieté*?

LII.

M. RUGOSUS. RUGGED.

LII.

Mytilus parvus rhomboideo-ovalis, subalbescens, rugosus. Rugosus.

Pholas noster, sive concha intra lapidem quendam cretaceum degens. List. H. An. Angl. p. 172. tit. 21. tab. 4. fig. 21. Id. App. H. An. Angl. p. 18. & App. H. An. Angl. in Goed. p. 30.

Chamæ pholas angusta, parva, sinu utrinque leviter muricato conspicua. List. H. Conch. tab. 426. fig. 267.—*Mytilus rugosus. Rugged.* Penn. Brit. Zool. No. 72. tab. 63. fig. 72.

Rugosus. Mytilus testa rhomboideo-ovali rugosa obtusa antiquata. Lin. S. N. p. 1156. No. 249.

This *species* is small, extremely thick, strong, and opake, of the *size* of a large filberd, or about an inch in *breadth*, for it is a broad and somewhat squarish shell, and about half an inch long, of an oval *shape*, intirely of a dirty or sullied white *colour*, and the *valves* are pretty deep or concave.

Outside, rugged, and transversely *wrinkled*. The *sides* dissimilar, one very lengthened, the other very short, but both extremes are rounded and obtuse. The *lower end*, or *bottom*, lies on near a strait line with the *upper* part, which gives it a somewhat squarish figure. The *beaks* large, prominent, turned inwards, level with the commissure, and are placed near the edge of the short side, like a *cuneus* or *purr*. The *margins* very thick, and plain.

Inside,

Cette *espece* est petite, extremement epaisse, forte & opaque, du *volume* d'une grande noisette, ou environ un pouce en *largeur*, car elle est une coquille large & un peu quarrée, & environ un demi pouce en *longueur*, d'une *forme* ovale, tout à fait d'une *couleur* blanche sale ou morne, & les *valves* sont asses profondes ou concaves.

L'*exterieur*, raboteux, & *ridé* transversalement. Les *cotés* inegaux, un fort allongé, l'autre tres court, mais les deux extremités sont arrondies & obtuses. Le *bas*, ou *bord inferieur*, court presque sur une ligne droite avec la partie *superieure*, qui fait qu'il a une forme quasi quarrée. Les *becs* grands, saillants, tournès en dedans, de niveau avec la commissure, & sont situés près du bord du coté court, comme le *cuneus*. Les *bords* sont epais, & unis.

Le

Inside, ſmooth, and no ways gloſſy. The *margins* thick and plain.

This *ſpecies* is not very *common* on our *coaſts*. Liſter notes it at *Hartlepool* in *Durham*; but on the coaſt of *Yorkſhire*, as about *Scarborough*, *Whitby*, &c. it is found in *incredible abundance*, *nitch'd* or *burrowed* in the *rocks* of lime-ſtone, alum-ſtone, &c.

Le *dedans*, liſſe, mais pas luſtré. Les *bords* epais & unis.

Cette *eſpece* n'eſt pas fort *commune* ſur nos *cotes*. Liſter note qu'elle ſe trouve à *Hartlepool*, au comté de *Durham*; mais ſur la cote du comté de *York*, comme à *Scarborough*, *Whitby*, &c. elle ſe trouve en *abondance incroyable*, *nichées* ou *enterrées* dans les *rochers* de pierre à chaux, mine d'alun, &c.

GENUS XII.

MYA.

Bivalves with *equal valves*, and shut close; generally *broad* shells. The *hinge* is set with large thick rugged teeth, and lateral ones, like plates, locking into one another.—The animal an *Ascidia*.

GENRE XII.

Bivalves à battans egaux, qui se ferment exactement; generalement des coquilles *larges*. La *charniere* à grandes dents, epaisses & raboteuses, & dents laterales, comme des lames, s'engrainant reciproquement.—L'animal est un *Ascidia*.

*FLUVIATILES. RIVER.

LIII.

M. MARGARITIFERA.
RIVER PEARL MUSCLE.

Tab. XV. fig. 3. 3.

*FLUVIATILES.

LIII.

Pl. XV. fig. 3. 3.

MYA *nigrescens crassa & ponderosa margaritifera. Margaritifera.* Tab. 15. fig. 3. 3.

Musculus niger omnium crassissima & ponderosissima testa. Conchæ longæ species Gesn. & Aldrov. List. App. H. An. Angl. p. 11. tit. 31. tab. 1. fig. 1. & App. H. An. Angl. in Goedart. p. 15. tit. 31. tab. 1. fig. 1.—*Musculus niger omnium longe crassissimus. Conchæ longæ species Gesn. & Aldrov.* Hist. Conch. tab. 149. fig. 4.

Musculi margaritiferi. Bede Hist. Ecclesiast. l. 1. c. 1. Platt's Jewel-house. Martin W. Isles, p. 7. &c.—*Pearl muscles.* Leigh Lancash. p. 134. Morris Charts of Wales. Survey of County of Down, p. 46.—*Mytulus major margaritiferus.* Wallis Northumb. p. 403. No. 42.—*Mya margaritifera. Pearl.* Penn. Brit. Zool. No. 18. tab. 43. fig. 18.

Chama-glycimeris margaritifera. Velschii Ephem. Nat. Curios. German. an. 3. obs. 36. Klein Ostrac. tab. 10. fig. 47.—*Concha testa oblonga, medio antice contracta.* Lin. F. Suec. I. p. 380. No. 1331. II. No. 2130. Et *Margaritifera. Mya testa ovata antice coarctata, cardinis dente primario conico, natibus decorticatis.* S. N. p. 1112. No. 29.

This *species* is extremely thick, strong, heavy, opake, and broad; from five to six inches in *breadth*, or from side to side, and from one and a half to three inches in *length*, or from the hinge to the opposite margin;

Cette *espece* est extremement epaisse, forte, pesante, opaque, & large; de cinq à six pouces en *largeur*, ou de coté à coté, & de un & demi à trois pouces en *longueur*, ou dès la charniere au bord opposé; d'une

margin; of an oblong-oval *ſhape*; and the *valves* are moderately deep or concave.

Outſide, black, wrought with numerous concentric tranſverſe *wrinkles*. The *ſides* very diſſimilar or unequal, one lengthened and oblique, the other ſhort and rounded. The *bottom margin*, or that oppoſite the hinge, runs on a ſerpentine or waved line. The *ſummit* round the beaks is always greatly worn down or decorticated, to the inner pearly coat. The *beaks* are large, prominent, blunt, level with the commiſſure, and placed near to the ſhort ſide. The *margins* are much thinner than the reſt of the ſhell, ſharp and plain.

Inſide, of *orient pearl*, with dazzling glows of fire, blue, roſy, and other bright and vivid colours. The *margins* plain and ſharp. All the *hinge* is very thick and ſinuous, furniſhed in one valve with a large rugged and notch'd conic tooth, like a canine tooth, placed juſt under the beak, which locks into a cavity between two other, like teeth, but ſmaller, in the other valve.

This *ſpecies* is found in our *great rivers*, eſpecially thoſe in the *northern* parts of *Great Britain*, as the *Tees*, *Alne*, *North* and *South Tyne*, *Tweed*, *Dee*, and *Don*, &c. alſo in the great rivers of *Ireland*; it generally inhabits the deeper parts, as gulphs, whirlpools, &c.

This ſhell has long been noted for producing quantities of *pearl*. *Suetonius* reports *(Vita Jul. Cæſ.)* that *Cæſar* was induced to undertake his *Britiſh* expedition for the ſake of our pearls, which were very large. The criſtalline ſpheres and hemiſpheres, now known as *Druid beads*, and mentioned

d'une *forme* oblongue-ovale; & les *valves* ſont moderement profondes ou concaves.

L'*exterieur*, noir, à *rides* nombreuſes concentriques & tranſverſales. Les *cotés* fort inegaux, un allongé & oblique, l'autre court & arrondi. Le *bord du bas*, ou celui oppoſé à la charniere, court en ligne ſerpentine ou ondée. Le *ſommet* autour des becs eſt toujours fort frotté, uſé, ou depouillé de ſa robe, juſques la couche interieure nacrée. Les *becs* ſont grands, saillants, obtus, de niveau avec la commiſſure, & ſitués près du coté court. Les *bords* ſont plus minces que le reſte de la coquille, tranchants & unis.

Le *dedans*, de *nacré* d'un *bel orient*, avec des teints eclatants de feu, bleu, de roſe, & autres couleurs brillantes & vives. Les *bords* unis & aigus. Toute la *charniere* eſt tres epaiſſe & ſinueuſe; dans l'une des valves elle eſt compoſée d'une grande dent conique, raboteuſe & entaillée, comme une dent canine, ſituée juſtement au deſſous du bec, qui s'engraine dans une cavité entre deux autres, pareilles dents, mais plus petites, dans l'autre valve.

Cette *eſpece* ſe trouve dans nos *grandes rivieres*, eſpecialement celles des parties *ſeptentrionales* de la *Grande Bretagne*, comme la *Tees*, l'*Alne*, la *Tyne Septentrionale* & *Meridionale*, la *Dee*, & la *Don*, &c. auſſi dans les grandes rivieres de l'*Irlande*, & generalement ſe trouve dans les eaux profondes, comme les golfes, gouffres, &c.

Cette coquille à eté depuis long tems renommée pour la quantité des *perles* qu'elle produit. *Suetone* rapporte *(Vita Jul. Cæſ.)* que *Cæſar* etoit excité à entreprendre ſon expedition *Britannique* par la conſideration de nos perles, qui etoient tres grandes. Les ſpheres & hemiſpheres de

mentioned in *old deeds* by the name of *mineral pearl (Woodward's Method of Fossils, lett. 3. p. 29. Phil. Transf. No. 201.)* might probably mislead him; however, we are told by *Pliny* and *Tacitus (Hist. Nat. l. 9. c. 35. & Vita Agricolæ)* that he dedicated, and hung up in the temple of *Venus Genetrix*, a buckler made of *British pearl.*

The venerable Bede takes notice of these pearls for their orient lustre and brightness. Pearl fisheries have been established, at different times, in several places. The *Conway* was noted for pearls in the days of *Cambden.* A notion (says Mr. *Pennant*) prevails that Sir Richard Wynne, of Gwydir, chamberlain to Catherine, consort of Charles II. presented that queen with a *pearl* of the *Conway,* which to this day is honoured with a place in the *regal crown. Cambden* also reports, that the famous circumnavigator, Sir John Hawkins, had a patent for fishing pearl muscles in the river *Irt,* in *Cumberland.* Sir Hugh Platt notes those of *Buckinghamshire,* and of the river *Clun* in *Shropshire;* and *Leigh* the fishery of the river *Wire,* in *Lancashire,* where they are called *Hambleton Hookins,* from the manner of taking them, and the place. Mr. *Pennant* also informs us of the valuable pearls of the rivers of *Tyrone* and *Donegal,* in *Ireland.*

Obs.—*Linné* has a genus called *Mya:* it is *toto cœlo* different from this. He *defines* his *myæ* to be *conchæ biantes,* or *gapers:* these, besides the difference of the *hinge, shut quite close,* though he has ranked

Hh them

de cristal, à present connues par le noms de *Druid beads,* ou *grains de Druides,* & mentionnées dans les *vieux actes* sous le nom de *perle minerale (Woodward Methode de Fossiles, lett. 3. p. 29. Transf. Philos. No. 201.)* pouvoit peutetre lui seduire; cependant, nous sommes informés par *Pline* & *Tacite (Hist. Nat. l. 9. c. 35. & Vita Agricolæ)* qu'il a dedié, & mis dans le temple de *Venus Genetrix,* un bouclier fait de *perles Britanniques.*

Le venerable Bede prend connoissance de ces perles pour leur bel orient & eclat. Des pesches des perles ont eté etablies en differents tems, & en differents lieux. La riviere *Conway* etoit fameuse pour les perles, au tems de *Cambden.* Une notion domine (dit M. *Pennant*) que le Chevalier Richard Wynne, de Gwydir, chambellan de Catherine, consorte de Charles II. presenta à cette reine une *perle* du *Conway,* qui même jusques à cet temps est honorée d'une place dans la *couronne royale. Cambden* rapporte aussi, que le fameux navigateur, le Chevalier Jean Hawkins, eut une patente pour pescher ces moules à perles dans la riviere *Irt,* au comté de *Cumberland.* Le Chevalier Hugh Platt donne notice de celles du comté de *Buckingham,* & de la riviere *Clun* au comté de *Salop;* & *Leigh* la pesche qui se fait dans la riviere *Wire,* au comté de *Lancaster,* ou elles sont nommées *Hambleton Hookins,* de la maniere de les pescher, & du lieu. M. *Pennant* nous avise aussi des perles de valeur des rivieres de *Tyrone* & *Donegal,* en *Irlande.*

Obs.—*Linné* a un genre nommé *Mya:* il est *tout à fait* different de ceci. Il donne la *definition* de ses *myæ* d'etre *conchæ biantes,* ou *beantes:* celles-ci, outre la difference en *charniere, se fermeat exactement,* quoiqu'il

les

them in his said genus. Pliny *(Hist. Nat. l. 9. c. 35.)* says, that pearls were bred in a shell called *mya*, but does not mention any marks to ascertain the species: it perhaps was this kind, from whence I have given the *genus* the name of *Mya*.

les à rangé dans son dit genre. Pline *(Hist. Nat. l. 9. c. 35.)* dit, que les perles etoient produites dans une coquille nommée *mya*, mais il ne donne aucunes marques à determiner l'espece: peutetre c'etoit de cette sorte, c'est pourquoi j'ay donné ce *genre* le nom de *Mya*.

LIV.

M. Pictorum. Painters.
Tab. XIV. fig. 4. 4.

LIV.

Pl. XIV. fig. 4. 4.

Mya minor ex flavo viridescens. Pictorum. Tab. 14. fig. 4. 4.

Musculus angustior, ex flavo viridescens, validus, umbonibus acutis, valvarum cardinibus velut pinnis donatis, sinuosis. List. H. An. Angl. p. 149. tit. 30. tab. 2. fig. 30.—*Musculus ex viridi pallidus, omnium angustissimus, cardinis altero denticulo, quasi continuo, serrato.* App. H. An. Angl. p. 13. tit. 33. tab. 1. fig. 4. & App. H. An. Angl. in Goedart. p. 20. tit. 33. tab. 1. fig. 4.—*Musculus angustus, subflavus, sive citrinus.* Hist. Conch. tab. 147. fig. 3.

Long thick horse muscle. Petiv. Gaz. tab. 93. fig. 9. Morton Northampt. p. 417.—*Mya pictorum. Painters.* Penn. Brit. Zool. No. 17. tab. 43. fig. 17.

Buon. Ricr. p. 161. fig. 41.—*Moule des rivieres.* Argenv. Conch. I. p. 372. pl. 27. fig. 10. No. 4. II. p. 329. pl. 31. fig. 10. No. 4. & p. 76. pl. 8. fig. 11.—*Mytulus, testa fusca, umbone prominente.* Geoffroy Coq. de Paris, p. 141. No. 2.

Musculus fluviatilis, striatus angustior, umbonibus acutis, valvarum cardinibus veluti pinnis donatis, sinuosis, ex flavo viridescens, intus argenteus. Gualt. I. Conch. tab. 7. fig. E.

Pictorum. Mya testa ovata, cardinis dente primario crenulato, laterali longitudinali: alterius duplicato. Lin. S. N. p. 1112. No. 28. & F. Suec. II. p. 2129.

A *broad shell*; strong, heavy, and thick, yet semitransparent; of an oblong-oval *shape*; about the *size* of a common muscle, or from two and a half to three inches in *breadth*, and very *short*, being only from one to one and a quarter inch *long*, or from the hinge to the opposite margin. The *valves* are moderately deep or concave.

Outside,

Une *coquille large*; forte, pesante, & epaisse, cependant demitransparente; d'une *forme* oblongue-ovale; environ la *grandeur* d'une moule commune, ou de deux & demi à trois pouces en *largeur*, & tres *court*, ayant seulement un à un & un quart de pouce en *longueur*, ou dès la charniere au bord opposé. Les *valves* sont moderement profondes ou concaves.

L'ex-

Outside, near smooth, being wrought only with some few concentric transverse wrinkles, of a *yellowish green*, and sometimes darker, or of an olive green *colour*. The *beaks* placed near the short side, are large, prominent, pointed inwards, and rise above the commissure, and generally around them the shell is much rubbed or worn down to the inner pearly coat, as in the preceding species, or *Margaritifera*. The *sides* very unequal, one lengthened on a strait line, and ending angular, like a fin, as the *M. Anatinus*, No. 47, *supra*, thence proceeds down the side in an oblique line to the bottom, which is rounded, and all this side slopes, and is narrow; the other side is much dilated or rounded; the *bottom* is also rounded, and the *margins* are sharp and plain.

Inside, very pearly, with glares of bluish in the *smaller*, and in the *large* of red or yellow. The *margins* sharp and plain. All the *hinge* is thick and sinuous, with a large thick notch'd tooth, and a lateral one, like a plate, with its groove, in one valve; the other has two like lateral teeth, with grooves, and a cavity, and all lock in together.

This species is common in our rivers and fresh waters. It sometimes has small or seed pearl.

L'*exterieur* presque lisse, etant seulement à quelques rides transversales & concentriques, *jaune verdatre*, & quelquefois plus foncè, ou d'une *couleur* verte d'olive. Les *becs* situés près du coté court, sont grands, saillants, pointus en dedans, & dejettent la commissure, & generalement autour d'eux la coquille est beaucoup usée & depouillée jusques la couche interieure nacrée, comme dans l'espece precedente, ou la *Margaritifera*. Les *cotés* fort inegaux, un allongé en ligne droite, & finissant angulaire, comme une nageoire, telle que la *M. Anatinus*, No. 47, *supra*, de la il descende le coté en ligne oblique jusques au bas, ou il est arrondi, & tout cet coté est en pente & etroit; l'autre coté est fort etendu ou arrondi; le *bas* est aussi arrondi, & les *bords* sont tranchants & unis.

Le *dedans*, fort nacré, avec des nuances bleuatres dans les *petites* coquilles, & dans les *grandes* de rouge ou jaune. Les *bords* tranchants & unis. Toute la *charniere* est epaisse & sinueuse; dans une valve elle est composée d'une grande dent, epaisse & entaillée, & une dent laterale, comme une lame, avec sa rainure; dans l'autre deux pareilles dents, avec leurs rainures, & une cavité, & toutes s'emboitent ensemble.

Cette espece est commune dans nos rivieres & eaux courantes. Elle produit quelquefois de petites perles.

PART III.

Bivalves with *equal valves*, and never *ʃhut cloʃe*, but are always *open* or *gaping* in ſome part.

TROISIEME PARTIE.

Bivalves à *battans egaux*, & jamais ne ſe *ferment exactement*, mais ſont toujours *ouvertes* ou *beantes* dans quelque partie.

GENUS XIII.

CHAMA. GAPER.

Bivalves of a round or oval *ʃhape*, or approaching thereto, and never *ʃhut cloʃe*. The *hinge* of different ſtructures.

GENRE XIII.

CHAMA.

Bivalves d'une *forme* arrondie ou ovale, ou approchante, & ne ſe *ferment* jamais *exactement*. La *charniere* diverſe en ſtructure.

* MARINÆ. SEA.

LV.

C. MAGNA. LARGE.

Tab. XVII. fig. 4.

* DE MER.

LV.

Pl. XVII. fig. 4.

CHAMA magna planior, craſſa, albeſcens. Magna. Tab. 17. fig. 4.

Concha longa lataque in mediis cardinibus cavitate quadam pyriformi inſignita. An *Chamæ glycymeris Rondeletii?* Liſt. H. An. Angl. p. 170. tit. 19. tab. 4. fig. 19.—*Chama fuſca lata planior.* Hiſt. Conch. tab. 415. fig. 259.

Wallace Orkn. p. 42.—*Chama lata & compreſſa noſtras.* Muſ. Petiv. p. 83. No. 810. —*The long and broad conch.* Wallis Northumb. p. 396. No. 10 & 11.—*Mactra lutraria; large.* Penn. Brit. Zool. No. 44. tab. 52. fig. 44.

Buon. Ricr. p. 154. fig. 19.—*Muſculus rugis tranſverſis inæqualibus ſignatus, ex fulvo & albido infectus.* Gualt. I. Conch. tab. 90. fig. A. *inferior.*

Lutraria. Mactra teſta ovali oblonga levi, dentibus lateralibus nullis. Lin. S. N. p. 1126. No. 101. F. Suec. II. No. 2128.—*Mya lutraria.* Muſ. Reg. p. 470. No. 9.

A *broad ſhell*, thick yet ſemitranſparent, ſtrong and heavy; generally from four inches and a half to five inches in *breadth*, or from ſide to ſide, and about two inches and a half in *length*, or from the hinge to the oppoſite margin; of an oblong-oval *ſhape*, ſinuous or waved in its contour, and patulous at the gaping or open end; and the *valves* are moderately concave.

Outſide, ſomewhat rough, with numerous ſtrong concentric tranſverſe *wrinkles*; *whitiſh*, variegated with pale reddiſh or cheſnut *colour*. The *beaks* ſmall, ſtrong, pointed inwards, juſt overtop the commiſſure, and are placed towards the ſhort ſide. The *ſides* very unequal, one greatly extended, the other ſhort and rounded: the *long* one is the gaping ſide, turns outwards, and is rounded, but the *other* alſo does not ſhut exactly cloſe. The *margins* are plain.

Inſide, ſmooth, ſomewhat gloſſy and white, and the *margins* are plain.

The *hinge*, a large triangular cavity on the long ſide; a ſmaller contiguous one on the ſhort ſide, with an erect ſtrong tooth, plate-like, and divided perpendicularly, that ſockets in the ſmaller cavities of each valve. It has no lateral teeth. The animal *probably* an *aſcidia*.

This *ſpecies* is found on many of our ſhores, and near the mouths of rivers; as in *Yorkſhire*, in the *oſtium* of the *Tees*, in great plenty, at *Scarborough*, &c.; in *Northumberland*, *Lancaſhire*, and other northern coaſts; in *Dorſetſhire*, *Cornwall*, and other weſtern coaſts; in the *Orkneys*, and other ſhores of *Scotland*.

Une *coquille large*, epaiſſe cependant demitranſparente, forte & peſante; generalement de quatre pouces & demi à cincq pouces en *largeur*, ou de coté à coté, & environ deux pouces & demi en *longueur*, ou dès la charniere au bord oppoſé; d'une *forme* oblongue-ovale, ſinueuſe ou ondée dans ſon contour, & debouchée ou etendue à l'extremité ouverte ou beante; & les *valves* ſont moderement concaves.

L'*exterieur*, un peu aſpre, par les *rides* nombreuſes fortes, concentriques & tranſverſales; *blanchatre*, bigarré d'une *couleur* de chataigne ou pale rougeatre. Les *becs* petits, forts, pointus en dedans, dejettent à point nommé la commiſſure, & ſont ſitués vers le coté court. Les *cotés* fort inegaux, un beaucoup etendu, l'autre court & arrondi: le coté *long* eſt celui qui eſt beant, il eſt retrouſſé & arrondi, mais l'*autre coté* ne ſe ferme exactement. Les *bords* ſont unis.

Le *dedans*, liſſe, un peu luſtré & blanc, & les *bords* ſont unis.

La *charniere*, une grande cavité triangulaire du coté long; une plus petite contigue du coté court, avec une dent forte, comme une Lame, erigée, & fendue à plomb, qui s'emboite dans les petites cavités de chaque valve. Elle n'a point de dents laterales. L'animal *vraiſemblablement* eſt un *aſcidia*.

Cette *eſpece* ſe trouve ſur pluſieurs de nos cotes, & près des embouchures des rivieres, comme au comté de *York*, à l'*embouchure* du *Tees*, en grande abondance, à *Scarborough*, &c.; aux comtés de *Northumberland* & *Lanceſter*, & autres cotes ſeptentrionales; aux comtés de *Dorſet*, *Cornwall*, & autres cotes occidentales; aux *Iſles Orcades*, & autres cotes de l'*Ecoſſe*.

LVI. | LVI.

C. Arenaria. Sand.

Chama media ovata fuſca. Arenaria.

Mya arenaria. Sand. Penn. Brit. Zool. No. 16. tab. 42. fig. 16.

Arenaria. Mya teſta ovata poſtice rotundata, cardinis dente antrorſum porrecto rotundato denticuloque laterali. Lin. S. N. p. 1112. No. 27. F. Suec. II. p. 2127.

A *broad ſhell*, pretty thick, ſtrong, and ſemitranſparent; of an oval *ſhape*; from three and a half to four inches *broad*, and from two to two one quarter inches *long*; and the *valves* are moderately concave or deep.

Outſide, ſtrongly and tranſverſely rugated, of a ruſty or cheſnut *colour*. The *beaks* are near central, ſmall, not very prominent, pointed, and that in the untoothed valve turns within the ledge of the commiſſure, and cloſes in a ſmall hole in the commiſſure of the other or toothed valve; the ledges or borders of the commiſſure alſo turn very much outward. The *ſides* are near equal; the *ſhorteſt* rounded and dilated; the *other*, or *gaping* ſide, is narrower, or has leſs breadth, and opens or turns outwards. The *margins* are plain.

Inſide, milk white and gloſſy. The *margins* plain. The *hinge* in *one valve* is a large hollow or cavity, which receives the large rounded erect thick plate-like *tooth*, depreſſed on the outſide, and with a ſlight convexity inwards of the other valve. It has no lateral teeth. The animal an *aſcidia*.

This *ſpecies* is *not common*. I have received it from the *Iſle of Wight*, near *Newport*, and from *Hearne Bay*, near *Faverſham*, in *Kent*.

Une *coquille large*, aſſes epaiſſe, forte, & demitranſparente; d'une *forme* ovale; de trois & demi à quatre pouces en *largeur*, & de deux à deux un quart pouces en *longueur*; & les *valves* ſont moderement concaves ou profondes.

L'*exterieur*, fortement & tranſverſalement ridé; d'une *couleur* chataigne, ou de rouille. Les *becs* preſque centrals, petits, pas fort ſaillants, & celui dans la valve ſans dent tourne en dedans du bord de la commiſſure, & s'emboite dans un petit trou dans la commiſſure de l'autre ou la valve à dent; les bords des commiſſures ſont auſſi fort retrouſſés. Les *cotés* ſont à peu pres egaux; le plus *court* eſt arrondi & dilaté; l'*autre*, ou le coté *beant*, eſt plus etroit, ou de moindre largeur, & s'ouvre & ſe retrouſſe. Les *bords* ſont unis.

Le *dedans*, blanc de lait & luſtré. Les *bords* unis. La *charniere* dans *une valve* à un grand enfoncement ou cavité, qui recoit une grande *dent* erigée, arrondie, epaiſſe, & comme une lame, comprimée a l'exterieur, & un peu convexe vers le dedans de l'autre valve. Il n'a point de dents laterales. L'animal eſt un *aſcidia*.

Cette *eſpece n'eſt pas commune*. Je l'ai reçu de l'*Iſle de Wight*, près de *Newport*, & de la *Baie de Hearne*, près de *Faverſham*, au comté de *Kent*.

LVII.

C. Truncata. Truncated.
Tab. XVI. fig. 1. 1.

LVII.

Pl. XVI. fig. 1. 1.

Chama subrotunda fusca rugosa, ex altera parte truncata. Truncata. Tab. 16. fig. 1. 1.

Concha lævis, altera tantum parte clusilis, apophysi admodum prominente lataque prædita. Lift. H. An. Angl. p. 191. tit. 36. tab. 5. fig. 36. Et *Chamæ pholas, latus, ex altera parte obtusus, scaber sive rugosus.* Hist. Conch. tab. 428. fig. 269.

Wallace Orkn. p. 45.—*Chamæ pholas latus; the broad pholade muscle.* Petiv. Gaz. tab. 79. fig. 12.—*Mya truncata; abrupt.* Penn. Brit. Zool. No. 14. tab. 41. fig. 14.

La came patagau. Argenv. Conch. II. p. 51. tab. 5. fig. C.

Truncata. Mya testa ovata postice truncata, cardinis dente antrorsum porrecto obtusissimo. Lin. S. N. p. 1112. No. 26. & F. Suec. II. No. 2126.

This *species* is pretty thick, strong, and heavy, yet semitransparent, of a roundish *shape*, about the *size* of a large pippin, or from one and a half to two inches *long*, and two to two and a half inches *broad*; of a very sinuous contour, on account of the gaping or open side; and the *valves* are concave or deep.

Outside, rough, by being thickly set with concentric transverse *wrinkles*, and of a light rust *colour*. The *beaks* small, strong, pointed, near central, and just overtop the commissure. The *closed side* rounded, and exceeding convex to the middle of the back, and from thence slopes to near the other or *open side*, where it rises again, and is turned greatly outwards, and *truncated*, or as if cut, and very *gaping*. The *bottom* is rounded, and the *margins* are plain.

Inside, milk white, and somewhat glossy. The *margins* plain. The *hinge* a large thick erect round plate in one valve, which locks in an answerable cavity in the other valve, like as in the *preceding species*,

Cette *espece* est asses epaisse, forte, & pesante, cependant demitransparente, d'une *forme* arrondie, environ la *grandeur* d'une grande pomme renette, ou de un & demi à deux pouces en *longueur*, & de deux à deux pouces & demi en *largeur*; d'un contour tres sinueux, causé par le coté ouvert ou beant; & les *valves* sont concaves ou profondes.

L'*exterieur*, aspre, causé par les *rides* concentriques transversales, tres serrées, & d'une *couleur* rouille claire. Les *becs* petits, forts, pointus, presque centrals, & dejettent à point nommé la commissure. Le *coté fermé* arrondi, & extremement convexe jusques au milieu du dos, & de la il va en pente jusques près de l'autre coté ou le *coté beant*, ou il s'eleve derechef, & est beaucoup retroussé, & *tronqué*, ou comme coupé, & tres *beant*. Le *bas* est arrondi, & les *bords* sont unis.

Le *dedans*, blanc de lait, & un peu lustré. Les *bords* unis. La *charniere* à grande lame arrondie, epaisse & erigée, dans une valve, qui s'emboite dans une cavité conforme dans l'autre valve, comme dans

species, or *Arenaria*. The animal is an *ascidia*.

This *shell* is found in *plenty* on many of our shores, as *Kent*, the *western* and *northern* coasts, the coasts of *Wales*, and the *Orkneys* in *Scotland*, &c.

dans l'*espece precedente*, ou *Arenaria*. L'*animal* est un *ascidia*.

Cette *coquille* se trouve en *abondance* sur plusiéurs de nos cotes, comme du comté de *Kent*, les cotes *occidentales* & *septentrionales*, les cotes de la principauté de *Galles*, & les *Isles Orcades* en *Ecosse*, &c.

LVIII.

LVIII.

C. Parva. Small.

Chama parva conoides, fragilis, albescens. Parva.
Mya dubia. Dubious. Penn. Brit. Zool. No. 19. tab. 44. fig. 19.

This *species* is of the *size* of a filberd or Pistachio nut, and like it in figure, or in *shape* of a blunted cone, thin and brittle, and the *valves* are pretty deep or concave.

Outside, whitish sullied with brown, set with fine concentric transverse *striæ*. The *beaks* prominent, pointed, and rise above the commissure. The *open* or *gaping part* is oval, very large and wide, and lies opposite to the hinge. The *margins* plain.

Inside, white and smooth. The *margins* plain. The *hinge* is set with a small tooth.

I have *as yet* only seen this shell from the coast at *Weymouth* in *Dorsetshire*.

Cette *espece* est de la *grandeur* d'une noisette ou Pistache, & ressemblante en figure, ou de la *forme* d'un cone obtus, mince & fragile, & les *valves* sont asses concaves ou profondes.

L'*exterieur*, blanchatre sali de brun, à *stries* fines concentriques & transversales. Les *becs* haussés, pointus, & dejettent la commissure. La *partie ouverte* ou *beante* est ovale, grande & ample, & vis à vis de la charniere. Les *bords* unis.

Le *dedans*, blanc & lisse. Les *bords* unis. La *charniere* avec une petite dent.

J'ai vu *jusques à present* cette coquille seulement du rivage de *Weymouth*, au comté de *Dorset*.

GENUS XIV.

SOLEN. SHEATH, or RAZOR SHELL.

Bivalves with *equal valves, open* or *gaping* at *both ends*; extremely *broad* and very *short shells*, of an *oblong shape*; the *hinge* with pointed teeth, like thorns.—The animal an *Ascidia*.

GENRE XIV.

SOLEN. LE MANCHE DE COUTEAU.

Bivalves à *battans egaux, ouvertes* ou *beantes* aux *deux bouts*; des *coquilles* extremement *larges* & tres *courtes*, d'une *forme oblongue*; la *charniere* à dents aigues, comme des epines.—L'animal est un *Ascidia*.

LIX.

S. SILIQUA. POD.

Tab. XVII. fig. 5.

LIX.

Pl. XVII. fig. 5.

SOLEN *major subfuscus rectus. Siliqua.* Tab. 17. fig. 5.

Concha fusca, longissima, angustissimaque, musculo ad cardinem aigeo, quibusdam solen dicta. Lift. H. An. Angl. p. 192. tit. 37. tab. 5. fig. 37. App. H. An. Angl. p. 19. App. H. An. Angl. in Goedart. p. 33.—*Solen major, subfuscus, rectus.* H. Conch. tab. 409. fig. 255.

Solen unguis; the sheath, razor, or spout fish. Grew Muf. p. 143. Merret Pin. p. 193. —*Solen sive concha tenuis longissimaque ab utraque parte naturaliter hians; the spout fish.* Wallace Orkn. p. 45.—*Spout fish.* Martin W. Isles, p. 6. 343. &c.—*Solen nostras vulgaris.* Muf. Petiv. p. 87. No. 844. Leigh Lancash. p. 120.—*Razor fish.* Rutty Dublin, p. 383. Smith Waterford, p. 272. Wallis Northumb. p. 395. No. 8.—*Solen Siliqua. Pod.* Penn. Brit. Zool. No. 20. tab. 45. fig. 20.

Rondel. Testac. l. 1. c. 43. p. 43. Bellon Aquat. tab. 414. fig. 2. Buon Ricr. p. 165. fig. 57. Argenv. Conch. I. p. 313. tab. 27. fig. L. II. p. 305. tab. 24. fig. L. & p. 39. tab. 6.—*Solen, lævis, albidus, candidus, ex fusco & subrosco colore variegatus & fasciatus.* Gualt. I. Conch. tab. 95. fig. C.—*Siliqua. Solen testa lineari recta, cardine altero bidentato.* Lin. S. N. p. 1113. No. 34. F. Suec. II. No. 2131. Muf. Reg. p. 473. No. 13.

This is a very *broad* and *short shell*, moderately thick and semitransparent, strait, and of an oblong *shape*, of equal dimensions, and flattens from the back to the sides; generally from six to seven inches, and sometimes to eight inches, *broad*, or from

Cette *coquille* est tres *large* & *courte*, moderement epaisse & demitransparente, droite, & d'une *forme* oblongue, egale dans toute sa dimension, & applatie du dos aux bords; generalement de six a sept pouces, & quelquefois de huit pouces, en *largeur*,

from end to end, and not above one and a quarter inch in *length*, or from the hinge to the opposite margin; and the *valves* are moderately deep or concave.

Outside, covered with a thin transparent yellowish brown *cuticle* or *epidermis*, like glue, which peels off soon after the fish is dead or exposed on the shore. Under this cuticle the shell is smooth, very glossy, and marked with many concentric transverse *wrinkles*, from the middle to one extreme; the other half is *striated lengthways*, of a yellowish-greenish, or brownish olive *colour*, variegated with white, with a *conoid mark*, well defined, of the same colour, that divides the shell *diagonally*. The *extremes* or *ends* gape, or are open; the edge of *one* is thick, and a little reflected outward, and near this the hinge is placed; the *other*, or *lower extreme*, has a sharp edge. All the *margins* are plain.

Inside, white and glossy. The *margins* plain. The *hinge* is placed at one extreme, and is set with two pointed erect teeth, with a fixed lateral one directed downwards in *one valve*; and a single erect sharp tooth with the lateral one in the *other valve*, which all lock in together.

This *species* is found in great *abundance* on many of the *English shores*, more especially our *northern* and *western coasts*; on the coasts of *Scotland*, and those of *Ireland*.

In *England* and *Scotland* it is mostly used for baits, and not for the table; in *Ireland* it is much eaten in Lent. The *antients* esteemed it a delicious food; and Dr. *Lister* informs us he thought it near as rich and pa-

largeur, ou de bout à bout, & pas plus de un pouce & un quart en *longueur*, ou dès la charniere au bord opposé; & les *valves* sont moderement profondes ou concaves.

L'*exterieur* couvert d'une *petite peau* ou *epiderme*, mince, transparente, jaunatre brune, comme la colle, qui se detache bientot apres que le poisson est mort ou exposé sur le rivage. Sous cette peau la coquille est lisse, fort lustré, & à plusieurs *rides* concentriques transversales, du milieu à un bout; l'autre moitie est *striée longitudinalement*, d'une *couleur* jaunatre-verdatre, ou olive brunatre, variée de blanc, avec une *marque conoide*, bien formée, de la même couleur, qui divise la coquille *diagonalement*. Les *extremités* ou *bouts* sont beants, ou ouverts; le bord d'*un* est epais, & un peu retroussé, & près de celui-ci la charniere est située; l'*autre*, ou l'*extremité inferieure*, est à bord tranchant. Touts les *bords* sont unis.

Le *dedans*, blanc & lustré. Les *bords* unis. La *charniere* est placée a une extremité, & composée de deux dents pointues erigées, avec une seule laterale fixée, & tournée en bas, dans *une valve*; & une seule dent pointue & erigée, avec la dent laterale, dans l'*autre valve*, qui s'emboitent toutes ensemble.

Cette *espece* se trouve en grande *abondance* sur plusieurs de nos *cotes Angloises*, specialement les *septentrionales* & *occidentales*; sur les cotes de l'*Ecosse*, & celles de l'*Irlande*.

En *Angleterre* & l'*Ecosse* elle est employée le plus generalement pour des amorces, & pas pour le table; en *Irlande* elle est beaucoup mangée en Careme. Les *anciens* l'ont conté une mangeaille tres delicieuse; &

palatable as the lobster. The *ſeaſon* of this ſhell-fiſh is the Spring.

& le Dr. *Liſter* nous dit qu'il l'a penſa auſſi riche & agreable au gout que l'ecreviſſe. La *ſaiſon* pour cet poiſſon eſt au Printems.

OBS.—Mr. Wallis, in his Hiſtory of Northumberland, p. 396, No. 9, notes a ſort of this ſhell he calls the *orange and white ſolen*, found in *Budle ſands* with this *common ſort*, and in *all reſpects like it, except in colour*, which is a deep orange and white in tranſverſe fillets, in alternate variegations. *Quere, if a diſtinct ſpecies, or only a variety?*

OBS.—M. Wallis, dans ſon Hiſtoire du comté de Northumberland, p. 396, No. 9, note une ſorte de cette coquille, qu'il nomme le *manche de couteau orange & blanc*, qui ſe trouve aux *rivages* de *Budle* avec cette *eſpece commune*, & en *toutes circonſtances reſſemblante, excepté en couleur*, qui eſt, en bandes tranſverſales & alternatives, oranges foncées & blanches. *Demande, ſi c'eſt une eſpece diſtincte, ou ſeulement une variete?*

LX.

S. ENSIS. SCYMETAR.

Solen ſubarcuatus. Enſis.

Solen alter curvus minor. Liſt. App. H. An. Angl. p. 20. App. H. An. Angl. in Goed. p. 36. tab. 2. fig. 9.—*Solen curvus.* Hiſt. Conch. tab. 411. fig. 257.—*Solen enſis. Scymetar.* Penn. Brit. Zool. No. 22. tab. 45. fig. 22.

Enſis. Solen teſta lineari ſubarcuata, cardine altero bidentato. Lin. S. N. p. 1114. No. 35. Muſ. Reg. p. 473. No. 14.

This *ſpecies* is rather of a ſmall *ſize*; not ſtrait, but *curved* or *bowed*; moſt generally from three and a half to about four and a half inches *broad*, or from end to end, and about half an inch *long*, or from the hinge to the oppoſite margin; thinner and more brittle than the preceding, but in *colour* and other reſpects like it.

Cette *eſpece* eſt plutot d'un petit *volume*; point droite, mais *courbée* ou *pliée*; le plus generalement de trois & demi à environ quatre & demi pouces en *largeur*, ou d'extremité à extremité, & environ un demi pouce en *longueur*, ou dès la charniere au bord oppoſé; plus mince & plus fragile que la precedente, mais en *couleur* & autres circonſtances eſt ſemblable.

This *kind* is not *common* on our coaſts. I have ſeen it from *Weymouth* in *Dorſetſhire*; and Dr. Liſter was informed it is found in plenty in the *æſtuary* of the *Severn*, on the *Welſh ſide*.

Cette *ſorte* n'eſt pas *commune* ſur nos cotes. Je l'ai vu de *Weymouth*, au comté de *Dorſet*; & Dr. Liſter fut informé qu'elle ſe trouve en abondance dans le *bras* de la riviere *Severn*, ſur la *cote* de *Galles*.

LXI.

S. Legumen. Peasecod.

Solen curtus subpellucidus, ad chamas quodammodo accedens. Legumen.

Chama subfusca, angustissima, ad solenes quodammodo accedens. List. H. Conch. tab. 420. fig. 264.—*Solen, legumen. Suboval.* Penn. Brit. Zool. No. 24. tab. 46. fig. 24.

Concha soleniformis, levis aut levissime striata, fragilis, pellucida, testa tenuissima cornea, subalbida, aliquando flavescens. Gualt. I. Conch. tab. 91. fig. A.

This *species* is strait and oblong; both ends are rounded, and one is somewhat broader than the other. It is very *thin*, *delicate* and *brittle*, *whitish* and near *transparent*, has a very thin glue-like glossy light-brown *cuticle* or *epidermis*, and is marked with extreme fine or capillary concentric transverse *striæ*. The usual *breadth* is about two and a half inches from end to end, and about three quarters of an inch *long*, or from the hinge to the opposite margin. The *inside* same as the outside; and all the *margins* are plain. The *hinge* is near central, and set with a single tooth in each valve.

This *kind* is very *rare* on our coasts. Mr. Pennant notes it to be found at *Red Wharf*, in *Anglesea*, *Wales*. I received it from the shore near *Christchurch*, in *Hampshire*.

LXI.

Cette *espece* est droite & oblongue; les deux extremités sont arrondies, & une est quelque peu plus large que l'autre. Elle est tres *mince*, *delicate* & *fragile*, *blanchatre* & presque *transparente*, à *petite peau* ou *epiderme* tres mince, de couleur brune claire, & comme de la colle, & à *stries* concentriques extremement fines. La *largeur* generale est environ deux pouces & demi, ou de bout à bout, & environ trois quarts d'un pouce en *longueur*, ou des la charniere au bord opposé. Le *dedans* de même que l'exterieur; & touts les *bords* sont unis. La *charniere* est presque centrale, & à une seule dent dans chaque valve.

Cette *sorte* est tres *rare* sur nos cotes. M. Pennant dit qu'elle se trouve à *Red Wharf* en *Anglesea*, en *Galles*. Je l'ai reçu de la cote du comté de *Hants*, près de *Christchurch*.

LXII.

Chama-Solen.

Solen. Testa ovali-oblonga subpellucida, sinuosa. Chama-Solen.

Chama angustior, ex altera parte sinuosa. List. H. Conch. tab. 421. fig. 265—*Solen cultellus. Kidney.* Penn. Brit. Zool. No. 25. tab. 46. fig. 25.

This *shell* is strait and broad, but much longer than the *genus* of *solens* commonly are,

LXII.

Cette *coquille* est droite & large, mais beaucoup plus longue que le *genre* de *solens* font.

are, and in that respect approaches to the *chama*; generally about two inches *broad*, or from end to end, and about one inch in *length*, or from the hinge to the opposite margin. It is thin, brittle, glossy, whitish, and near transparent, with extreme fine concentric transverse *striæ*, and has a very thin brownish *epidermis*; of an oblong-oval *shape*; the ends rounded and gaping, but the bottom margin is somewhat sinuous or serpentine. The *hinge* is near central, with a pointed tooth in each valve, that lock together. All the *margins* are plain, and the *inside* is smooth and whitish.

This *species borders* on the *chama*, and connects the *genera*, for which *reason* I have named it *Chama-Solen*. It is very *rare* on the *English coasts*. Mr. Pennant notes it from *Weymouth*, in *Dorsetshire*. I received it from the shores of *Dorsetshire* and *Hampshire*.

Obs.—Said *author* very *erroneously quotes* the *Cultellus*, *Lin.* S. N. p. 1114. No. 37. for this *species*.

sont ordinairement, & en cela elle approche aux *cames*; generalement environ deux pouces de *large*, ou d'extremité à extremité, & environ un pouce en *longueur*, ou dès la charniere au bord opposé. Elle est mince, fragile, lustrée, blanchatre, & presque transparente, à *stries* extremement fines, transversales & concentriques, & à *epiderme* tres mince & brunatre; d'une *forme* oblongue-ovale; les extremites arrondies & beantes, mais le bord du bas est quelque peu sinueux ou ondé. La *charniere* est presque central, avec une dent pointue dans chaque valve, qui s'emboitent ensemble. Touts les *bords* sont unis, & le *dedans* est lisse & blanchatre.

Cette *espece approche* aux *cames*, & lie les *genres*, pour laquelle *raison* je l'ai nommé le *Chama-Solen*. Elle est tres *rare* sur les *cotes Angloises*. M. Pennant dit qu'elle se trouve à *Weymouth*, au comté de *Dorset*. Je l'ai reçu dès cotes des comtés de *Dorset* & *Hants*.

Obs.—Le dit *auteur* cite *tres erronnement* le *Cultellus*, *Lin.* S. N. p. 1114. No. 37. pour *cette espece*.

GENUS

GENRE

GENUS XV.

PINNA, SEA HAM OR WING.

Bivalves with *equal valves*, of a triangular *shape*, and *open* or *gape* at the bottom; the *hinge* inarticulate or toothless, and placed on one side.—The animal a *Slug*.

GENRE XV.

PINNE MARINE, JAMBON & JAMBONNEAU.

Bivalves à *battans egaux*, d'une *forme* triangulaire, & *ouverte* ou *beante* au bas. La *charniere* sans dent ou articulation, & située d'un coté.—L'animal est une *Limace*.

* MARINÆ. SEA.

LXIII.

P. MURICATA. THORNY.

Tab. XVI. fig. 3.

* DE MER.

LXIII.

Pl. XVI. fig. 3.

PINNA *tenuis costis longitudinalibus muricatis. Muricata, Thorny.* Tab. 16. fig. 3.

Pinna tenuis, striata, muricata. List. H. Conch. tab. 370. fig. 210.—*Pinna, commonly called a Skare.* Rutty Dublin, p. 387.—*Pinna fragilis. Brittle.* Penn. Brit. Zool. No. 80. tab. 59. fig. 80.

Pinna recta transversim & directe striata & rugose, striis in summitate aculeis exasperatis, ex fusco rubro nigricans. Gualt. I. Conch. tab. 79. fig. D.

Muricata. Pinna testa striata, squamis concavis ovatis acutis. Lin. S. N. p. 1160. No. 266. Mus. Reg. p. 545. No. 143.

A very thin brittle *shell*, of a triangular *shape*, *narrow* or pointed at the *top*, and *wide* at the *bottom*, which part always *gapes* or is *open*; rather *flatted*, for the *valves* are but moderately concave. The largest that our coasts afford are between six and seven inches *long*, and three to three and a half *broad* at the bottom, or the broadest part, of a horn *colour*, and semitransparent.

Outside, marked with slight longitudinal *ribs*, that are roughened with rows of small *prickles*, not very prominent, broad or plate-like, and concave, the concavity turning

Une *coquille* tres mince, fragile, d'une *forme* triangulaire, *etroite* ou pointue au *sommet*, & *large* au *bas*, laquelle partie est toujours *beante* ou *ouverte*; plutot *comprimée*, car les *valves* ne sont que moderement concaves. Les plus grandes que fournit nos cotes sont entre six & sept pouces en *longueur*, & trois à trois & demi en *largeur* au bas, ou la partie plus large, d'une *couleur* cornée, & demitransparente.

L'*exterieur* à *cotes* longitudinales legeres, qui sont aspres ou rudes par des rangées des petites *epines*, pas fort saillantes, larges & comme des lames concaves, la

turning towards the bottom; and in the *young shells* both *ribs* and *prickles* are near obsolete. The *whole shell* is besides thickly set with very fine or capillary transverse *striæ*.

Inside, smooth, of same colour, but towards the top of a *pearly* lustre.

This *species* has been fished at *Weymouth* in *Dorsetshire*, and I have seen a very small one from the coasts of *Wales*.

Dr. Rutty mentions a *pinna* ten inches *long* and five *broad*, caught near the *Skerries* in *Ireland*; and Mr. Pennant saw vast pinnæ, found among the *farther Hebrides*, or *Western Islands of Scotland*, in the collection of Dr. Walker, at Moffat; but the said *authors* do not *ascertain* whether of *this* or *other species*.

la concavité tournant toujours vers le bas; & dans les *jeunes coquilles* tant à l'egard des *cotes* que des *epines*, elles sont presque usées ou frottées. *Toute* la *coquille* est outre cela à *stries* fort fines ou capillaires transversales.

Le *dedans* lisse, de la méme couleur, mais vers le sommet d'un lustre tres *nacré*.

Cette *espece* à eté pesché à *Weymouth*, au comté de *Dorset*, & j'ai vu une tres petite de la cote de *Galles*.

Le Dr. Rutty fait mention d'une *pinne* dix pouces de *long* & cincq en *largeur*, peschée pres de *Skerries* en *Irlande*; & M. Pennant à vu des pinnes fort grandes, trouvées entre les *Hebrides eloignées*, ou *Isles Occidentales d'Ecosse*, dans la collection du Dr. Walker, à Moffat: mais les dits *auteurs* ne *designent* si elles sont de *cette* ou d'*autre espece*.

PART QUATRIEME

PART IV.

MULTIVALVES.

Shells composed of *many pieces* or *valves.*

QUATRIEME PARTIE.

MULTIVALVES.

Coquilles composées de *plusieurs pieces* ou *battans.*

GENUS XVI.

PHOLAS. PIDDOCKS.

Shells composed of *two large valves,* and *small accessorial ones* near to or at the hinge, and are *open* or *gape* at both *ends.* The *hinge* generally folded back, and under it, in the *inside,* is a long *curved tooth* or *spur.*—The animal an *Ascidia.*

GENRE XVI.

PHOLAS.

Coquilles composées de *deux grands battens,* & de quelques autres *petits accessoires* pres dè ou à la charniere, & sont *ouvertes* ou *beantes* aux deux *extremités.* La *charniere* generalement repliée, & en dessous, dans le *dedans,* à longue *dent* ou *eperon courbé.*—L'animal est un *Ascidia.*

LXIV.

P. BIFRONS. DOUBLE-FRONT.

Tab. XVI. fig. 4. 4.

LXIV.

Pl. XVI. fig. 4. 4.

PHOLAS *ovalis, parte dimidia striis undatim crispatis, altera lævis; dens longus angustus curvus. Bifrons.* Tab. 16. fig. 4. 4.

Concha altera parte dimidia striis undatim crispatis donata, altera lævis, apophysi longâ, angusta, recurva, dentiformi. An è peloridibus antiquorum? Lift. H. An. Angl. p. 192. tit. 38. tab. 5. fig. 38.—*Pholas angulosus, nobis olim, concha altera, &c.* Tit. 38. App. H. An. Angl. in Goedart. p. 36. tab. 2. fig. 7.—*Pholas latus rugosus ex dimidio dorso & asper.* Hist. Conch. tab. 279. fig. 436.

Concha ex dimidia pene margine profunde striata. Merret Pin. p. 194.—*Chamæ pholas bifrons. Furrow-ribbed pholas muscle.* Petiv. Gaz. tab. 79. fig. 13.—*Pholas crispatus. Curled.* Penn. Brit. Zool. No. 12. tab. 40. fig. 12.

Pitaut

Pitaut ou Dail Pholade. Argenv. Conch. I. p. 365. pl. 30. fig. K. M. II. p. 322. pl. 26. fig. H.—*Crispata.* *Pholas testa ovali hinc obtusiore crispato-striata, cardinis dente curvo.* Lin. S. N. p. 1111. No. 25. F. Suec. II. No. 2125. Mus. Reg. p. 469. No. 8.

This *shell* is thick, strong, and semitransparent; of a suboval *shape*; sinuous or uneven in its contour and appearance; from one and a half to two inches *long*, or from the hinge to the opposite margin, and from two and a half to three inches and a half *broad*, or from end to end; the *valves* concave or deep, and *gape*, or are *open* at both *ends*; of a *white*, with a great part of a *rust colour*.

Outside. Down the middle runs a broad, oblique, and transversely wrinkled *furrow*, that divides the shell into nearly two equal areas or compartments, *one* whereof is rough, and wrought with transverse concentric *wrinkles*, the *other* with very prominent concentric and undulated *wrinkles*, imbricated or partly lying over each other; their edges are finely scolloped or indented, and their extremes end sharp, like thorns, on the edge of the side. The *beaks* slope greatly inwards, in a spiral manner. The *hinge* is folded back on the body, on the imbricated side, and spreads into a very thick, broad and smooth border to the sides, and between this border and the back is a deep long hollow. The whole back is very convex, but slopes greatly at the top, the bottom, and the imbricated side. The *sides* are dissimilar; the wrinkled one lengthened, rounded and spread; the *imbricated* one, with the folded edge of the hinge, runs into an acute angle, and thence is cut or truncated

Cette *coquille* est epaisse, forte, & demitransparente; d'une *forme* quasi ovale; sinueuse ou inegale dans son contour & apparence; de un & demi à deux pouces en *longueur*, ou des la charniere au bord opposé, & de deux & demi à trois pouces & demi en *largeur*, ou d'extremité à extremité; les *valves* concaves ou profondes, & *beantes* ou *ouvertes* aux deux *bouts*; *blanche*, avec beaucoup de *couleur roussatre*.

L'exterieur. Le long du milieu court un *sillon* large, oblique, transversale & ridé, qui divise la coquille en presque deux compartiments ou aires egaux, *un* desquels est aspre, & à *rides* concentriques & transversales, l'*autre* à *rides* saillantes concentriques & ondées, tuilées ou mises en partie l'une sur l'autre; leur bords sont finement dechiquetés ou decoupés à languettes, & à leur fins ils finissent en epines aux cotés. Les *becs* penchent fort en dedans, d'une maniere spirale. La *charniere* est repliée sur le corps, sur le coté tuilé, & s'etend dans une bordure tres epaisse, lisse & large, jusques aux cotés, & entre cette bordure & le dos il se trouve un creux profond & long. Tout le dos est tres convexe, mais penche beaucoup au sommet, au bas, & le coté tuilé. Les *cotés* sont inegaux; celui qui est ridé allongé, arrondi & etendu; l'*autre* tuilé, avec le bord retroussé de la charniere, court en un angle aigu, & de la est coupé ou tronqué jusques au bas, dans un angle

to the bottom, in a very oblique angle. The bottom is finuous or waved. The *margins* are plain.

Infide, white and fmooth, but not gloffy. The *beaks* are turned quite inwards, and have a very broad fmooth edge, which folds on the back. In the cavity under the beaks rifes a long, broad, flat curved *tooth* or *fpur*, very remarkable, and about half an inch long. The *margins* are plain.

It is a *trivalve fhell*, having a fmall valve on the hinge.

This *fpecies* is found in *great abundance* on many of our fhores, nitched or burrowed in rocks and ftones, as *Cornwall*, *Lincolnfhire*, *Yorkfhire*, *Wales*, &c.

tres oblique. Le bas eft finueux ou ondé. Les *bords* font unis.

Le *dedans*, blanc & liffe, mais point luftré. Les *becs* font entierement retrouffés en dedans, & ont un bord tres large & liffe, qui fe replie fur le dos. Dans la cavité au deffous des becs s'eleve une *dent* ou *eperon*, long, large, plat & courbe, tres remarquable, & environ un demi pouce en longueur. Les *bords* font unis.

C'eft une *coquille trivalve*, ayant une petite valve à la charniere.

Cette *efpece* fe trouve en *grande abondance* fur plufieurs de nos cotes, enfevelie ou nichée dans les rochers ou pierres, comme aux comtés de *Cornwall*, *Lincoln*, *York*, la principauté de *Galles*, &c.

LXV.

P. Muricatus. Prickly.

Tab. XVI. fig. 2. 2.

LXV.

Pl. XVI. fig. 2. 2.

Pholas anguftius ftriatus & veluti aculeatus. Muricatus. Tab. 16. fig. 2. 2.

Pholas roftratus major diepenfis vulgò Gallice Piteau dictus. App. H. An. Angl. in Goedart. p. 37. tab. 2. fig. 3.—*Pholas ftriatus, finuatus ex altera parte.* Hift. Conch. tab. 433. fig. 276.—*Pholas alte ftriatus, ex altera parte finuatus, eadem mucronatus, Hift. noft. Conch. Anglice Piddocks, Gallice Pitau; earumque pifcatores pitauquieres.* Exercit. Anat. 3. p. 88. tab. 7. fig. 1. 2.

Pholas anguftius; long Pierce ftone or Pholade. Petiv. Gaz. tab. 79. fig. 10.—*Piddocks.* Dale Harw. p. 389. Borlafe Cornw. p. 278. tab. 28. fig. 31. Wallis Northumb. p. 398. No. 21. Rutty Dublin, p. 388.—*Pholas dactylus. Dactyle.* Penn. Brit. Zool. No. 10. tab. 39. fig. 10.

Ballano o pholades. Buon. Ricr. p. 157. fig. 26. 27.—*Reaumur fur les Dails.* Acad. de Paris, 1712.—*Pholas major ftriis cancellatis exafperata, fubalbida.* Gualt. I. Conch. tab. 105. fig. A. B. C. D.—*Dactylus tefta oblonga hinc reticulato-ftriata.* Lin. S. N. p. 1110. No. 20. F. Succ. II. No. 2124.

The *shell* is entirely *milk white*; rather thin, brittle, and semitransparent; from three to four inches *broad*, or from end to end, and about one to one and a half inch *long*, or from the hinge to the opposite margin, but much shorter in the other parts of the shell. Of an oblong-oval *shape*. The *sides* very unequal; *one* greatly extended, narrow, and rounded at the end; the *other* runs into a very pointed short narrow neck, and thence goes obliquely down, or as if truncated, to the bottom. The *bottom margin* is sinuous or winding on the short side, and is all along scolloped or notched into small spines or prickles. The *valves* are moderately concave; and it is a *trivalve*, the *small accessorial valve* lying over the hinge.

La *coquille* est totalement *blanche de lait*, plutot mince, fragile, & demitransparente; de trois à quatre pouces en *largeur*, ou de bout à bout, & environ un à un & demi pouce en *longueur*, ou dès la charniere au bord opposé, mais beaucoup plus courte dans les autres parties de la coquille. D'une *forme* oblongue-ovale. Les *cotés* fort inegaux; *un* extremement etendu, etroit, & arrondi au bout; l'*autre* passe dans une langue courte, fort pointue & etroite, & de la va obliquement, ou comme tronqué, au bas. Le *bord du bas* sinueux ou tortueux du coté court, & dentelé ou entaillé son long en petits piquants ou epines. Les *valves* sont moderement concaves, & c'est une *trivalve*; la *petite valve* etant située par dessus la charniere.

Outside, finely striated in concentric transverse *striæ* till near the middle, but from thence the striæ become very rough by notches or minute prickles, and are crossed by other very slight *longitudinal* striæ. All these *striæ* are stronger, more rough and prickly, as they approach the short side, and the *margins* there are also more notched or prickly.

L'*exterieur*, strié tres finement à *stries* concentriques transversales à presque au milieu, mais de la les stries deviennent fort aspres par des entailleures ou epines menus, qui sont traversées par d'autres stries *longitudinales* tres legeres. Toutes ces *stries* sont fortes, plus aspres & piquantes, comme elles approchent au coté court, & les *bords* sont aussi plus entaillés ou epineux.

Inside, smooth and glossy. The *hinge* is situated quite near the short side; it *reflects* or *folds* very much, on the back, into a fine broad smooth border or plate, which is wrought underneath into a pretty cancellated or cavernous work. Under the hinge rises a long flatted curved *tooth* or *spur*, as in the preceding species.

Le *dedans*, lisse & lustré. La *charniere* tout près du coté court; elle est *retroussée* ou beaucoup *tournée en arriere*, sur le dos, dans une bordure ou lame, fine, large & lisse, que en dessous est travaillée dans un joli ouvrage lamelleux ou caverneux. Au dessous de la charniere s'eleve une *dent* ou *eperon*, long, applati & courbé, comme dans l'espece precedente.

This species is very frequent on many of the shores of *Great Britain* and *Ireland*, and

Cette espece est tres frequent sur plusieurs des cotes de la *Grande Bretagne* & l'*Irlande*;

and works itſelf, or perforates and burrows in rocks and other ſtones. The Spring is the ſeaſon for it, and it is ſaid to be a very excellent and dainty food.

l'*Irlande*; elles travaillent ou percent & ſe nichent dans les rochers & autres pierres. Le Printems eſt ſa ſaiſon, & on dit que c'eſt une mangeaille tres excellente & de bon gout.

LXVI.

P. CANDIDUS. WHITE.

LXVI.

Pholas tenuis candidus ovatus decuſſatim ſtriatus. Candidus.

Concha candida, dupliciter ſtriata & veluti aculeata. Liſt. H. An. Angl. p. 193. tit. 39. tab. 5. fig. 39.—*Pholas alter.* App. H. An. Angl. in Goedart. p. 37. tab. 2. fig. 4. & 5. —*Pholas parvus aſper.* H. Conch. tab. 435. fig. 278.

Pholas latus; ſhort Pierce-ſtone or Pholade. Petiv. Gaz. tab. 79. fig. 11.—*Pholas candidus. White.* Penn. Brit. Zool. No. 11. tab. 39. fig. 11.

Pholas teſta tenuiſſima, ſtriis minoribus cancellatis ſignata. Gualt. I. Conch. tab. 105. fig. E.—*Candidus. Pholas teſta oblonga undique ſtriis decuſſatis muricata.* Lin. S. N. p. 1111. No. 23. Muſ. Reg. p. 469. No. 7.

This *ſpecies differs* from the laſt in not being of one quarter of its *ſize*, and of a more oval *ſhape*, by not having the narrow prolonged neck on the ſhort ſide, for both ends are equally rounded. In all other *reſpects* it entirely *reſembles* it.

This *kind* is found on the ſame ſhores, but is more *rare*. *May it not be only a variety in growth, inſtead of a diſtinct ſpecies?*

Cette *eſpece differe* de la derniere en ce qu'elle n'eſt pas un quart de ſa *grandeur*, & d'une *forme* plus ovale, n'ayant pas cette langue allongée & etroite du coté court, car les deux cotés ſont egalement arrondis. En toutes autres *circonſtances* elle la *reſſemble* parfaitement.

Cette *eſpece* ſe trouve ſur les mêmes cotes, mais eſt plus *rare*. *Eſt ce ſeulement une varieté d'etat ou age, plutot qu'une eſpece diſtincte?*

LXVII.

LXVII.

LXVII.

P. Parvus. Small.

Pholas parvus simillima tota structura Pholade Bifronte. Parvus.

Pholas parvus. Little. Penn. Brit. Zool. No. 13. tab. 40. fig. 13.

This *shell*, in *every particular except size*, so exactly resembles the *Pholas Bifrons*, No. 64, *supra*, that I doubt whether it be a *distinct species*, or otherwise than a *young shell* of that *kind*.

The *shell* is not much above the *size* of a hazel nut: it is found in *great quantities* on the *same coasts* as the other kinds, nitched in the rocks, stones, &c. There is an *amazing abundance* at *Scarborough* and *Whitby*, in *Yorkshire*, nitched in the alum and other stones.

LXVII.

Cette *coquille*, dans *chaque circonstance excepte la grandeur*, ressemble si exactement la *Pholas Bifrons*, No. 64, *supra*, que je doute si elle est une *espece distincte*, ou autrement qu'une *jeune coquille* de cette *espece*.

La *coquille* ne surpasse pas beaucoup la *grandeur* d'une noisette: elle se trouve en *grandes quantités* sur les *mêmes cotes* que les autres especes, nichées dans les rochers, pierres, &c. Une *abondance etonnante* se trouve à *Scarborough* & *Whitby*, au comté de *York*, nichées dans la mine d'alun & autres pierres.

GENUS

GENRE

GENUS XVII.

BALANUS. ACORN.

Multivalve shells, most generally of a conic *shape*, composed of *several unequal parallel* and *perpendicular valves*, or *immoveable parts*, without *hinges* or *joints*. The *aperture* on the top or summit. The *base* flat, and always *affixed* to shells, stones, and other bodies, and are never found *detached* or *free*.—The animal a *Triton*.

GENRE XVII.

GLAND DE MER.

Coquilles multivalves, le plus generalement d'une *forme* conique, composées de *plusieurs valves* ou *pieces fixes*, ou sans *jointures* ou *charnieres*, *inegales*, *paralleles* & *perpendiculaires*. L'*ouverture* sur le haut ou sommet. La *base* applatie, & toujours *affichée* à des coquilles, pierres, & autres corps, & ne se trouvent jamais *detachées* ou *isolées*.—L'animal est un *Triton*.

* MARINÆ. SEA.

LXVIII.

B. VULGARIS. COMMON.

Tab. XVII. fig. 7.

* DE MER.

LXVIII.

Pl. XVII. fig. 7.

B*ALANUS parvus conicus é senis laminis compositus, vertice operculo bifido rhomboide occluso. Vulgaris.* Tab. 17. fig. 7.

Balanus cinereus, velut é senis laminis striatis compositus, ipso vertice altera testa, bifida, rhomboide occluso. Balani parva species. Lift. H. An. Angl. p. 196. tit. 41. tab. 5. fig. 41. —*Balanus parvus striatus.* Hist. Conch. tab. 444. fig. 287. Exercit. Anat. 3. p. 96. tab. 7 & 8.

Spunge centre shell.—Grew Mus. p. 148.—*Balani sive polycipides. Center shells.* Merret Pin. p. 193.—*Balanus noster parvus vulgaris.* Mus. Petiv. p. 82. No. 804. Wallace Orkn. p. 45.—*The acorn fish.* Dale Harw. p. 390. Wallis Northumb. p. 402.—*Lepas balanoides. Sulcated.* Penn. Brit. Zool. No. 5. tab. 37. fig. 5. & 5 A.

Petits glands de mer. Argenv. Conch. I. p. 322. tab. 26. fig. D. II. p. 364. tab. 30. fig. D.—*Lepas testa conica truncata; operculo obtuso.* Lin. F. Suec. I. p. 385. No. 1348. II. No. 2123.—*Balanoides. Lepas testa conica truncata Levi fixa, operculis obtusis.* S. N. p. 1108. No. 11.

This *species*, as well as the other *kinds* of *balani*, are *affixed* strongly by their *base* to shells, stones, and other bodies, and leave the vestiges or impressions of their bases on the things they adhere to, for they have no *progressive motion*.

The

Cette *espece*, aussi bien que toutes les autres *sortes* de *glands de mer*, sont fortement *affichées* par leur *bases* à des coquilles, des pierres, & autres corps, & laissent les vestiges ou empreintes de leur bases sur les corps auxquels elles s'affichent, car elles n'ont point de *mouvement progressif*.

Les

The *shells* are generally tender, brittle, and whitish, of a conic *shape*, and from a very small *size* to above that of a *pea*. It is composed of six *pieces* or *valves*, distinguished from each other by deep furrows: the other parts are near smooth. The *summit* or *top* is open, and within it is seen *two small valves*, closed together into a very sharp ridge at top.

When not *affixed* on *flat*, but *uneven surfaces*, they *sometimes*, but *rarely*, extend down into a pretty long rugged tubular *stalk* or *root*, which is the *variety noted* and *well figured* by *Mr. Pennant*, *fig.* 5 *A*.

This *species* is found in *great abundance* on all the *British shores*, sticking to the rocks, shells, &c.

Dr. *Lister* informs us he eat of the *fish cooked*, not *raw*: on first taste it was very palatable, but after some time proved hot, like pepper, and burnt his palate

Les *coquilles* sont generalement delicates, fragiles, & blanchatres, d'une *forme* conique, & de tres petit *volume* jusques à surpasser la *grandeur* d'un *pois*. Elle est composée de *six pieces* ou *valves*, distinguées l'une de l'autre par des sillons profonds: les autres parts sont presque lisses. Le *sommet* ou *haut* est ouvert, & en dedans se voit *deux petites valves*, fermées ensemble dans un faite tres tranchant au haut.

Quand point *affichée* sur des *surfaces applaties*, mais *inegales*, *quelquefois*, mais *rarement*, elles s'etendent ou croissent en bas dans une *tige* ou *racine*, asses, longue, raboteuse & tubulaire, qui est la *varieté* que *M. Pennant* à *noté* & *tres bien figuré*, *fig.* 5 *A*.

Cette *espece* se trouve en *grande abondance* sur toutes les *cotes* de la *Grande Bretagne*, affichée aux rochers, coquilles, &c.

Le Dr. *Lister* nous dit qu'il à mangé le *poisson apprefté*, non pas *crud*: au premier gout il le trouva fort agreable, mais apres quelque tems il eprouva qu'il etoit chaud, comme le poivre, & brula son palais.

LXIX.

B. PORCATUS. RIDGED.

Balanus majusculus valvis porcatis. Porcatus.

Balanus Anglica vulgaris ore patulo. Phil. Transf. P. II. 1758, tab. 34. fig. 17.—*Lepas balanus. Common.* Penn. Brit. Zool. No. 4. tab. 37. fig. 4.

Balanus. Lepas testa conica fulcata fixa, operculis acuminatis. Lin. S. N. p. 1107. No. 10. F. Suec. I. p. 385. No. 1349. II. No. 2122. Muf. Reg. p. 466. No. 2.

This is a larger and much stronger *species* than the preceding, seldom less than the *size* of a filberd, the *colour* whitish, the *shape* more conic. The *summit* has a wide opening; of a *rugged appearance*, for the *sides* are wrought with very prominent and strong longitudinal *ribs*.

This

Cette *espece* est plus grande & beaucoup plus forte que la precedente, rarement de moindre *volume* qu'une noisette, de *couleur* blanchatre, la *forme* plus conique. Le *sommet* à ouverture large; d'une *apparence raboteuse*, car les *cotés* sont à *cotes* longitudinales, tres fortes & saillantes.

Cette

This *kind* is found very *frequently* on the *British coasts*, adhering to rocks, shells, and other bodies, like the preceding species.

Cette *sorte* se trouve tres *frequemment* sur les *cotes Britanniques*, affichée aux rochers, coquilles, & autres corps, comme l'espece precedente.

B.

B.

Balanus striatus. Striated.

Lepas striata. Striated. Penn. Brit. Zool. No. 7. tab. 38. fig. 7.

Lepas with the *shells lapping* over each other, and obliquely *striated*; from the coast of *Weymouth*, in *Dorsetshire*.

Gland de mer avec les *valves pliées* l'une sur l'autre; & *striées* obliquement; de la cote à *Weymouth*, au comté de *Dorset*.

LXX.

B. TINTINNABULUM. BELL.

LXX.

GLAND DE MER CLOCHETTE.

Balanus major purpurascens, conicus, angustus, tintinnabuliformis, apertura valde patente. Tintinnabulum.

Balanus major angustus purpurascens, capitis apertura valde patente. List. H. Conch. tab. 433. fig. 285.—*Balanus major. The conick centre shell.* Grew. Mus. p. 148.—*Balanus maximus ore patulo.* Mus. Petiv. p. 82. No. 803.—*B. tintinnabuliformis & B. calyciformis orientalis.* Phil. Trans. 1758, P. II. tab. 34. fig. 8 & 9.—*B. ore biante magnus.* Borlase Cornw. p. 278.—*Balanus. Tintinnabulum. Bell.* Penn. Brit. Zool. No. 8.

Rumph. Mus. tab. 41. fig. A.—Klein Ostrac. p. 176. spec. 2. No. 2. tab. 12. fig. 97. —Argenv. Conch. I. p. 322. tab. 26. fig. A. II. p. 364. tab. 30. fig. A.

Tintinnabulum. Lepas testa conica rugosa obtusa fixa. Lin. S. N. p. 1108. No. 12. Mus. Reg. p. 466. No. 3.—*Gland de mer clochette.* D'Avila Cab. p. 404. No. 922.

This *species* is rather to be accounted an *exotic*, and imported to us from *hot climates*, as the *West Indies*, *Guinea*, &c.

It is large and deep, somewhat above the *size* of a walnut, or about two inches in *height* or *length*, and an inch and a half *cross* the *base*, or *broad*; of a bell-like *shape*, but not expanded outwards at the *aperture*, which is large; of a purple *colour*, and slightly *striated* the length. The *interstices* between the valves are depressed below the surface, striated transversely with extreme fine *striæ*, and form *cones* with

Cette *espece* doit plutot etre reputé une *exotique*, & transporté chez nous des *climats chauds*, comme les *Indes Occidentales*, *la Guinée*, &c.

Elle est grande & profonde, quelque chose au dela la *grandeur* d'une noix, ou environ deux pouces en *hauteur* ou *longueur*, & un pouce & demi *à travers* la *base*, ou *large*; d'une *forme* comme une clochette, mais pas deployée exterieurement à l'*ouverture*, qui est grande; d'une *couleur* de pourpre, & legerement *striée* sa longueur. Les *intervalles* entre les valves sont enfoncés au dessous de la surface, stri-

with their *points downwards*; the *reverse* of the very *valves*, whose conic points lie *upwards*.

ſtriés tranſverſalement à *ſtries* extremement fines, & forment des *cones* dont les *pointes tournent en bas*; l'*oppoſé* des *valves* mêmes, dont les pointes coniques tournent *par haut*.

This *kind* is *frequently* found adhering to the *bottoms* of *ſhips*, in great cluſters.

Cette *ſorte* ſe trouve *frequemment* affichée aux *fonds* des *navires*, en grands amas.

LXXI.
B. Balæna. Whale.
Tab. XVII. fig. 2. 2. 2.

LXXI.
Pl. XVII. fig. 2. 2. 2.

Balanus hemiſphericus ſexlobatus. Balænæ. Tab. 17. fig. 2. 2. 2.

Balanus balænæ cuidam Oceani Septentrionalis adhærens. Liſt. H. Conch. tab. 445. fig. 288.—*Pediculus ceti.* Phil. Tranſ. No. 222. p. 323. Epitome Tranſ. Soc. R. Angl. Vol. V. p. 381. tab. 17. fig. 2.—*Pediculus ceti, vel Lepas nuda carnoſa aurita.* Idem, 1758, Vol. 50. P. II. tab. 34. fig. 1. & fig. 7. Martin W. Iſles, p. 162 & 166. Edward's Gleanings, Vol. II. p. 162. tab. 264. c. 76. fig. a. b. c.

Pediculus ceti. Bocone Rech. & Obſ. Nat. p. 287.—*Quarta ſpecies echini plani.* Rumph. Muſ. tab. 14. fig. H.—*Balanus balænaris.* Klein Oſtrac. p. 176. tab. 12. fig. 98. Gualt. I. Conch. tab. 106. fig. Q.—*Grand pou de Baleine.* D'Av. Cab. p. 404.—*Diadema. Lepas teſta ſubrotunda ſex lobata ſulcata fixa.* Lin. S. N. p. 1198. No. 15.

The *top figure* is the full view of the *upper ſide*, the *middle figure* a *ſide view*, and the *lower figure* is a view of the *under ſide* or *bottom*.

This *ſpecies* is of a compreſſed hemiſpherical *ſhape*, from the bigneſs of a walnut to above double that *ſize*, of a ſtrong bone-like *ſubſtance*, and *white*. It is ſlightly *hexangular*, or divided into ſix parts; the *angles*, however, are not acute, but rather obtuſely defined. The *top* has a very deep ſmooth hexangular cavity. All down the *ſides* it is regularly divided into *ſix parts*, each *part* conſiſting of three, and ſometimes four, parallel prominent rounded *ribs*, and an intervening broad *plane*, which all

La *figure d'enhaut* repreſente le *ſommet* ou la *partie ſuperieure* de la coquille, la *figure* du *milieu* la repreſente de *coté*, & l'*inferieure* repreſente la *baſe*.

Cette *eſpece* eſt d'une *forme* hemiſpherique comprimée, de la grandeur d'une noix à double cet *volume*, d'une *ſubſtance* forte quaſi oſſeuſe, & *blanche*. Elle eſt un peu *hexangulaire*, ou diviſée en ſix parties; les *angles* deſquelles, ne ſont pas aigus, mais plutot marqués d'une maniere obtuſe. Le *ſommet* ou *haut* à une cavité hexangulaire, tres profonde & liſſe. Le long des *cotés* eſt regulierement diviſé en *ſix parties*, chaque *partie* compoſée de trois, & quelquefois de quatre *cotes* paralleles, ſaillantes

all widen to the bottom. The *ribs* open at top into ſome plates or cells, then cloſe, and are tranſverſely ſet with very fine *ſtriæ*; but at bottom the *ſtriæ* are very ſtrong, rough, and like ſhagreen. The *intervals* or *planes* are alſo tranſverſely and finely ſtriated. The *baſe* is cavernous, and ſet with a central deep cell, from which a great number of narrow cells, bounded by plates, radiate to the circumference.

The *animal* is finely *figured* by Mr. Ellis, in the *Phil. Tranſ.* for 1758, and reſembles a *cluſter of ſmall hooded and eared ſerpents* iſſuing from the *top cavity*, and from the *openings* or *cells* on the tops of the ribs. When *alive*, the *baſe*, by which it *affixes* itſelf, is *covered* with a coriaceous or ſtrong ſkin.

This *ſhell* is found in the *Northern Seas*, and always *adheres* to the *whale*, *from whence its name*. In the ſea round *Scotland* many are found. Mr. Martin tells us, that at *Fladda Chuan*, one of the inferior *Weſtern Iſlands*, near the *Iſle of Skie*, where are big whales, the natives diſtinguiſh one kind above all others for its great ſize, and told him that they had *big limpets* growing on their backs. This *ſpecies* is alſo *common* on the whales about *Newfoundland*.

ſaillantes & arrondies, & un *plan* large intermediat, qui touts elargiſſent au bas. Les *cotes* s'ouvrent au haut en quelques lames ou cellules, & de la elles ſont couvertes, & à *ſtries* tres fines tranſverſales; mais au bas les *ſtries* ſont tres fortes, aſpres, & comme la peau chagrin. Les *intervalles* ou *plans* ſont auſſi finement ſtriés à travers. La *baſe* eſt caverneuſe, & à cellule central profonde, de laquelle un grand nombre de cellules etroites, bornées par des lames, ſe repandent d'une maniere rayonnnante à la circonference.

L'*animal* eſt parfaitement bien *figuré* par M. Ellis, dans les *Tranſactions Philoſophiques* pour 1758, & reſſemble un *amas de petits ſerpents coiffés & à oreilles*, ſortant de la *cavité* du *ſommet*, & des *ouvertures* ou *cellules* des ſommets des cotes. Quand *vivant*, la *baſe*, par laquelle il s'*affiche*, eſt *couverte* d'une peau forte & coriacée.

La *coquille* ſe trouve dans les *Mers Septentrionales*, & toujours *adherente* à la *baleine*, *d'ou elle à pris ſon nom*. Dans la mer environ l'*Ecoſſe* pluſieurs ſe trouvent. M. Martin nous raconte, que à *Fladda Chuan*, une des *Iſles Occidentales* inferieures d'*Ecoſſe*, près de l'*Iſle de Skie*, ou il y a des grandes baleines, les habitans diſtinguent une ſorte plus que les autres pour ſon grand volume, & lui ont dit qu'elles avoient de *grands lepas* ou *patelles* qui croiſſoient ſur leurs dos. Cette *eſpece* eſt auſſi *commune* ſur les baleines à la *Terre Neuve*.

LXXII.

LXXII.

LXXII.

B. ANATIFERUS. BARNACLE.
Tab. XVII. fig. 3.

LXXII.

CONQUE ANATIFERE.
Pl. XVII. fig. 3.

Balanus compressus quinquevalvis lævis, tubo seu colle membranaceo insidente. Anatifera. Tab. 17. fig. 3.

Concha anatifera margine lævi. List. H. Conch. tab. 440. fig. 283. & Exercit. Anat. 3. p. 94. tab. 7. fig. 4. 5. 9.

Gerard Herbal, p. 1587.—*Conchâ quinquevalvis compressa, tubulo quodam lignis aut algæ marinæ adhærens; animal sui generis multis cirrhis instructum continens, falso dicta anatifera.* Sibbald Mus. p. 170. No. 2. & *Concha anatifera.* Prodr. Hist. Nat. Scotiæ, p. 2. l. 3. c. 112. tab. 18. fig. 1. 2. 3.—*Barnacle shell, or Concha anatifera.* Merret Pin. p. 194. Phil. Transf. Abridg. vol. 2. p. 853.—*Balanus compressa. Flat centre shell.* Grew Mus. p. 148. Wallace Orkn. p. 45. fig. 1. Mus. Petiv. p. 82. No. 802. Leigh Lancash. p. 157. Dale Harw. p. 389.

Concha anatifera vulgaris. Phil. Transf. 1758, vol. 50. p. 11. tab. 34. fig. 5. Smith Cork, p. 318.—*Lepas anatifera. Anatiferous.* Penn. Brit. Zool. No. 9. tab. 38. fig. 9.

Anates conchiferæ vel anatiferæ falso dictæ aut tellinæ & balani. Bauh Hist. Plant. 3. p. 803.—*Conchæ anatiferæ ex arbore dependentes.* Aldrovand Exsang. p. 543. Ornithol, c. 20. fig. 548.—*Conchæ anatiferæ Britannicæ.* Lobel Icon 2. p. 250.—*Arbores conchiferæ & anatiferæ dictæ.* Chabræi Sciagraph. p. 580. fig. 3. 4.

Concha anatifera. Worm. Mus. p. 256.—*Concha pedata.* Imperat. Hist. Nat. p. 901. Lat. & Ital. p. 683.—*Telline pedate.* Buon. Ricr. p. 148. fig. 2. Klein Ostrac. p. 175. No. 5. Argenv. Conch. I. p. 364. tab. 30. fig. F. tab. 26. fig. D. II. p. 322. tab. 26. fig. E. & p. 67. tab. 7. fig. I.—*Tellina cancellifera major, fere lævis, quinque portionibus testaceis composita subalbida.* Gualt. I. Conch. tab. 106. fig. A.—*Lepas testa compressa, basi membrana cylindratea.* Lin. F. Suec. I. p. 385. No. 1350. II. No. 212.—*Anatifera. Lepas testa compressa quinquevalvi lævi, intestino insidente.* S. N. p. 1109. No. 18. Mus. Reg. p. 468. No. 6.

The *general name* of *Anatifera*, or *Barnacle*, given to the *species*, is owing to the *fabulous story* of its *breeding* or *becoming Barnacle geese*, being first *observed* where that *sea fowl breeds* and *resorts* in great numbers. The *animal*, a *triton*, having many *cirrhi*, or fine *feather-like tentacula*, with also a produced neck, to which it hung, gave rise to this *strange conceit* and

Ll 2 vulgar

Le *nom general* de *Anatifera*, ou *Bernacle*, donné à cette *espece*, est du à l'*histoire fabuleuse* de cette coquille *engendrant* ou *devenant* des *oyes Barnacles*, etant premierement *observé* ou cet *oiseau anatique couve* & *frequente* en tres grands nombres. L'*animal*, un *triton*, ayant plusieurs *cirrhi*, ou *tentacula fines* & comme des *plumes*, avec aussi un col allongé, auquel il pendoit, à donné

origine

vulgar opinion, believed not only by the *commonalty*, but even by the *learned naturalists* and *botanists*. One *instance* of this credulity is our countryman the *botanist Gerard:* He firmly believes it by facts, *says he*, in his knowledge; and, after reciting the story in a circumstantial manner, he *gravely* ends his narration with the following words: *for the truth hereof, if any doubt, may it please them to repaire unto me, and I shall satisfie them by the testimonie of good witnesses.*

origine à cette *etrange imagination* & *opinion vulgaire*, cru non seulement par *le peuple*, mais même par les *naturalistes* & *botanistes sçavans*. Un *exemple* de cette credulité est notre compatriote le *botaniste Gerard*: Il l'a cru fermement par des faits, *dit il*, de son sçavoir; &, apres avoir recité l'histoire d'une maniere circonstanciée, il conclut sa narration tres *serieusement* par les paroles suivantes: *pour la verité de laquelle, si personne en doute, qu'il leur plaisent de se rendre ches moi, & je les satisferai par le temoignage de bons temoins.*

This *shell* is *milk white*, smooth, thin, and semitransparent. It is composed of *five subtriangular valves*, depressed or not very concave, laid flatways. The *two upper* ones are near treble as large as the *two bottom* ones that join them, and the *fifth valve* is narrow, pointed, sharp, and like a spur or keel, and runs *longitudinally* down one side, or in a *contrary position* to the *others*, and connects them close. All the valves, both within and without, have plain *margins*. The *top* is open, and from it proceeds a pretty long, cylindric, membranous wrinkled *tube*, *neck*, or *pedicle*, of a *blood red colour*; and by this neck it *affixes* itself pendent to the bottom of *ships*, *planks*, *logs*, and other pieces of *wood*.

Cette *coquille* est *blanche de lait*, lisse, mince, & demitransparente. Elle est composée de *cinq valves*, *à peu pres triangulaires*, comprimées, ou point fort concaves, posées horizontalement. Les *deux valves superieures* sont presque triple la grandeur des *deux inferieures*, qui les joignent, & la *cinquieme valve* est etroite, pointue, tranchante, & comme un eperon ou carene, & posée *longitudinalement* le long d'un coté, ou dans une *situation contraire* aux *autres*, & les lient serrement ensemble. Toutes les valves, tant en dedans que en dehors, ont les *bords* unis. Le *sommet* est ouvert, & de la s'eleve un *tube*, *col*, ou *pedicule*, asses, long, cylindrique, membraneux & ridé, d'une *couleur rouge de sang*; & par cet col elle s'*affiche* pendant au fond des *vaissaux*, *planches*, *troncs*, & autres pieces de *bois*.

It is *frequently* found on the coasts of *Scotland*, as also on the coasts of *England* and *Ireland*.

Elle est *frequente* sur les cotes d'*Ecosse*, comme aussi sur celles d'*Angleterre* & d'*Irlande*.

END OF THE BIVALVES AND MULTIVALVES.

FINIS. FIN.

INDEX,

In which the species of *Linné* and *Pennant* are also enumerated by the capitals L. and P.

TABLE DES MATIERES,

Dans laquelle les espece de *Linné* & *Pennant* sont aussi marquées par les lettres capitales de L. & P.

Chama

Mactra

INDEX of the FIGURES quoted in LISTER's HIST. CONCHYLIORUM.

The alternate columns mark the numbers of *Lister's* tables and figures, and the pages where they are described in this work.

INDICE des FIGURES citées de l'HISTORIA CONCHYLIORUM de LISTER.

Les colomnes alternatives marquent les nombres des planches & figures de *Lister*, & les pages ou elles se trouvent decrites dans cet ouvrage.

Tab.	*fig.*	*Page*	*Tab.*	*fig.*	*Page*	*Tab.*	*fig.*	*Page*	*Tab.*	*fig.*	*Page*
19	14	90	180	17	151	370	210	240	585	43	98
27	5	86	181	18	151	375	216	205	588	51	115
41	30	107	189	23	151	376	218	} 205	590	54 ?	110
48	46	67	190	27	144		219		590	55 ?	112
49	47	72	191	28	144	376	217	207	591	57 ?	112
56	53	75	193	30	} 154	394	241	209	591	59 ?	110
57	54	76	194	31		405	250	} 210	611	2	15
59	56	79	202	36	} 162		251		616	1	37
69	68	55	203	37		409	255	235	616	2	40
71	—	53 80	204	38	165	411	257	237	621	8	41
78	78	53	235	69	171	415	259	230	630	16	38
79	79	53	247	82	168	420	264	238	631	17	38
123	23	92	250	84?	199	421	265	238	632	19 ?	38
123	21	93	251	85	198	423	271	202	633	21	42
123	22	95	253	87	197	426	267	223	641	31	43
126	26	81	253	88	200	428	269	233	697	39	50
132	32	91	272	108	183	433	276	244		a44	
134	34	96	279	116	190	433	285	250	707	57	33
136	40	60	279	436	242	435	278	246	714	70	31
137	41	60	281	119	188	440	283	253	714	71	26
138	42	63	282	120	188	444	287	248	835	58	101
138	43	65	284	122	185	445	288	251	865	20	136
141	39	1	290	126	187	533	5	22	913	4	120
141	38	48	291	127	187	535	14	6	913	5	124
147	3	228	292	128	187	542	26	8	942	38 ?	133
149	4	225	293	129	187	543	27	7	962	14	122
153	8	215	299	136	187	543	28	11	962	15 a	122
156	11	214	324	161	176	547	2	24	965	18. 19	125
159	14	173	332	169	178	553	—	} 22	966	21	131
163	1	140	334	171	180	561	8	83	1055	6	81
165	2. 3	143	359	198	219	568	19?	83	1058	11	57
172	9	148	362	200	216		20?				

E X-

EXPLANATIONS OF PLATES,

And References to the Pages where they are described.

ER-

E R R A T A.

E. English, F. French Text

Page 1. Order I. add *Univalves.*

P. 3. F. l. 3. for *Pl. II.* read *Pl. I.*—l. 19. add *Pl. I.*

P. 6. E. l. 16. *Orkneys*, r. *Western Isles.*—F. l. 17. r. *Isles Occidentales.*

P. 22. F. l. 9. *Exterieur*, r. *exterieure.*—l. 15. *entremelées dans*, r *entremelées dans.*

P. 23. F. *partout les*, add *partout sur les.*

P. 64. F. l. 3. *est*, r. *et.*

P. 66. l. 8. *p.* 2177, r. *No.* 2177.

P. 77. E. l. 25. *an* dele—l. 36. *that* dele—F. l. 25. *il fut nous*, r. *il nous fut.*

P. 85. E. l. 30. *them*, r. *it*—F. l. 14. *Continent*, r. *contient*—l. 29. *l'a*, read *la.*

P. 89. E. l. 2. *Cylindraceous*, r. *Cylindraceus.*

P. 96. E. l. 25. *Neritædes*, r. *Neritoides.*

P. 98. F. l. 5. *cellui*, r. *celui.*

P. 105. E. l. 23. *fuscus*, add *ulva.*

P. 118. F. l. 9. *en trecoupent*, r. *entrecoupent.*

P. 128. F l. 8. *de*, r. *le.*

P. 135. F. l. 8. *de*, r. *à.*

P. 149. E. l. 31. *concave*, add *or.*

P. 152. F. l. 10. *le*, r. *les*; *lustrées*, r. *lustrés*—l. 11. *blanches*, r. *blancs.*

P. 160. F. l. 15. *l'eau sortir*, r. *sortir l'eau.*

P. 174. F. l. 5. *sont*, add *egaux.*

P. 178. E. l. 32. *incisis*, r. *incisis.*

P. 185. F. l. 28. *dit*, add *que*—l. 29. *Asiatque*, r. *Asiatique.*

P. 200. E. l. 9. *side*, r. *half*—F. l. 11. *un coté*, r. *une moitié.*

P. 216. E. l. 32. *Mytulis*, r. *Mytilus.*

P. 218. F. l. 10. *de Bretagne*, add *la Grande.*

P. 226. F. l. 31. add *la Tweed, la Don.*

P. 228. E. l. 9. *Tab. XIV.* r. *Tab. XV.*

What other *Errata* occur, the candid Reader is desired to correct, especially the tex in the quotations of the figures of Pl. III. which correct according to the Index c Explanations of Plates, p. vii.

SUBSCRIBERS Omitted.

Mr. Stanesby Alchorne, *coloured*
Mr. William Hudson, F. R. S. *coloured*
Mr. John Hunter, F. R. S. *coloured*
Walter Synnot, Esq; *coloured.*

The NAMES of the SUBSCRIBERS.

N. B. *Pl.* stands for uncoloured or plain copies, and *col.* for coloured ones.

A.

JOHN Adams, Esq; *painted*
Hon. Baron Aguilar, *col.*
Benjamin d'Aguilar, Esq; *pl.*
Mr. John Anderson, Prof. Hist. Nat. Glasgow, F. R. S. *col.*
Mr James Arden, *col.*

B.

Mr. Matthew Boulton, Birmingham, *col.*
Gustavus Brander, Esq; Mus. Brit. Cur. F R. S. *pl.*
George Browne, Esq; *pl.*
John Smith Budgen, Esq; *col.*

C.

Joseph Capadose, Esq; *col.*
Daniel de Castro, Esq; *col.*
Mrs. de Castro, *painted*
Samuel de Castro, Esq; *pl.*
Hon. Mrs. Cavendish, *col.*
Anthony Champion, Esq; *col.*
Thomas Chowne, Esq; *col.*
Mr. —— Church, Islington, *col.*
Monf. J. P. Cobre, Augsberg, Germany, *col.*
John Cordley, Esq; York, *col.*
Hananel Mendes da Costa, Esq; *col.*
Joshua Mendes da Costa, Esq; *col.*
Philip de la Cour, M. D. Bath, *col.*
Rev. Sir John Cullum, Bart. *pl.*
William Cuming, M. D. Dorchester, *col.*
John Cuthbert, Esq; F. R. S. *col.*

D.

John Debonnaire, jun. Esq; *col.*
James Dewar, Esq; *col.*
Mr. Dru Drury, *col.*

E.

George Edwards, Esq; *col.*
John Ellis, Esq; F. R. S. *col.*
Miss Ellis, *col.*

F.

Mr. Fenn, *col.*
Martin Fonnereau, Esq; *painted*
Mrs. Fordyce, *col.*
Mrs. Forster, *pl.*
John Reinhold Forster, LL.D. F.R.S. *col.*
Mr. Ingham Foster, 3 copies, *painted, col.* and *pl.*
Raphael Franco, Esq; *col.*
John Fothergill, M.D. F.R.S. 2 copies, *col.*

G.

Rev. G. Gretton, A. B. Trinity College, Cambridge, *col.*
Counsellor Thomas Griffin, *col.*
George Griffin, Esq; *col.*
John Grimstone, Esq; Killwick, near Beverley, Yorkshire, *painted*

H.

Anthony Hamilton, D. D. *col.*
Arthur Holdsworth, Esq; Governor of Dartmouth Castle, *col.*
John Holland, Esq; of Tierdan, Denbighshire, *col.*
Mr. George Humphrey, *col.*

I.

Mr. Edward Jacobs, Feversham, *col.*
John Ibbetson, Esq; F. R. S. *col.*
Mrs. Jones, *col.*
Mr. Thomas Jones, Printer, *col.*

Richard

K.

Richard Kaye, LL. D. Muf. Brit. Cur. F. R. S. *col.*

George Keate, Efq; F. R. S. *col.*

Mr. Thomas Knowlton, Loanfborough Caftle, Yorkfhire, *col.*

L.

Mr. Abraham Lara, *pl.*

Charlwood Lawton, Efq; *pl.*

John Coakley Lettfom, M.D. F.R.S. *col.*

Sir Afhton Lever, Knt. 4 copies, 2 *col.* 2 *pl.*

Mofes Ifaac Levi, Efq; *pl.*

Gomperts Levifon, M. D. *pl.*

Mifs Lewis, Wimbledon, *col.*

Michael Lort, B.D. F.R.S. Trinity College, Cambridge, *pl.*

John Lloyd, Efq; F.R.S. *col.*

M.

Edward Maefe, Efq; *pl.*

Mr. Thomas Martyn, *col.*

Lady Mill, *col.*

Mr. John Millan, Bookfeller, 6 copies, 1 *col.* 5 *pl.*

Mr. Ifaac Minors, *pl.*

Hans Winthrop Mortimer, Efq; M. P. for Shaftefbury, *col.*

Charles Morton, M.D. Principal Librarian Brit. Muf. F.R.S. *col.*

Charles Mofs, Efq; Wells, *pl.*

William Myddelton, Efq; *pl.*

N.

Mr. Ifaac Naffo, *col.*

John Nicols, Efq; *col.*

O.

Humphrey Owen, M. A. Warrington, Lancafhire, *col.*

P.

Her Grace the Duchefs Dowager of Portland, *col.*

Mr. Samuel Platt, *col.*

Thomas Pennant, Efq; F.R.S. 5 copies, *pl.*

Mrs. Pierfon, *col.*

Mr. Pointer, *col.*

Richard Pulteney, M.D. F.R.S. Blandford

R.

David Alvez Rabello, Efq; *pl.*

Mr. Cater Rand, Lewes, Suffex, *col.*

Philip Rafhleigh, Efq; M.P. for Fowey, *col.*

John Rafhleigh, Efq; Penguite, Cornwall, *col.*

—— Rawlinfon, M. D. *col.*

Mr John Van Rymfdyk, *col.*

S.

Right Hon. Earl of Seaforth

Jofeph Salvador, Efq; F. R. S. *col.*

George Scott, LL.D. F.R.S. Woolftonhall, Effex, *pl.*

Phineas Serra, Efq; *pl.*

Henry Seymer, Efq; Handford, Dorfet, *pl.*

Mr. Robert Shore, Snitterton, Derbyfhire, *col.*

Ifaac Siqueira, M. D. *col.*

Sir John Smith, Bart. Sydling St. Nicholas, Dorfet, *col.*

James Stuart, Efq; F. R. S. *col.*

T.

Edward Thomas, A.M. F.R.S. *col.*

Henry Tuckfeild, Efq; *painted*

Marmaduke Tunftal, Efq; F. R. S. *painted*

W.

Richard Hill Waring, Efq; F. R. S. Leefwood, Flintfhire, 2 copies, *col.* & *pl.*

Mifs Waring, *pl.*

Mrs. Watfon, Bath, *col.*

Mr. John White, *col.*

Mr. John Whitehurft, *col.*

Jofeph Williams, Efq; Glanravon, Carnarvonfhire, *col.*

Andrew Wilfon, M. D. *pl*

Mr. John Wooller, *pl.*

Pl. I.

page			page
3	1.2.81.	Patella vulgata Gm.	2697
		Lin. Trans. 8 p. 229. Montag	475
14	3.	Patella græca	3728
		Lin. Tr. 236 – Mont.	492
11	4.	Patella Fissura	3728
		Lin. Tr. 230 – Mont.	490
7	5.6.	Patella pellucida	3717
		Lin. Tr. 233 – Mont.	477
12	7.7.	Patella Ungarica	3700
		Lin. Tr. 230 – Mont.	486
26	9.9.	Bulla lignaria	3425
		Lin. Tr. 125 – Mont.	205
28	10.10.	Bulla Hydatis	3422
		Lin. Tr. 123 – Mont.	217

page Pl. II.

15 — 1. 2. Haliotis tuberculata 3687
Lin. Tr. 227 — M. 473.

30 — 3. Bulla aperta — 3424
Lin. Tr. 131 — M. Sup. 92.

40 — 4. Trochus erythroleucos — 3581
Tr. exiguus L. Tr. 156 — M. 277.

18 — 5. Serpula vermicularis — 3748
L. Tr. 243 — M. 509.

38 — 6. Cypraea Pediculus — 3418
L. Tr. 120 — M. 201.

31 — 7. 7. Voluta pallida — 3444
L. Tr. 132 — M. 232

1 — 8. 8. Patella lacustris — 3710
L. Tr. 232 — Mont. 482.

20 — 9. Serpula triquetra — 3740
L. Tr. 244 — M. 511

24 — 10. Dentalium Entalis — 3736
L. Tr. 237 — M. 474

23 — 11. Serpula spiralis — 3740
Lin. Tr. 244. Mont. 498

Plate II

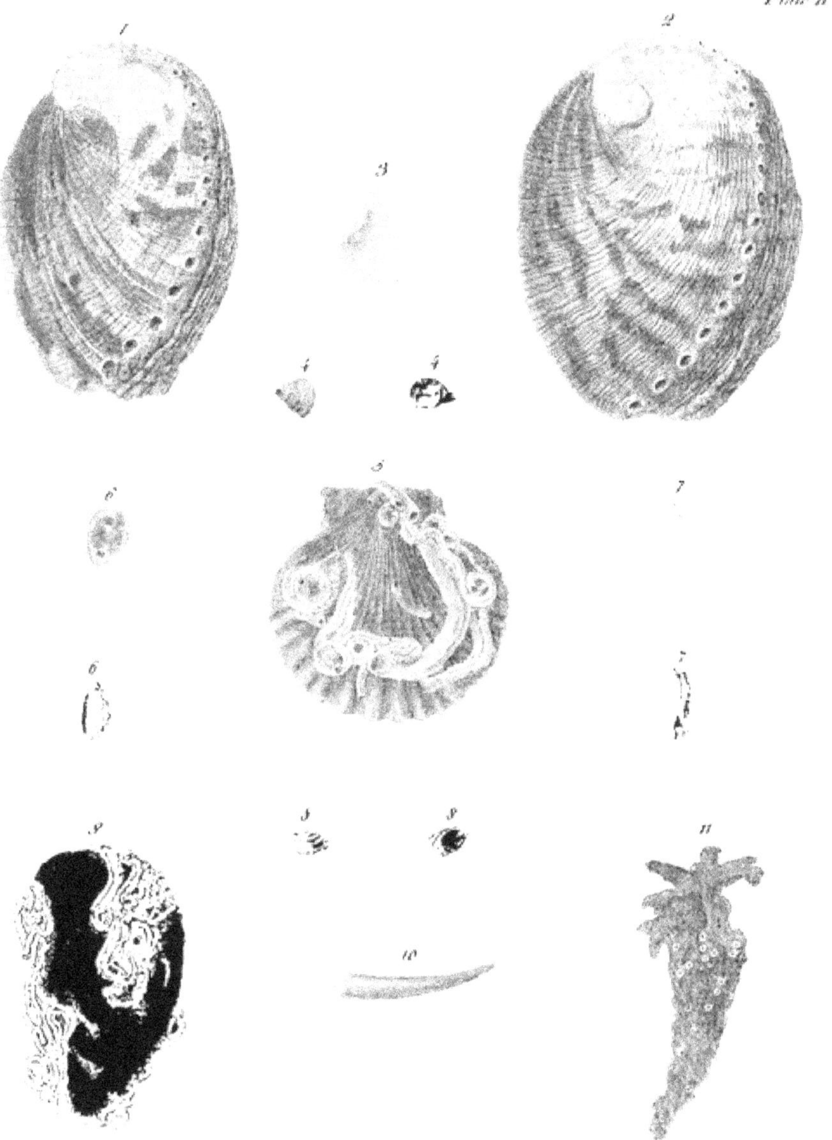

Pl. III.

page

44 — 1. 2. Trochus Magus — Gm. 3567
Lin. Tr. 191 — Montag. 283

37 — 3. 4. Tro: Ziziphinus — 3579
L. Tr. 155 — Mont. 274

38 — 5. 6. Tro. papillosus Lin. Tr. 155
Tro. tenuis Mont. 275
an. Tro. Granatum Gmel. 3584

46 — 7. 8. Tr. umbilicatus — Lin. Tr. 153
Montag. 286

42 — 9. 10. Tro. cinereus — Lin. Tr. 152
Montag — 289

43 — 11. 12. Tro. lineatus — L. Tr. 152
Montag. 284

48 — 13 — 16. Nerita littoralis Gm. 3677
L. Tr. 226 — Mont. 467
see t. 4. f. 23.

50 — 17. 18. Nerita fluviatilis — 3676
L. Tr. 225 — Mont. 470.

PL. III

Pl. IV.

72 — 1. 1. Helix aspersa — — Gmel. 3677
H. hortensis. Lin. Tr. 200 — M. 407

50 — 2. 3. Nerita litteralis vid. t. 2. — 3677
Lin. Tr. 227 — M. 467

51 — 4. 5. Nerita pallidula — Lin. Tr. 226
Montag. Test. Britan. — 468

80 — 6. Hel. Turtarum — — Gm. 3634
— — Sericea — — 3617
Lin. Tr. 196. 15 — Mont. 420

14 — 7. Hel. Pisana — — 3621
— — media — — 3640
H. virgata, L. Tr. 195 — Mont. 415

58 — 8. Helix Ericetorum — — 3632
Lin. Tr. 191 — M. 437

55 — 9. Hel. lapicida — — 3613
Lin. Tr. 187 Mont. 435

53 — 10. Hel. complanata — — 3617
Lin. Tr. 188 — Mont. 450

66 — 11. Helix contorta — — 3624
Lin. Tr. 191 — Mont. 457

65 — 12. Helix Vortex — — 3620
Lin. Tr. 189 — Mont. 451.

50 — 13. Hel. cornea — — 3622
L. Tr. 190 — Mont. 448

07 — 14. Hel. Pomatia — — 3627
L. Tr. 201 — M. 405

37 — 15. 16. Hel. rotundata — — 3633
H. radiata L. Tr. 199 — Mont. 432. 3.22.3

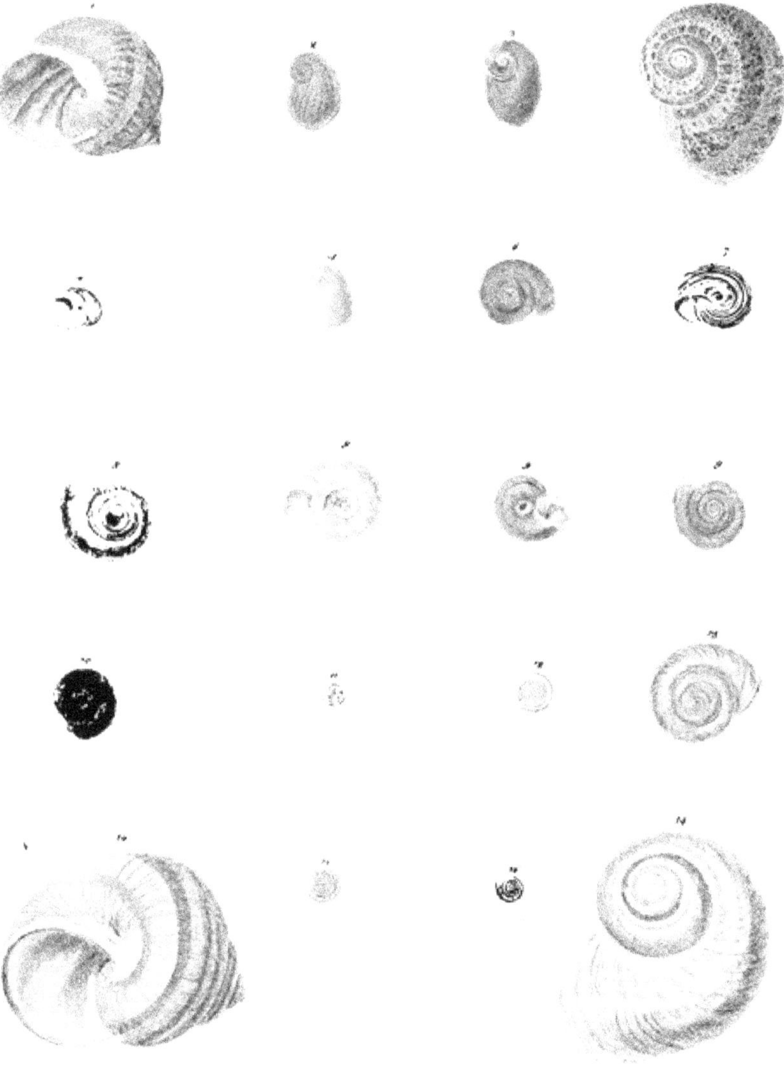

4.5.14.15 — ——— ♂
Lin. Tr. 308 — Montag. 407

96–6. Bulla fontinalis ——— 3427
Lin. Tr. 126 — Montag. 226

83–7. Nerita glaucina ——— 3671
Lin. Tr. 224 — Mont. 269

86–9. Turbo elegans ——— 3606
Lin. Tr. 164 — Mont. 342

58–10. Helix hispida ——— 3625
Lin. Tr. 198 — Mont. 423

93–11. Helix stagnalis ——— 3657
Lin. Tr. 214 — Mont. 367

91–12. Helix tentaculata ——— 3662
Lin. Tr. 220 — Mont. 389

92–13. Helix putris ——— 3659
Lin. Tr. 219 — Mont. 376.

14.19. Helix nemoralis ♂ — 3647
4.4.5 — L. Tr. 207. — Mont. 411.

15. Turbo perversus ——— 3609
Lin. Tr. 219 — Mont. 376

89–16. Turbo Muscorum — 3611
Lin. Tr. 182 — Mont. 334

17. Helix auricularia — 3662
Lin. Tr. 221 — Montag. 375

18. Helix lubrica ——— 3661
Lin. Tr. 313 — Montag. 390

Pl. VI

page

98 – 1. 1. Turbo littoreus ——— Gm. 3588
Lin. Tr. 150 – Mont. 301

81 – 2. 2. Helix vivipara ——— 3646
L. Tr. 209 – Mont. 386 –

110. 3. Turbo duplicatus ——— 3607.
L. Tr. 175

120 – 4. Murex antiquus ——— 3546
Gm. M. despectus ——— 3547
Lin. Tr. 145

134. 5. Murex corneus ——— 3552
Lin. Tr. 147 – Montag. 258

132. 6. Buccin. undatum ——— 3492
Lin. Tr. 137 – Montag. 237*

100 – 7. Trochus crassus – Lin. Tr. 164
Montag. 281.

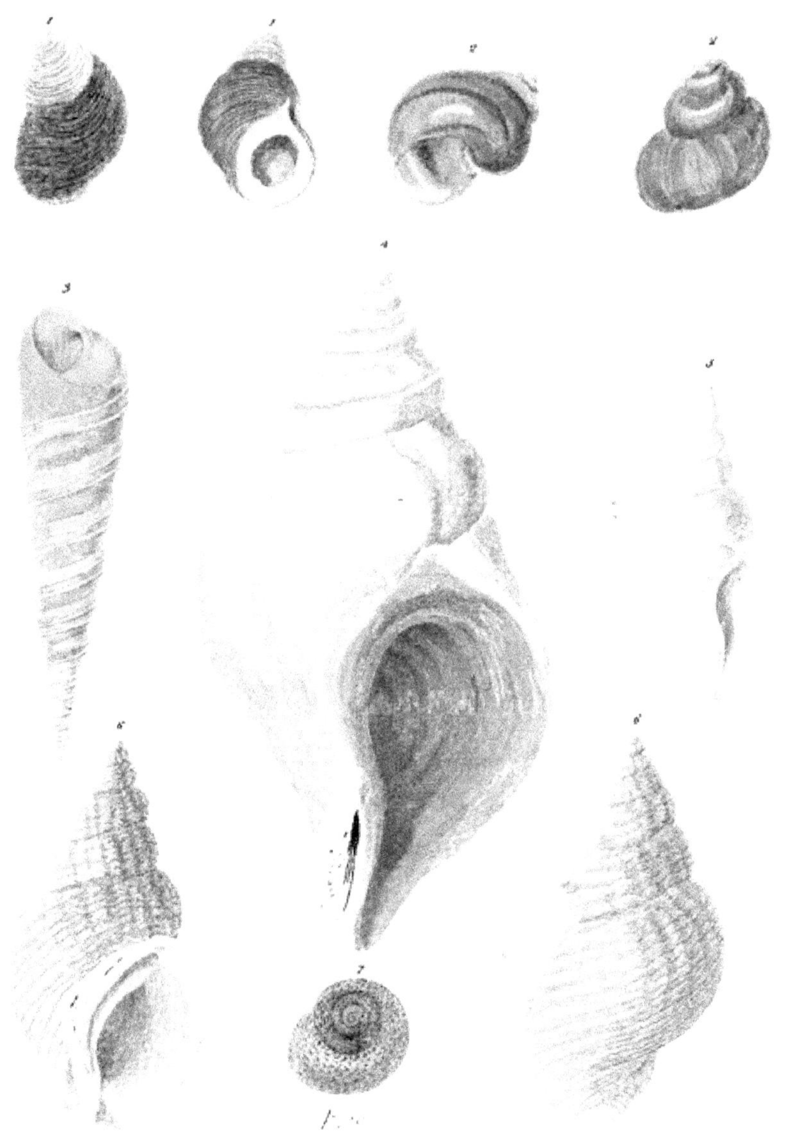

Pl. VII.

page
125 — 1. 4. 9. 12. Bucc: Lapillus Gm. 3488
Lin. Tr. 135 — Montag 239
112 — 5. 6. Turbo Terebra — 3608
Lin. Tr. 176 — Mont. 293
136 — 7. Stromb: Pes Pelecani — 3507
Lin. Tr. 141 — M. 253
114 — 8. Turbo exoletus — 3607
Lin. Tr. 176 — M. 295
121 — 10. Bucc. reticulatum 3495
Lin. Tr. 37 — M. 240
115 — 11. Turbo Clathrus — 3603
L. Tr. 170 — Mont. 296

1
2
3
4
5
6
7
8
9
10
11
12

Pl. VIII.

103 — 1 — 3 Turbo Pullus — 3589
Lin. Tr. V. 8. 162 — Mont. 319
101 — 2. Voluta tornatilis 3437
Lin. Tr. 129 — M. 231
4. Murex costatus Lin. Tr. 144
18. Mont. Test. Brit. 265
130 — 5. Bucc. Lineatum L. Tr. 136
Montag. — 245
104 — 6. 9 Turbo Cimex — Gm. 3589
Lin. Tr. 161 — M. 315.
102 — 7. Murex erinaceus — 3530
Lin. Tr. 161 — M. 315
63 — 8. Hel. complanata — 3617
L. Tr. 100 — Mont. 450
102 — 10. Turbo striatulus 3604
L. Tr. 172 — M. 316. t. 10. f. 5
7 — 11. Patella virginea — 3711
L. Tr. 238 — M. 400
315 — 12. Helix parva Da Costa 85
117 — 13. Hel. reticulatus D. C. 117

144–1. 2. 4. 5. Ostrea opercularis. Gm. 3325
Lin. Tr. 901 – Mont. 145

3. 3. Ostrea maxima ——— 3315
Lin. Tr. 96 – Mont. 143.

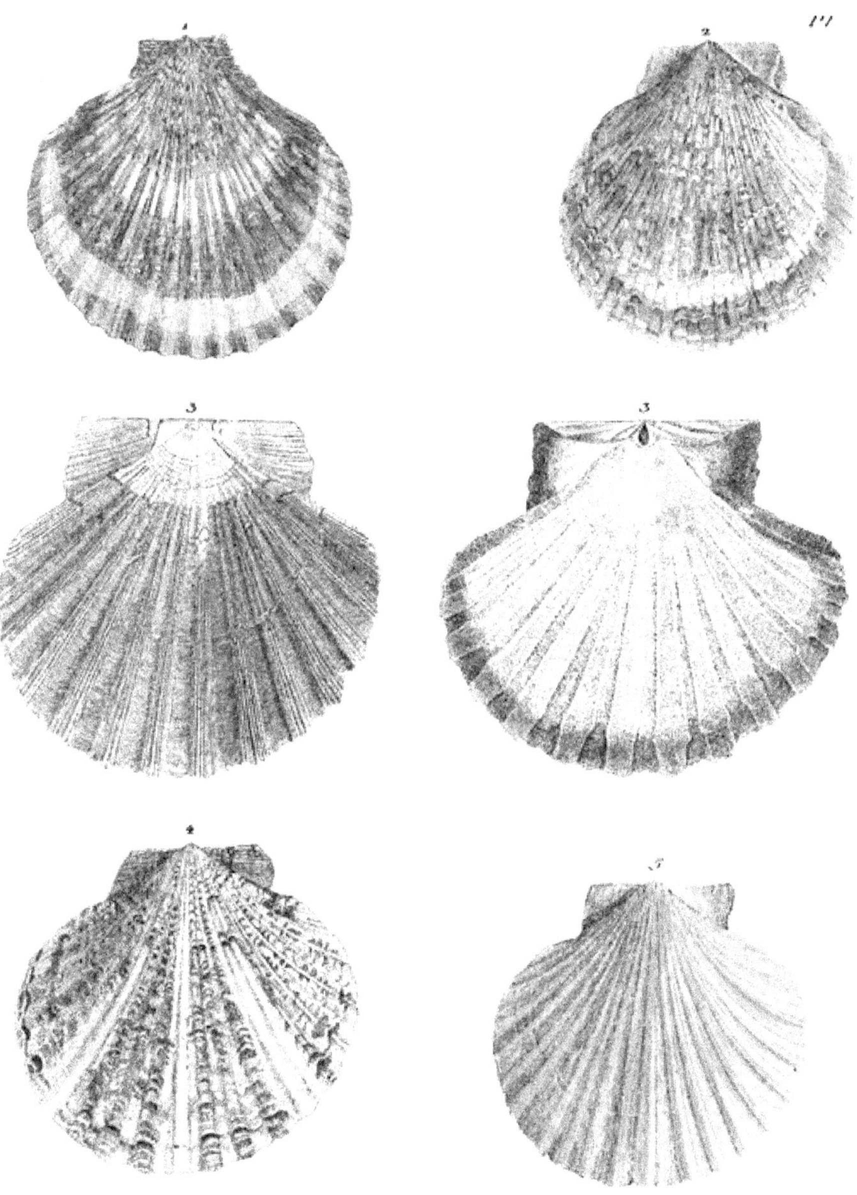
1
2
3
3
4
5

Pl. X.

page

151 — 1.2.4.5.7.9. Ostrea varia — Gm. 3324
Lin. Tr. 97 — Mont. 146

3.6. Ostrea sinuosa ——— 3319
Lin. Tr. 99. — Mont. 148 (distorta)

8. Ostrea lineata — Lin. Tr. 99
Pecten lineatus — Montag 147.

1
2
3
4
5
6
7
8
9

Pl. XIII.

200 1.1. Mactra Listeri —3261
—— compressa Lin. Tr. 71
Martyn. 96 b

173 2.2. Tellina cornea —3241
Lin. Tr. 8. 59 — Mart. 86

1001 3. Venus fasciata Lin. Tr. 80
—— Paphia Mart. 110

193}
1059} 4. left. Ven. verrucosa Gm. 3269
Lin. Tr. 701 — Mart. 112. 574

194 4. right Tellina crassa 3280
V. crassa Lin. Tr. 55. Mart. 65

1001 5.5. Venus Paphia —3266
Lin. Tr. 810 — Mart. 110

1701 6.6. Card. laevigatum —3251
Lin. Tr. Mart. 86

Pl. III

Pl. XII.

105 – 1. 1. Venus verrucosa. Gm. 3260
Lin. Tr. 8. 78 — M. 112 & 574.

191 - 2. 2. Venus Gallina — 3270
Lin. Tr. 82 — M. 113

196 – 3. 3 Mactra Stultorum – 3258
Lin. Tr. 69 — M. 94

211 – 4. Tellina incarnata – 3234
Tel. solidula L. Tr. 58 – M. 63

147 – 5. 5. Venus exoleta — 3284
Lin. Tr. 87 – Mont. 116.

Pl. XI.

200 – 1. 1. Cardium edule — 3252
Lin. Tr. – 6 — Mont. 76

168 – 2. 2. Arca pilosa — 3314
Lin. Tr. 8. p. 99 — Mont. 136

163 – 3. Anomia Ephippium — 3340
Lin. Tr. 102 — Mont. 155

162 – 4. 4. Ostrea undulata — 3346
Lin. Tr. 103 — Mont. 157

171 – 5. 5. Arca lactea — 3309
Lin. Tr. 92 — Mont. 130

154 6. Ostrea edulis — 3334
Lin. Tr. 101 — Mont. 151.

Pl. XI.

page

Pl. XIV.

209—1. Tellina Fervensis — Gm. 3235
Lin. Tr. 8. p. 49 — Montag. 55

176—2. Cardium echinatum — 3247
Lin. Tr. 63 — Mont. 78

207—3. Donax Trunculus — 3263
Lin. Tr. 74 — Mont. 103

162—4. Venus decussata — 3294
Lin. Tr. 88 — Mont. 124

183—5. Venus Islandica — 3271
Lin. Tr. 83 — M. 14

197—6. Mactra solida — 3259
Lin. Tr. 70 — M. 92

184—7. Venus Chione — 3272 (sup. t. 19. f. 1)
Lin. Tr. 84 — M. 115.

Pl. XV.

page

1.1. Mactra solida — 3259
Lin. Tr. 8. p. 70 — Mont. 92

214 – 2. Mytilus cygneus — 3375
——— anatinus 3355
Lin. Tr. 109 — Montag. 170

219 – 3. Mya margaritifera 3219
Lin. Tr. 40 — Mont. 33

228 – 4. Mya Pictorum — 3218
Lin. Tr. 28 — Mont 34

219 – 5. Mytilus edulis — 3343
L. Tr. 105 — Mont 163

219 – 5. (right) Myt. Modiolus – 3354
L. Tr. 95. Mont. 141

170 6. (right) arca Nucleus – 3314
L. Tr 95 — Mont. 141

204 6. (left) Donax Irus – 3265
L. Tr. 77 — Mont. 100

Pl. XV

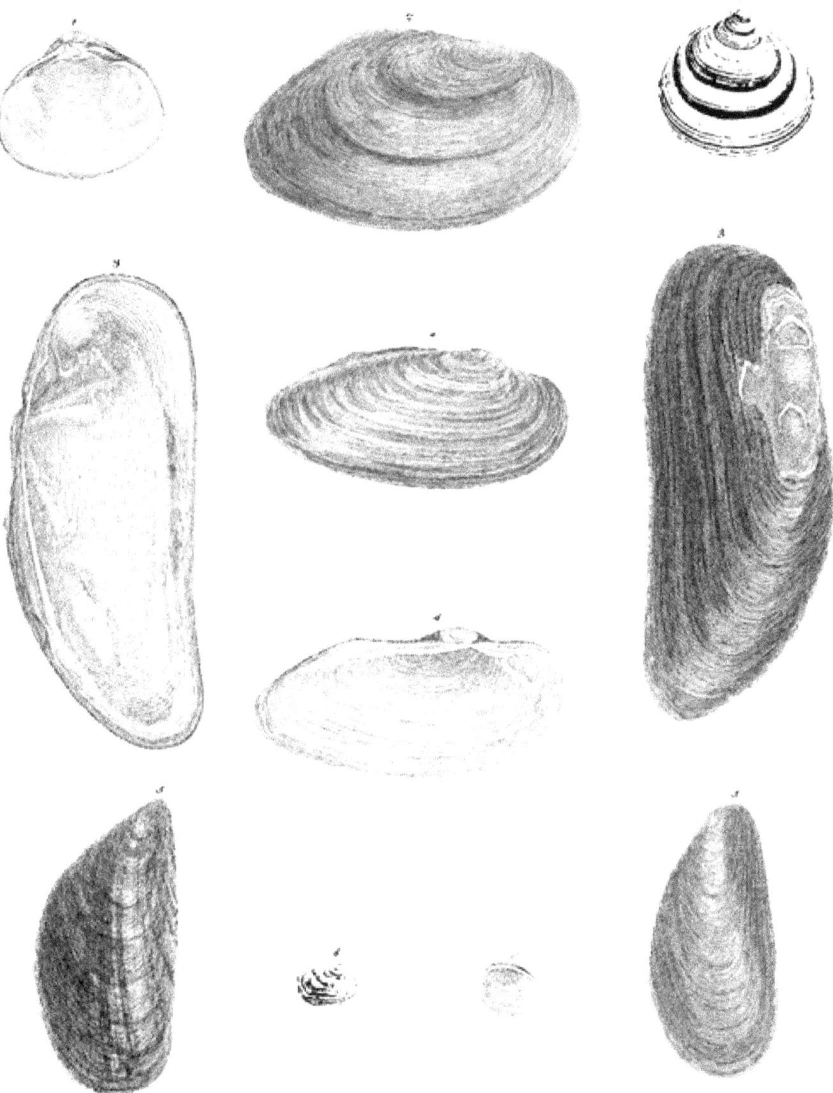

Pl. XVI.

Page

233 – 1. 1. *Mya truncata* Gm. 3217
Lin. Tr. 35 – Montag. 32

144 – 2. 2. *Pholas Dactylus* — 3214
Lin. Tr. 30 — Montag. 20

240 – 3 *Pinna pectinata* — 3363
Lin. Tr. 113 — Mont. 170

4. 4. *Pholas crispata* – 3216
Lin. Tr. 32 — Mont. 23

Pl. XVI.

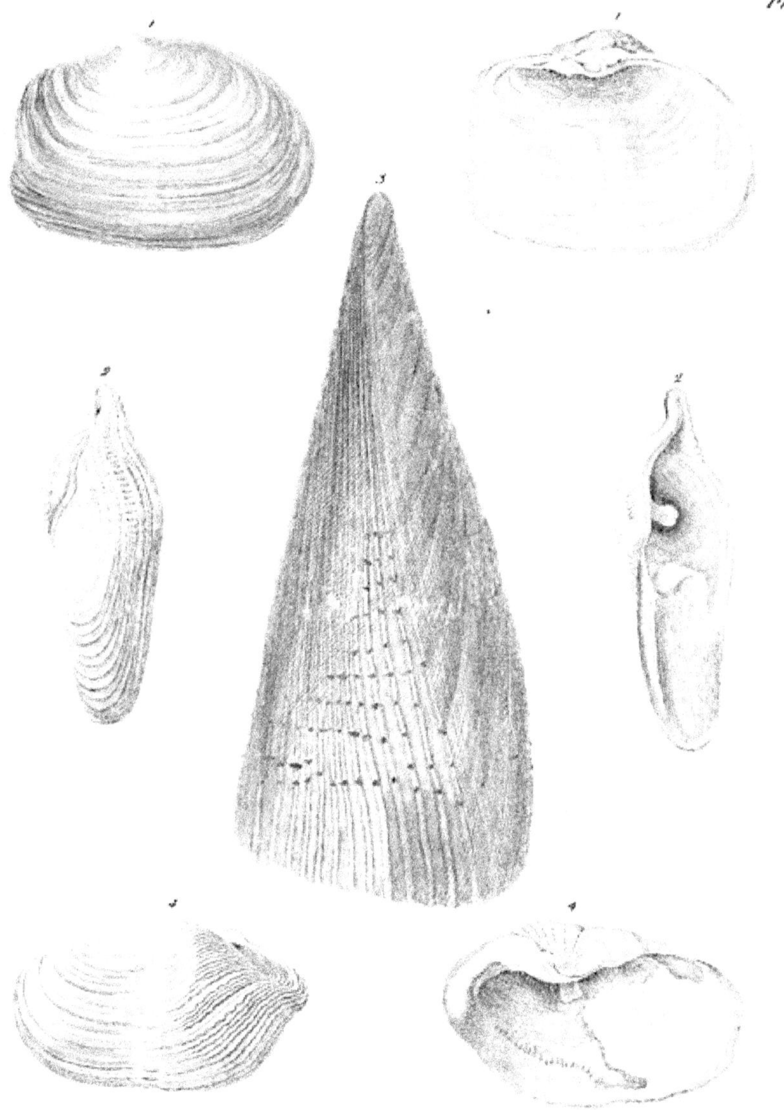

Pl. XVII.

221—1. Mytilus discors —— 3356
Lin. Tr. 8. p. 111 — Montag. 167

251—2. 2. Lepas Diadema — 3208
Lin. Tr. 27 —— Mont. 13

3. Lepas anatifera —— 3211
Lin. Tr. 201 — Mont. 15

4. Mya oblonga —— 3221
Mactra lutraria — 3259
Lin. Tr. 74 — Mont. 101

5. Solen Siliqua —— 3223
Lin. Tr. 43 — Mont. 46

75—6. Helix Arbustorum — 3630
Lin. Tr. 202 — Mont. 413.

7. Lepas balanoides — 3207
Lin. Tr. 23 — Mont. p. 7

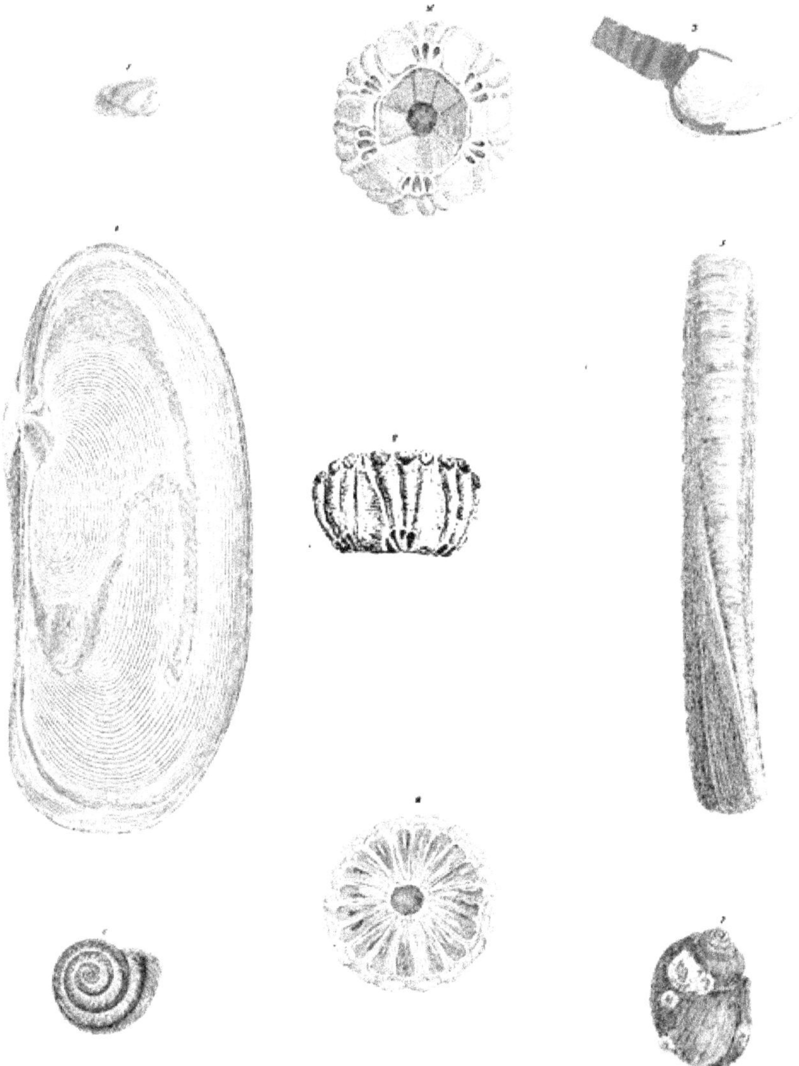

www.ingramcontent.com/pod-product-compliance
Lightning Source LLC
Chambersburg PA
CBHW021504210326
41599CB00012B/1127

* 9 7 8 3 7 4 4 6 5 7 3 6 5 *